Graduate Texts in Mathematics **223**

Graduate Texts in Mathematics

1 TAKEUTI/ZARING. Introduction to Axiomatic Set Theory. 2nd ed.
2 OXTOBY. Measure and Category. 2nd ed.
3 SCHAEFER. Topological Vector Spaces. 2nd ed.
4 HILTON/STAMMBACH. A Course in Homological Algebra. 2nd ed.
5 MAC LANE. Categories for the Working Mathematician. 2nd ed.
6 HUGHES/PIPER. Projective Planes.
7 J.-P. SERRE. A Course in Arithmetic.
8 TAKEUTI/ZARING. Axiomatic Set Theory.
9 HUMPHREYS. Introduction to Lie Algebras and Representation Theory.
10 COHEN. A Course in Simple Homotopy Theory.
11 CONWAY. Functions of One Complex Variable I. 2nd ed.
12 BEALS. Advanced Mathematical Analysis.
13 ANDERSON/FULLER. Rings and Categories of Modules. 2nd ed.
14 GOLUBITSKY/GUILLEMIN. Stable Mappings and Their Singularities.
15 BERBERIAN. Lectures in Functional Analysis and Operator Theory.
16 WINTER. The Structure of Fields.
17 ROSENBLATT. Random Processes. 2nd ed.
18 HALMOS. Measure Theory.
19 HALMOS. A Hilbert Space Problem Book. 2nd ed.
20 HUSEMOLLER. Fibre Bundles. 3rd ed.
21 HUMPHREYS. Linear Algebraic Groups.
22 BARNES/MACK. An Algebraic Introduction to Mathematical Logic.
23 GREUB. Linear Algebra. 4th ed.
24 HOLMES. Geometric Functional Analysis and Its Applications.
25 HEWITT/STROMBERG. Real and Abstract Analysis.
26 MANES. Algebraic Theories.
27 KELLEY. General Topology.
28 ZARISKI/SAMUEL. Commutative Algebra. Vol.I.
29 ZARISKI/SAMUEL. Commutative Algebra. Vol.II.
30 JACOBSON. Lectures in Abstract Algebra I. Basic Concepts.
31 JACOBSON. Lectures in Abstract Algebra II. Linear Algebra.
32 JACOBSON. Lectures in Abstract Algebra III. Theory of Fields and Galois Theory.
33 HIRSCH. Differential Topology.
34 SPITZER. Principles of Random Walk. 2nd ed.
35 ALEXANDER/WERMER. Several Complex Variables and Banach Algebras. 3rd ed.
36 KELLEY/NAMIOKA et al. Linear Topological Spaces.
37 MONK. Mathematical Logic.
38 GRAUERT/FRITZSCHE. Several Complex Variables.
39 ARVESON. An Invitation to C^*-Algebras.
40 KEMENY/SNELL/KNAPP. Denumerable Markov Chains. 2nd ed.
41 APOSTOL. Modular Functions and Dirichlet Series in Number Theory. 2nd ed.
42 J.-P. SERRE. Linear Representations of Finite Groups.
43 GILLMAN/JERISON. Rings of Continuous Functions.
44 KENDIG. Elementary Algebraic Geometry.
45 LOÈVE. Probability Theory I. 4th ed.
46 LOÈVE. Probability Theory II. 4th ed.
47 MOISE. Geometric Topology in Dimensions 2 and 3.
48 SACHS/WU. General Relativity for Mathematicians.
49 GRUENBERG/WEIR. Linear Geometry. 2nd ed.
50 EDWARDS. Fermat's Last Theorem.
51 KLINGENBERG. A Course in Differential Geometry.
52 HARTSHORNE. Algebraic Geometry.
53 MANIN. A Course in Mathematical Logic.
54 GRAVER/WATKINS. Combinatorics with Emphasis on the Theory of Graphs.
55 BROWN/PEARCY. Introduction to Operator Theory I: Elements of Functional Analysis.
56 MASSEY. Algebraic Topology: An Introduction.
57 CROWELL/FOX. Introduction to Knot Theory.
58 KOBLITZ. p-adic Numbers, p-adic Analysis, and Zeta-Functions. 2nd ed.
59 LANG. Cyclotomic Fields.
60 ARNOLD. Mathematical Methods in Classical Mechanics. 2nd ed.
61 WHITEHEAD. Elements of Homotopy Theory.
62 KARGAPOLOV/MERLZJAKOV. Fundamentals of the Theory of Groups.
63 BOLLOBAS. Graph Theory.

(continued after index)

Anders Vretblad

Fourier Analysis and Its Applications

Anders Vretblad
Department of Mathematics
Uppsala University
Box 480
SE-751 06 Uppsala
Sweden
anders.vretblad@math.uu.se

Mathematics Subject Classification (2000): 42-01

Library of Congress Cataloging-in-Publication Data
Vretblad, Anders.
Fourier analysis and its applications / Anders Vretblad.
p. cm.
Includes bibliographical references and index.

1. Fourier analysis. I. Title.
QA403.5. V74 2003
515'2433—dc21 2003044941

ISBN 978-1-4419-1841-3 e-ISBN 978-0-387-21723-9

Printed in the United States of America.

9 8 7 6 5 4 3 2 Corrected second printing, 2005

springeronline.com

To

YNGVE DOMAR,

my teacher, mentor, and friend

Preface

The classical theory of Fourier series and integrals, as well as Laplace transforms, is of great importance for physical and technical applications, and its mathematical beauty makes it an interesting study for pure mathematicians as well. I have taught courses on these subjects for decades to civil engineering students, and also mathematics majors, and the present volume can be regarded as my collected experiences from this work.

There is, of course, an unsurpassable book on Fourier analysis, the treatise by Katznelson from 1970. That book is, however, aimed at mathematically very mature students and can hardly be used in engineering courses. On the other end of the scale, there are a number of more-or-less cookbook-styled books, where the emphasis is almost entirely on applications. I have felt the need for an alternative in between these extremes: a text for the ambitious and interested student, who on the other hand does not aspire to become an expert in the field. There do exist a few texts that fulfill these requirements (see the literature list at the end of the book), but they do not include all the topics I like to cover in my courses, such as Laplace transforms and the simplest facts about distributions.

The reader is assumed to have studied real calculus and linear algebra and to be familiar with complex numbers and uniform convergence. On the other hand, we do not require the Lebesgue integral. Of course, this somewhat restricts the scope of some of the results proved in the text, but the reader who *does* master Lebesgue integrals can probably extrapolate the theorems. Our ambition has been to prove as much as possible within these restrictions.

Some knowledge of the simplest distributions, such as point masses and dipoles, is essential for applications. I have chosen to approach this matter in two separate ways: first, in an intuitive way that may be sufficient for engineering students, in star-marked sections of Chapter 2 and subsequent chapters; secondly, in a more strict way, in Chapter 8, where at least the fundaments are given in a mathematically correct way. Only the one-dimensional case is treated. This is not intended to be more than the merest introduction, to whet the reader's appetite.

Acknowledgements. In my work I have, of course, been inspired by existing literature. In particular, I want to mention a book by Arne Broman, *Introduction to Partial Differential Equations...* (Addison–Wesley, 1970), a compendium by Jan Petersson of the Chalmers Institute of Technology in Gothenburg, and also a compendium from the Royal Institute of Technology in Stockholm, by Jockum Aniansson, Michael Benedicks, and Karim Daho. I am grateful to my colleagues and friends in Uppsala. First of all Professor Yngve Domar, who has been my teacher and mentor, and who introduced me to the field. The book is dedicated to him. I am also particularly indebted to Gunnar Berg, Christer O. Kiselman, Anders Källström, Lars-Åke Lindahl, and Lennart Salling. Bengt Carlsson has helped with ideas for the applications to control theory. The problems have been worked and re-worked by Jonas Bjermo and Daniel Domert. If any incorrect answers still remain, the blame is mine.

Finally, special thanks go to three former students at Uppsala University, Mikael Nilsson, Matthias Palmér, and Magnus Sandberg. They used an early version of the text and presented me with very constructive criticism. This actually prompted me to pursue my work on the text, and to translate it into English.

Uppsala, Sweden — Anders Vretblad
January 2003

Contents

1 Introduction

1.1 The classical partial differential equations

In this introductory chapter, we give a brief survey of three main types of partial differential equations that occur in classical physics. We begin by establishing some convenient notation.

Let Ω be a domain (an open and connected set) in three-dimensional space $\mathbf{R}^3$, and let T be an open interval on the time axis. By $C^k(\Omega)$, resp. $C^k(\Omega \times T)$, we mean the set of all real-valued functions $u(x, y, z)$, resp. $u(x, y, z, t)$, with all their partial derivatives of order up to and including k defined and continuous in the respective regions. It is often practical to collect the three spatial coordinates (x, y, z) in a vector $\mathbf{x}$ and describe the functions as $u(\mathbf{x})$, resp. $u(\mathbf{x}, t)$. By Δ we mean the LAPLACE operator

$$\Delta = \nabla^2 := \frac{\partial^2}{\partial x^2} + \frac{\partial^2}{\partial y^2} + \frac{\partial^2}{\partial z^2}.$$

Partial derivatives will mostly be indicated by subscripts, e.g.,

$$u_t = \frac{\partial u}{\partial t}, \qquad u_{yx} = \frac{\partial^2 u}{\partial x \partial y}.$$

The first equation to be considered is called the *heat equation* or the *diffusion equation:*

$$\Delta u = \frac{1}{a^2}\frac{\partial u}{\partial t}, \qquad (\mathbf{x}, t) \in \Omega \times T.$$

As the name indicates, this equation describes conduction of heat in a homogeneous medium. The temperature at the point $\mathbf{x}$ at time t is given by $u(\mathbf{x}, t)$, and a is a constant that depends on the conducting properties of the medium. The equation can also be used to describe various processes of diffusion, e.g., the diffusion of a dissolved substance in the solvent liquid, neutrons in a nuclear reactor, BROWNian motion, etc.

The equation represents a category of second-order partial differential equations that is traditionally categorized as *parabolic.* Characteristically, these equations describe *non-reversible* processes, and their solutions are highly regular functions (of class C^∞).

In this book, we shall solve some special problems for the heat equation. We shall be dealing with situations where the spatial variable can be regarded as one-dimensional: heat conduction in a homogeneous rod, completely isolated from the exterior (except possibly at the ends of the rod). In this case, the equation reduces to

$$u_{xx} = \frac{1}{a^2} u_t .$$

The *wave equation* has the form

$$\Delta u = \frac{1}{c^2} \frac{\partial^2 u}{\partial t^2}, \qquad (\mathbf{x}, t) \in \Omega \times T.$$

where c is a constant. This equation describes vibrations in a homogeneous medium. The value $u(\mathbf{x}, t)$ is interpreted as the deviation at time t from the position at rest of the point with rest position given by $\mathbf{x}$.

The equation is a case of *hyperbolic* equations. Equations of this category typically describe reversible processes (the past can be deduced from the present and future by "reversion of time"). Sometimes it is even suitable to allow solutions for which the partial derivatives involved in the equation do not exist in the usual sense. (Think of shock waves such as the sonic bangs that occur when an aeroplane goes supersonic.) We shall be studying the *one-dimensional* wave equation later on in the book. This case can, for instance, describe the motion of a vibrating string.

Finally we consider an equation that does not involve time. It is called the *Laplace* equation and it looks simply like this:

$$\Delta u = 0.$$

It occurs in a number of physical situations: as a special case of the heat equation, when one considers a stationary situation, a *steady state*, that does not depend on time (so that $u_t = 0$); as an equation satisfied by the potential of a conservative force; and as an object of considerable purely mathematical interest. Together with the closely related POISSON equation, $\Delta u(\mathbf{x}) = F(\mathbf{x})$, where F is a known function, it is typical of equations classified as *elliptic.* The solutions of the Laplace equation are very regular functions: not only do they have derivatives of all orders, there are even certain possibilities to reconstruct the whole function from its local behaviour near a single point. (If the reader is familiar with analytic functions, this should come as no news in the two-dimensional case: then the solutions are harmonic functions that can be interpreted (locally) as real parts of analytic functions.)

The names *elliptic, parabolic,* and *hyperbolic* are due to superficial similarities in the appearance of the differential equations and the equations of conics in the plane. The precise definitions of the different types are as follows: The unknown function is $u = u(\mathbf{x}) = u(x_1, x_2, \dots, x_m)$. The equations considered are *linear;* i.e., they can be written as a sum of terms equal to a known function (which can be identically zero), where each term in the sum consists of a coefficient (constant or variable) times some derivative of u, or u itself. The derivatives are of degree at most 2. By changing variables (possibly locally around each point in the domain), one can then write the equation so that no mixed derivatives occur (this is analogous to the diagonalization of quadratic forms). It then reduces to the form

$$a_1u_{11} + a_2u_{22} + \cdots + a_mu_{mm} + \{\text{terms containing } u_j \text{ and } u\} = f(\mathbf{x}),$$

where $u_j = \partial u/\partial x_j$ etc. If all the a_j have the same sign, the equation is elliptic; if at least one of them is zero, the equation is parabolic; and if there exist a_j's of opposite signs, it is hyperbolic.

An equation can belong to different categories in different parts of the domain, as, for example, the TRICOMI equation $u_{xx} + xu_{yy} = 0$ (where $u = u(x, y)$), which is elliptic in the right-hand half-plane and hyperbolic in the left-hand half-plane. Another example occurs in the study of the so-called velocity potential $u(x, y)$ for planar laminary fluid flow. Consider, for instance, an aeroplane wing in a streaming medium. In the case of *ideal* flow one has $\Delta u = 0$. Otherwise, when there is friction (air resistance), the equation looks something like $(1-M^2)u_{xx}+u_{yy} = 0$, with $M = v/v_0$, where v is the speed of the flowing medium and v_0 is the velocity of sound in the medium. This equation is elliptic, with nice solutions, as long as $v < v_0$, while it is hyperbolic if $v > v_0$ and then has solutions that represent shock waves (sonic bangs). Something quite complicated happens when the speed of sound is surpassed.

1.2 Well-posed problems

A *problem* for a differential equation consists of the equation together with some further conditions such as initial or boundary conditions of some form. In order that a problem be "nice" to handle it is often desirable that it have certain properties:

1. There *exists* a solution to the problem.
2. There exists *only one* solution (i.e., the solution is uniquely determined).
3. The solution is *stable*, i.e., small changes in the given data give rise to small changes in the appearance of the solution.

A problem having these properties (the third condition must be made precise in some way or other) is traditionally said to be *well posed*. It is, however, far from true that all physically relevant problems are well posed. The third condition, in particular, has caught the attention of mathematicians in recent years, since it has become apparent that it is often very hard to satisfy it. The study of these matters is part of what is popularly labeled chaos research.

To satisfy the reader's curiosity, we shall give some examples to illuminate the concept of well-posedness.

Example 1.1. It can be shown that for suitably chosen functions $f \in C^\infty$, the equation $u_x + u_y + (x + 2iy)u_t = f$ has no solution $u = u(x, y, t)$ at all (in the class of complex-valued functions) (Hans Lewy, 1957). Thus, in this case, condition 1 fails. □

Example 1.2. A natural problem for the heat equation (in one spatial dimension) is this one:

$$u_{xx}(x,t) = u_t(x,t),\ x > 0,\ t > 0; \quad u(x,0) = 0,\ x > 0; \quad u(0,t) = 0,\ t > 0.$$

This is a mathematical model for the temperature in a semi-infinite rod, represented by the positive x-axis, in the situation when at time 0 the rod is at temperature 0, and the end point $x = 0$ is kept at temperature 0 the whole time $t > 0$. The obvious and intuitive solution is, of course, that the rod will remain at temperature 0, i.e., $u(x,t) = 0$ for all $x > 0$, $t > 0$. But the mathematical problem has additional solutions: let

$$u(x,t) = \frac{x}{t^{3/2}}\, e^{-x^2/(4t)}, \quad x > 0,\ t > 0.$$

It is a simple exercise in partial differentiation to show that this function satisfies the heat equation; it is obvious that $u(0,t) = 0$, and it is an easy exercise in limits to check that $\lim_{t \searrow 0} u(x,t) = 0$. The function must be considered a solution of the problem, as the formulation stands. Thus, the problem fails to have property 2.

The disturbing solution has a rather peculiar feature: it could be said to represent a certain (finite) amount of heat, located at the end point of the rod at time 0. The value of $u(\sqrt{2t}, t)$ is $\sqrt{(2/e)}/t$, which tends to $+\infty$ as $t \searrow 0$. One way of excluding it as a solution is adding some condition to the formulation of the problem; as an example it is actually sufficient to demand that a solution must be bounded. (We do not prove here that this does solve the dilemma.) □

Example 1.3. A simple example of instability is exhibited by an ordinary differential equation such as $y''(t) + y(t) = f(t)$ with initial conditions $y(0) = 1$, $y'(0) = 0$. If, for example, we take $f(t) = 1$, the solution is $y(t) = 1$. If we introduce a small perturbation in the right-hand member by taking $f(t) = 1 + \varepsilon \cos t$, where $\varepsilon \neq 0$, the solution is given by $y(t) = 1 + \frac{1}{2}\varepsilon t \sin t$. As time goes by, this expression will oscillate with increasing amplitude and "explode". The phenomenon is called *resonance.* □

1.3 The one-dimensional wave equation

We shall attempt to find *all* solutions of class C^2 of the one-dimensional wave equation

$$c^2 u_{xx} = u_{tt}.$$

Initially, we consider solutions defined in the open half-plane $t > 0$.

Introduce new coordinates (ξ, η), defined by

$$\xi = x - ct, \quad \eta = x + ct.$$

It is an easy exercise in applying the chain rule to show that

$$u_{xx} = \frac{\partial^2 u}{\partial x^2} = \frac{\partial^2 u}{\partial \xi^2} + 2\frac{\partial^2 u}{\partial \xi\, \partial \eta} + \frac{\partial^2 u}{\partial \eta^2}$$

$$u_{tt} = \frac{\partial^2 u}{\partial t^2} = c^2 \left(\frac{\partial^2 u}{\partial \xi^2} - 2\frac{\partial^2 u}{\partial \xi\, \partial \eta} + \frac{\partial^2 u}{\partial \eta^2} \right).$$

Inserting these expressions in the equation and simplifying we obtain

$$c^2 \cdot 4 \frac{\partial^2 u}{\partial \xi\, \partial \eta} = 0 \quad \Longleftrightarrow \quad \frac{\partial}{\partial \xi}\left(\frac{\partial u}{\partial \eta}\right) = 0.$$

Now we can integrate step by step. First we see that $\partial u/\partial \eta$ must be a function of only η, say, $\partial u/\partial \eta = h(\eta)$. If ψ is an antiderivative of h, another integration yields $u = \varphi(\xi) + \psi(\eta)$, where φ is a new arbitrary function. Returning to the original variables (x, t), we have found that

$$u(x,t) = \varphi(x - ct) + \psi(x + ct). \tag{1.1}$$

In this expression, φ and ψ are more-or-less arbitrary functions of one variable. If the solution u really is supposed to be of class C^2, we must demand that φ and ψ have continuous second derivatives.

It is illuminating to take a closer look at the significance of the two terms in the solution. First, assume that $\psi(s) = 0$ for all s, so that $u(x,t) =$

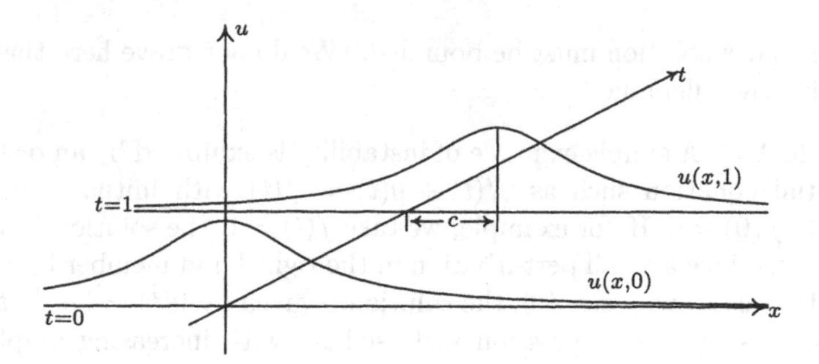

FIGURE 1.1.

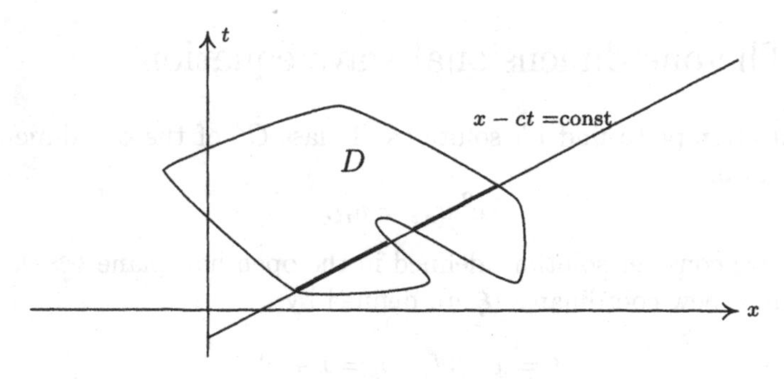

FIGURE 1.2.

$\varphi(x - ct)$. For $t = 0$, the graph of the function $x \mapsto u(x, 0)$ looks just like the graph of φ itself. At a later moment, the graph of $x \mapsto u(x, t)$ will have the same shape as that of φ, but it is pushed ct units of length to the right. Thus, the term $\varphi(x - ct)$ represents a *wave moving to the right along the x-axis* with constant speed equal to c. See Figure 1.1! In an analogous manner, the term $\psi(x + ct)$ describes a wave moving to the left with the same speed. The general solution of the one-dimensional wave equation thus consists of a superposition of two waves, moving along the x-axis in opposite directions.

The lines $x \pm ct =$ constant, passing through the half-plane $t > 0$, constitute a net of level curves for the two terms in the solution. These lines are called the *characteristic curves* or simply *characteristics* of the equation. If, instead of the half-plane, we study solutions in some other region D, the derivation of the general solution works in the same way as above, as long as the characteristics run unbroken through D. In a region such as that shown in Figure 1.2, the function φ need not take on the same value on the two indicated sections that do lie on the same line but are not connected inside D. In such a case, the general solution must be described in a more complicated way. But if the region is *convex*, the formula (1.1) gives the general solution.

Remark. In a way, the general behavior of the solution is similar also in higher spatial dimensions. For example, the two-dimensional wave equation

$$\frac{\partial^2 u}{\partial x^2} + \frac{\partial^2 u}{\partial y^2} = \frac{1}{c^2}\frac{\partial^2 u}{\partial t^2}$$

has solutions that represent wave-shapes passing the plane in all directions, and the general solution can be seen as a sort of superposition of such solutions. But here the directions are infinite in number, and there are both planar and circular wave-fronts to consider. The superposition cannot be realized as a sum — one has to use integrals. It is, however, usually of little interest to exhibit the general solution of the equation. It is much more valuable to be able to pick out some particular solution that is of importance for a concrete situation. □

Let us now solve a natural *initial value problem* for the wave equation in one spatial dimension. Let $f(x)$ and $g(x)$ be given functions on $\mathbf{R}$. We want to find all functions $u(x,t)$ that satisfy

$$\text{(P)} \qquad \begin{cases} c^2 u_{xx} = u_{tt}, & -\infty < x < \infty, \quad t > 0; \\ u(x,0) = f(x), \quad u_t(x,0) = g(x), & -\infty < x < \infty. \end{cases}$$

(The initial conditions assert that we know the shape of the solution at $t = 0$, and also its rate of change at the same time.) By our previous calculations, we know that the solution must have the form (1.1), and so our task is to determine the functions φ and ψ so that

$$f(x) = u(x,0) = \varphi(x) + \psi(x), \quad g(x) = u_t(x,0) = -c\,\varphi'(x) + c\,\psi'(x). \tag{1.2}$$

An antiderivative of g is given by $G(x) = \int_0^x g(y)\,dy$, and the second formula can then be integrated to

$$-\varphi(x) + \psi(x) = \frac{1}{c}\,G(x) + K,$$

where K is the integration constant. Combining this with the first formula of (1.2), we can solve for φ and ψ:

$$\varphi(x) = \frac{1}{2}\left(f(x) - \frac{1}{c}\,G(x) - K\right), \qquad \psi(x) = \frac{1}{2}\left(f(x) + \frac{1}{c}\,G(x) + K\right).$$

Substitution now gives

$$\begin{aligned} u(x,t) &= \varphi(x-ct) + \psi(x+ct) \\ &= \frac{1}{2}\left(f(x-ct) - \frac{1}{c}\,G(x-ct) - K + f(x+ct) + \frac{1}{c}\,G(x+ct) + K\right) \\ &= \frac{f(x-ct) + f(x+ct)}{2} + \frac{G(x+ct) - G(x-ct)}{2c} \\ &= \frac{f(x-ct) + f(x+ct)}{2} + \frac{1}{2c}\int_{x-ct}^{x+ct} g(y)\,dy. \end{aligned} \tag{1.3}$$

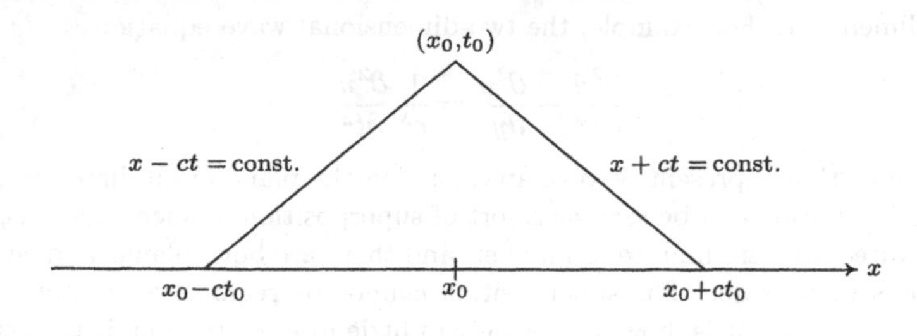

FIGURE 1.3.

The final result is called D'ALEMBERT'S formula. It is something as rare as an explicit (and unique) solution of a problem for a partial differential equation.

Remark. If we want to compute the value of the solution $u(x,t)$ at a particular point (x_0, t_0), d'Alembert's formula tells us that it is sufficient to know the initial values on the interval $[x_0 - ct_0, x_0 + ct_0]$: this is again a manifestation of the fact that the "waves" propagate with speed c. Conversely, the initial values taken on $[x_0 - ct_0, x_0 + ct_0]$ are sufficient to determine the solution in the isosceles triangle with base equal to this interval and having its other sides along characteristics. See Figure 1.3. □

In a similar way one can solve suitably formulated problems in other regions. We give an example for a semi-infinite spatial interval.

Example 1.4. Find all solutions $u(x,t)$ of $u_{xx} = u_{tt}$ for $x > 0$, $t > 0$, that satisfy $u(x,0) = 2x$ and $u_t(x,0) = 1$ for $x > 0$ and, in addition, $u(0,t) = 2t$ for $t > 0$.

Solution. Since the first quadrant of the xt-plane is convex, all solutions of the equation must have the appearance

$$u(x,t) = \varphi(x-t) + \psi(x+t), \qquad x > 0,\ t > 0.$$

Our task is to determine what the functions φ and ψ look like. We need information about $\psi(s)$ when s is a positive number, and we must find out what $\varphi(s)$ is for all real s.

If $t = 0$ we get $2x = u(x,0) = \varphi(x) + \psi(x)$ and $1 = u_t(x,0) = -\varphi'(x) + \psi'(x)$; and for $x = 0$ we must have $2t = \varphi(-t) + \psi(t)$. To liberate ourselves from the magic of letters, we neutralize the name of the variable and call it s. The three conditions then look like this, collected together:

$$\begin{cases} 2s = \varphi(s) + \psi(s) \\ \ \ 1 = -\varphi'(s) + \psi'(s) \\ 2s = \varphi(-s) + \psi(s) \end{cases} \qquad s > 0.$$

The second condition can be integrated to $-\varphi(s) + \psi(s) = s + C$, and combining this with the first condition we get

$$\varphi(s) = \tfrac{1}{2}\,s - \tfrac{1}{2}\,C, \quad \psi(s) = \tfrac{3}{2}\,s + \tfrac{1}{2}\,C \quad \text{for } s > 0.$$

The third condition then yields $\varphi(-s) = 2s - \psi(s) = \frac{1}{2}\,s - \frac{1}{2}\,C, \quad s > 0$, where we switch the sign of s to get

$$\varphi(s) = -\tfrac{1}{2}\,s - \tfrac{1}{2}\,C \quad \text{for } s < 0.$$

Now we put the solution together:

$$u(x,t) = \varphi(x-t) + \psi(x+t) = \begin{cases} \frac{1}{2}(x-t) + \frac{3}{2}(x+t) = 2x + t, x > t > 0, \\ \frac{1}{2}(t-x) + \frac{3}{2}(x+t) = x + 2t, 0 < x < t. \end{cases}$$

Evidently, there is just one solution of the given problem.

A closer look shows that this function is continuous along the line $x = t$, but it is in fact not differentiable there. It represents an "angular" wave. It seems a trifle fastidious to reject it as a solution of the wave equation, just because it is not of class C^2. One way to solve this conflict is furnished by the theory of ***distributions***, which generalizes the notion of functions in such a way that even "angular" functions are assigned a sort of derivative. □

Exercise

1.1 Find the solution of the problem (P), when $f(x) = e^{-x^2}$, $g(x) = \dfrac{1}{1+x^2}$.

1.4 Fourier's method

We shall give a sketch of an idea that was tried by JEAN-BAPTISTE JOSEPH FOURIER in his famous treatise of 1822, *Théorie analytique de la chaleur*. It constitutes an attempt at solving a problem for the one-dimensional heat equation. If the physical units for heat conductivity, etc., are suitably chosen, this equation can be written as

$$u_{xx} = u_t\,,$$

where $u = u(x,t)$ is the temperature at the point x on a thin rod at time t. We assume the rod to be isolated from its surroundings, so that no exchange of heat takes place, except possibly at the ends of the rod. Let us now assume the length of the rod to be π, so that it can be identified with the interval $[0,\pi]$ of the x-axis. In the situation considered by Fourier, both ends of the rod are kept at temperature 0 from the moment when $t = 0$, and the temperature of the rod at the initial moment is assumed to be equal to a known function $f(x)$. It is then physically reasonable that we should be able to find the temperature $u(x,t)$ at any point x and at any time $t > 0$. The problem can be summarized thus:

$$\begin{cases} \text{(E)}\ \ u_{xx} = u_t, & 0 < x < \pi, \quad t > 0; \\ \text{(B)}\ \ u(0,t) = u(\pi,t) = 0, & t > 0; \\ \text{(I)}\ \ u(x,0) = f(x), & 0 < x < \pi. \end{cases} \tag{1.4}$$

The letters on the left stand for *equation, boundary conditions,* and *initial condition*, respectively. The conditions (E) and (B) share a specific property: if they are satisfied by two functions u and v, then all linear combinations $\alpha u + \beta v$ of them also satisfy the same conditions. This property is traditionally expressed by saying that the conditions (E) and (B) are *homogeneous.* Fourier's idea was to try to find solutions to the partial problem consisting of just these conditions, disregarding (I) for a while.

It is evident that the function $u(x,t) = 0$ for all (x,t) is a solution of the homogeneous conditions. It is regarded as a trivial and uninteresting solution. Let us instead look for solutions that are not identically zero. Fourier chose, possibly for no other reason than the fact that it turned out to be fruitful, to look for solutions having the particular form $u(x,t) = X(x)T(t)$, where the functions $X(x)$ and $T(t)$ depend each on just one of the variables.

Substituting this expression for u into the equation (E), we get

$$X''(x)T(t) = X(x)T'(t), \qquad 0 < x < \pi, \quad t > 0.$$

If we divide this by the product $X(x)T(t)$ (consciously ignoring the risk that the denominator might be zero somewhere), we get

$$\frac{X''(x)}{X(x)} = \frac{T'(t)}{T(t)}, \qquad 0 < x < \pi, \quad t > 0. \tag{1.5}$$

This equality has a peculiar property. If we change the value of the variable t, this does not affect the left-hand member, which implies that the right-hand member must also be unchanged. But this member is a function of only t; it must then be constant. Similarly, if x is changed, this does not affect the right-hand member and thus not the left-hand member, either. Indeed, we get that both sides of the equality are constant for all the values of x and t that are being considered. This constant value we denote (by tradition) by $-\lambda$. This means that we can split the formula (1.5) into two formulae, each being an ordinary differential equation:

$$X''(x) + \lambda X(x) = 0, \quad 0 < x < \pi; \qquad\qquad T'(t) + \lambda T(t) = 0, \quad t > 0.$$

One usually says that one has *separated the variables*, and the whole method is also called the method of *separation of variables.*

We shall also include the boundary condition (B). Inserting the expression $u(x,t) = X(x)\,T(t)$, we get

$$X(0)\,T(t) = X(\pi)\,T(t) = 0, \quad t > 0.$$

Now if, for example, $X(0) \neq 0$, this would force us to have $T(t) = 0$ for $t > 0$, which would give us the trivial solution $u(x,t) \equiv 0$. If we want to find interesting solutions we must thus demand that $X(0) = 0$; for the same reason we must have $X(\pi) = 0$. This gives rise to the following *boundary value problem* for X:

$$X''(x) + \lambda X(x) = 0, \quad 0 < x < \pi; \qquad X(0) = X(\pi) = 0. \tag{1.6}$$

In order to find nontrivial solutions of this, we consider the different possible cases, depending on the value of λ.

$\lambda < 0$: Then we can write $\lambda = -\alpha^2$, where we can just as well assume that $\alpha > 0$. The general solution of the differential equation is then $X(x) = Ae^{\alpha x} + Be^{-\alpha x}$. The boundary conditions become

$$\begin{cases} 0 = X(0) = A + B, \\ 0 = X(\pi) = Ae^{\alpha\pi} + Be^{-\alpha\pi}. \end{cases}$$

This can be seen as a homogeneous linear system of equations with A and B as unknowns and determinant $e^{-\alpha\pi} - e^{\alpha\pi} = -2\sinh \alpha\pi \neq 0$. It has thus a unique solution $A = B = 0$, but this leads to an uninteresting function X.

$\lambda = 0$: In this case the differential equation reduces to $X''(x) = 0$ with solutions $X(x) = Ax + B$, and the boundary conditions imply, as in the previous case, that $A = B = 0$, and we find no interesting solution.

$\lambda > 0$: Now let $\lambda = \omega^2$, where we can assume that $\omega > 0$. The general solution is given by $X(x) = A\cos\omega x + B\sin\omega x$. The first boundary condition gives $0 = X(0) = A$, which leaves us with $X(x) = B\sin\omega x$. The second boundary condition then gives

$$0 = X(\pi) = B\sin\omega\pi. \tag{1.7}$$

If here $B = 0$, we are yet again left with an uninteresting solution. But, happily, (1.7) can hold without B having to be zero. Instead, we can arrange it so that ω is chosen such that $\sin\omega\pi = 0$, and this happens precisely if ω is an integer. Since we assumed that $\omega > 0$ this means that ω is one of the numbers $1, 2, 3, \ldots$.

Thus we have found that the problem (1.6) has a nontrivial solution exactly if λ has the form $\lambda = n^2$, where n is a positive integer, and then the solution is of the form $X(x) = X_n(x) = B_n \sin nx$, where B_n is a constant.

For these values of λ, let us also solve the problem $T'(t) + \lambda T(t) = 0$ or $T'(t) = -n^2 T(t)$, which has the general solution $T(t) = T_n(t) = C_n e^{-n^2 t}$.

If we let $B_n C_n = b_n$, we have thus arrived at the following result: *The homogeneous problem* (E)+(B) *has the solutions*

$$u(x,t) = u_n(x,t) = b_n\, e^{-n^2 t} \sin nx, \quad n = 1, 2, 3, \ldots.$$

Because of the homogeneity, all sums of such expressions are also solutions of the same problem. Thus, the homogeneous sub-problem of the original problem (1.4) certainly has the solutions

$$u(x,t) = \sum_{n=1}^{N} b_n\, e^{-n^2 t} \sin nx, \tag{1.8}$$

where N is any positive integer and the b_n are arbitrary real numbers. The great question now is the following: among all these functions, can we find one that satisfies the *non-homogeneous* condition (I): $u(x,0) = f(x) =$ a known function?

Substitution in (1.8) gives the relation

$$f(x) = u(x,0) = \sum_{n=1}^{N} b_n \sin nx, \quad 0 < x < \pi. \tag{1.9}$$

If the function f happens to be a linear combination of sine functions of this kind, we can consider the problem as solved. Otherwise, it is rather natural to pose a couple of questions:

1. Can we permit the sum in (1.8) to consist of an *infinity* of terms?
2. Is it possible to approximate a (more or less) arbitrary function f using sums like the one in (1.9)?

The first of these questions can be given a partial answer using the theory of *uniform convergence.* The second question will be answered (in a rather positive way) later on in this book. We shall return to our heat conduction problem in Chapter 6.

Exercise

1.2 Find a solution of the problem treated in the text if the initial condition (I) is $u(x,0) = \sin 2x + 2 \sin 5x$.

Historical notes

The partial differential equations mentioned in this section evolved during the eighteenth century for the description of various physical phenomena. The Laplace operator occurs, as its name indicates, in the works of PIERRE SIMON DE LAPLACE, French astronomer and mathematician (1749–1827). In the theory of analytic functions, however, it had surely been known to EULER before it was given its name.

The wave equation was established in the middle of the eighteenth century and studied by several famous mathematicans, such as J. L. R. D'ALEMBERT (1717–83), LEONHARD EULER (1707–83) and DANIEL BERNOULLI (1700–82).

The heat equation came into focus at the beginning of the following century. The most important name in its early history is JOSEPH FOURIER (1768–1830). Much of the contents of this book has its origins in the treatise *Théorie analytique de la chaleur*. We shall return to Fourier in the historical notes to Chapter 4.

2

Preparations

2.1 Complex exponentials

Complex numbers are assumed to be familiar to the reader. The set of all complex numbers will be denoted by **C**. The reader has probably come across complex exponentials at some occasion previously, but, to be on the safe side, we include a short introduction to this subject here.

It was discovered by EULER during the eighteenth century that a close connection exists between the exponential function e^z and the trigonometric functions cos and sin. One way of seeing this is by considering the Maclaurin expansions of these functions. The exponential function can be described by

$$e^z = 1 + z + \frac{z^2}{2!} + \frac{z^3}{3!} + \frac{z^4}{4!} + \cdots = \sum_{n=0}^{\infty} \frac{z^n}{n!},$$

where the series is nicely convergent for all real values of z. Euler had the idea of letting z be a *complex* number in this formula. In particular, if z is purely imaginary, $z = iy$ with real y, the series can be rewritten as

$$\begin{aligned} e^{iy} &= 1 + iy + \frac{(iy)^2}{2!} + \frac{(iy)^3}{3!} + \frac{(iy)^4}{4!} + \cdots \\ &= 1 + iy - \frac{y^2}{2!} - i\frac{y^3}{3!} + \frac{y^4}{4!} + i\frac{y^5}{5!} - \cdots \\ &= \left(1 - \frac{y^2}{2!} + \frac{y^4}{4!} - \frac{y^6}{6!} + \cdots\right) + i\left(y - \frac{y^3}{3!} + \frac{y^5}{5!} - \frac{y^7}{7!} + \cdots\right). \end{aligned}$$

In the brackets we recognize the well-known expansions of cos and sin:

$$\cos y = 1 - \frac{y^2}{2!} + \frac{y^4}{4!} - \frac{y^6}{6!} + \cdots,$$
$$\sin y = y - \frac{y^3}{3!} + \frac{y^5}{5!} - \frac{y^7}{7!} + \cdots.$$

Accordingly, we define

$$e^{iy} = \cos y + i \sin y. \tag{2.1}$$

This is one of the so-called Eulerian formulae. The somewhat adventurous motivation through our manipulation of a series can be completely justified, which is best done in the context of *complex analysis.* For this book we shall be satisfied that the formula is true and can be used.

What is more, one can define exponentials with general complex arguments:

$$e^{x+iy} = e^x e^{iy} = e^x(\cos y + i \sin y) \quad \text{if } x \text{ and } y \text{ are real.}$$

The function thus obtained obeys most of the well-known rules for the real exponential function. Notably, we have these rules:

$$e^z e^w = e^{z+w}, \quad \frac{1}{e^z} = e^{-z}, \quad \frac{e^z}{e^w} = e^{z-w}.$$

It is also true that $e^z \neq 0$ for all z, but it need no longer be true that $e^z > 0$.

Example 2.1. $e^{i\pi} = \cos \pi + i \sin \pi = -1 + i \cdot 0 = -1$. Also, $e^{ni\pi} = (-1)^n$ if n is an integer (positive, negative, or zero). Furthermore, $e^{i\pi/2} = i$ is not even real. Indeed, the range of the function e^z for $z \in \mathbf{C}$ contains all complex numbers except 0. □

Example 2.2. The modulus of a complex number $z = x + iy$ is defined as $|z| = \sqrt{z\bar{z}} = \sqrt{x^2 + y^2}$. As a consequence,

$$|e^z| = |e^{x+iy}| = |e^x \cdot e^{iy}| = e^x |\cos y + i \sin y| = e^x \sqrt{\cos^2 y + \sin^2 y} = e^x.$$

In particular, if $z = iy$ is a purely imaginary number, then $|e^z| = |e^{iy}| = 1$. □

Example 2.3. Let us start from the formula $e^{ix} e^{iy} = e^{i(x+y)}$ and rewrite both sides of this, using (2.1). On the one hand we have

$$\begin{aligned} e^{ix} e^{iy} &= (\cos x + i \sin x)(\cos y + i \sin y) \\ &= \cos x \cos y - \sin x \sin y + i(\cos x \sin y + \sin x \cos y), \end{aligned}$$

and on the other hand,

$$e^{i(x+y)} = \cos(x + y) + i \sin(x + y).$$

If we identify the real and imaginary parts of the trigonometric expressions, we see that

$$\cos(x+y) = \cos x \cos y - \sin x \sin y, \quad \sin(x+y) = \cos x \sin y + \sin x \cos y.$$

Thus the addition theorems for cos and sin are contained in a well-known exponential law! □

By changing the sign of y in (2.1) and then manipulating the formulae obtained, we find the following set of equations:

$$e^{iy} = \cos y + i \sin y \qquad \cos y = \frac{e^{iy} + e^{-iy}}{2}$$
$$e^{-iy} = \cos y - i \sin y \qquad \sin y = \frac{e^{iy} - e^{-iy}}{2i}$$

These are the "complete" set of Euler's formulae. They show how one can pass back and forth between trigonometric expressions and exponentials.

Particularly in Chapters 4 and 7, but also in other chapters, we shall use the exponential expressions quite a lot. For this reason, the reader should become adept at using them by doing the exercises at the end of this section. If these things *are* quite new, the reader is also advised to find more exercises in textbooks where complex numbers are treated.

Exercises

2.1 Compute the numbers $e^{i\pi/2}$, $e^{-i\pi/4}$, $e^{5\pi i/6}$, $e^{\ln 2 - i\pi/6}$.

2.2 Prove that the function $f(z) = e^z$ has period $2\pi i$, i.e., that $f(z+2\pi i) = f(z)$ for all z.

2.3 Find a formula for $\cos 3t$, expressed in $\cos t$, by manipulating the identity $e^{3it} = \left(e^{it}\right)^3$.

2.4 Prove the formula $\sin^3 t = \frac{3}{4}\sin t - \frac{1}{4}\sin 3t$.

2.5 Show that if $|e^z| = 1$, then z is purely imaginary.

2.6 Prove the DE MOIVRE formula:

$$(\cos t + i \sin t)^n = \cos nt + i \sin nt, \quad n \text{ integer}.$$

2.2 Complex-valued functions of a real variable

In order to perform calculus on complex-valued functions, we should define limits of such objects. As long as the domain of definition lies on the real axis, this is quite simple and straightforward. One can use similar formulations as in the all-real case, but now modulus signs stand for moduli of complex numbers. For example: if we state that

$$\lim_{t\to\infty} f(t) = A,$$

then we are asserting the following: for every positive number ε, there exists a number R such that as soon as $t > R$ we are assured that $|f(t) - A| < \varepsilon$.

If we split $f(t)$ into real and imaginary parts,

$$f(t) = u(t) + iv(t), \quad u(t) \text{ and } v(t) \text{ real},$$

the following inequalities hold:

$$|u(t)| \leq |f(t)|, \quad |v(t)| \leq |f(t)|; \qquad |f(t)| \leq |u(t)| + |v(t)|. \tag{2.2}$$

This should make it rather clear that convergence in a complex-valued setting is equivalent to the simultaneous convergence of real and imaginary parts. Indeed, if the latter are both small, then the complex expression is small; and if the complex expression is small, then both its real and imaginary parts must be small. In practice this means that passing to a limit can be done in the real and imaginary parts, which reduces the complex-valued situation to the real-valued case.

Thus, if we want to define the derivative of a complex-valued function $f(t) = u(t) + iv(t)$, we can go about it in two ways. Either we define

$$f'(t) = \lim_{h \to 0} \frac{f(t+h) - f(t)}{h},$$

which stands for an ε-δ notion involving complex numbers, or we can just say that

$$f'(t) = u'(t) + iv'(t). \tag{2.3}$$

These definitions are indeed equivalent. The derivative of a complex-valued function of a real variable t exists if and only if the real and imaginary parts of f both have derivatives, and in this case we also have the formula (2.3). The following example shows the most frequent case of this, at least in this book.

Example 2.4. If $f(t) = e^{ct}$ with a complex coefficient $c = \alpha + i\beta$, we can find the derivative, according to (2.3), like this:

$$\begin{aligned} f'(t) &= \frac{d}{dt}\left(e^{\alpha t}(\cos \beta t + i \sin \beta t)\right) = \frac{d}{dt}\left(e^{\alpha t} \cos \beta t\right) + i\frac{d}{dt}\left(e^{\alpha t} \sin \beta t\right) \\ &= \alpha e^{\alpha t} \cos \beta t - e^{\alpha t} \beta \sin \beta t + i\left(\alpha e^{\alpha t} \sin \beta t + e^{\alpha t} \beta \cos \beta t\right) \\ &= e^{\alpha t}(\alpha + i\beta)(\cos \beta t + i \sin \beta t) = ce^{ct}. \end{aligned}$$

□

Similarly, integration can be defined by splitting into real and imaginary parts. If I is an interval, bounded or unbounded,

$$\int_I f(t)\, dt = \int_I (u(t) + iv(t))\, dt = \int_I u(t)\, dt + i \int_I v(t)\, dt.$$

If the interval is infinite, the convergence of the integral on the left is equivalent to the simultaneous convergence of the two integrals on the right.

A number of familiar rules of computation for differentiation and integration can easily be shown to hold also for complex-valued functions, with virtually unchanged proofs. This is true for, among others, the differentiation of products and quotients, and also for integration by parts. The chain rule for derivatives of composite functions also holds true for an expression such as $f(g(t))$, when g is real-valued but f may take complex values.

Absolute convergence of improper integrals follows the same pattern. From (2.2) it follows, by the comparison test for generalized integrals, that $\int f$ is absolutely convergent if and only if $\int u$ and $\int v$ are both absolutely convergent.

The fundamental theorem of calculus holds true also for integrals of complex-valued functions:

$$\frac{d}{dx}\int_a^x f(t)\,dt = f(x).$$

Example 2.5. Let c be a non-zero real number. To compute the integral of e^{ct} over an interval $[a, b]$, we can use the fact that e^{ct} is the derivative of a known function, by Example 2.4:

$$\int_a^b e^{ct}\,dt = \left[\frac{e^{ct}}{c}\right]_{t=a}^{t=b} = \frac{e^{cb} - e^{ca}}{c}.$$

□

When estimating the size of an integral the following relation is often useful:

$$\left|\int_a^b f(t)\,dt\right| \le \int_a^b |f(t)|\,dt.$$

Here the limits a and b can be finite or infinite. This is rather trivial if f is real-valued, so that the integral of f can be interpreted as the difference of two areas; but it actually holds also when f is complex-valued. A proof of this runs like this: The value of $\int_a^b f(t)\,dt$ is a complex number I, which can be written in polar form as $|I|e^{i\alpha}$ for some angle α. Then we can write as follows:

$$\begin{aligned}\left|\int_a^b f(t)\,dt\right| &= |I| = e^{-i\alpha}\int_a^b f(t)\,dt = \int_a^b e^{-i\alpha} f(t)\,dt = \operatorname{Re}\int_a^b e^{-i\alpha} f(t)\,dt \\ &= \int_a^b \operatorname{Re}\left\{e^{-i\alpha} f(t)\right\}dt \le \int_a^b \left|e^{-i\alpha} f(t)\right|dt = \int_a^b |f(t)|\,dt.\end{aligned}$$

Here we used that the left-hand member is real and thus equal to its own real part.

Exercises

2.7 Compute the derivative of $f(t) = e^{it^2}$ by separating into real and imaginary parts. Compare the result with that obtained by using the chain rule, as if everything were real.

2.8 Show that the chain rule holds for the expression $f(g(t))$, where g is real-valued and f is complex-valued, and t is a real variable.

2.9 Compute the integral

$$\int_{-\pi}^{\pi} e^{int}\,dt,$$

where n is an arbitrary integer (positive, negative, or zero).

2.3 Cesàro summation of series

We shall study a method that makes it possible to assign a sort of "sum value" to certain divergent series. For a convergent series, the new method yields the ordinary sum; but, as will be seen in Chapter 4, the method is really valuable when studying a series which may or may not be convergent.

Let a_k be terms (real or complex numbers), and define the partial sums s_n and *the arithmetic means* σ_n *of the partial sums* like this:

$$s_n = \sum_{k=1}^{n} a_k, \qquad \sigma_n = \frac{s_1 + s_2 + \cdots + s_n}{n} = \frac{1}{n}\sum_{k=1}^{n} s_k. \tag{2.4}$$

Lemma 2.1 *Suppose that the series* $\sum_{k=1}^{\infty} a_k$ *is convergent with the sum* s. *Then also*

$$\lim_{n\to\infty} \sigma_n = s.$$

Proof. Let $\varepsilon > 0$ be given. The assumption is that $s_n \to s$ as $n \to \infty$. This means that there exists an integer N such that $|s_n - s| < \varepsilon/2$ for all $n > N$. For these n we can write

$$|\sigma_n - s| = \left|\frac{s_1 + s_2 + \cdots + s_n - ns}{n}\right|$$

$$= \frac{1}{n}\big|(s_1 - s) + \cdots + (s_N - s) + (s_{N+1} - s) + \cdots + (s_n - s)\big|$$

$$\le \frac{1}{n}\sum_{k=1}^{N}|s_k - s| + \frac{1}{n}\sum_{k=N+1}^{n}|s_k - s| \le \frac{1}{n}\cdot C + \frac{1}{n}\cdot(n-N)\frac{\varepsilon}{2} \le \frac{C}{n} + \frac{\varepsilon}{2}.$$

Here, C is a non-negative constant (that does not depend on n), and so, if $n > 2C/\varepsilon$, the first term in the last member is also less than $\varepsilon/2$. Put $n_0 = \max(N, 2C/\varepsilon)$. For all $n > n_0$ we have then $|\sigma_n - s| < \varepsilon$, which is the assertion. □

Definition 2.1 *Let s_n and σ_n be defined as in* (2.4). *We say that the series $\sum_{k=1}^{\infty} a_k$ is* summable according to Cesàro *or* Cesàro summable *or* summable $(C, 1)$ *to the value, or "sum", s, if $\lim_{n\to\infty} \sigma_n = s$.*

We write

$$\sum_{k=1}^{\infty} a_k = s \quad (C, 1).$$

The lemma above states that if a series is convergent in the usual sense, then it is also summable $(C, 1)$, and the Cesàro sum coincides with the ordinary sum.

Example 2.6. Let $a_k = (-1)^{k-1}$, $k = 1, 2, 3, \ldots$, which means that we have the series $1 - 1 + 1 - 1 + 1 - 1 + \cdots$. Then $s_n = 0$ if n is even and $s_n = 1$ if n is odd. The means σ_n are

$$\sigma_n = \frac{1}{2} \quad \text{if } n \text{ is even}, \qquad \sigma_n = \frac{\frac{1}{2}(n+1)}{n} = \frac{n+1}{2n} \quad \text{if } n \text{ is odd}.$$

Thus we have $\sigma_n \to \frac{1}{2}$ as $n \to \infty$. This divergent series is indeed summable $(C, 1)$ with sum $\frac{1}{2}$. □

The reason for the notation $(C, 1)$ is that it is possible to iterate the process. If the σ_n do not converge, we can form the means $\tau_n = (\sigma_1 + \cdots + \sigma_n)/n$. If the τ_n converge to a number s one says that the original series is $(C, 2)$-summable to s, and so on.

These methods can be efficient if the terms in the series have different signs or are complex numbers. A *positive* divergent series cannot be summed to anything but $+\infty$, no matter how many means you try.

Exercises

2.10 Study the series $1 + 0 - 1 + 1 + 0 - 1 + 1 + 0 - \cdots$, i.e., the series $\sum_{k=1}^{\infty} a_k$, where $a_{3k+1} = 1$, $a_{3k+2} = 0$ and $a_{3k+3} = -1$. Compute the Cesàro means σ_n and show that the series has the Cesàro sum $\frac{2}{3}$.

2.11 The results of Example 2.6 and the previous exercise can be generalized as follows. Assume that the sequence of partial sums s_n is periodic, i.e., that there is a positive integer p such that $s_{n+p} = s_n$ for all n. Then the series is summable $(C, 1)$ to the sum $\sigma = (s_1 + s_2 + \cdots + s_p)/p$. Prove this!

2.12 Show that if $\sum a_k$ has a finite $(C, 1)$ value, then

$$\lim_{n\to\infty} \frac{s_n}{n} = 0.$$

What can be said about $\lim_{k\to\infty} a_k/k$?

2.13 Prove that if $a_k \geq 0$ and $\sum a_k$ is $(C, 1)$-summable, then the series is convergent in the usual sense. (Assume the contrary – what does that entail for a positive series?)

2.14 Show that the series $\sum_{k=1}^{\infty}(-1)^k\, k$ is not summable $(C,1)$. Also show that it *is* summable $(C,2)$. Show that the $(C,2)$ sum is equal to $-\frac{1}{4}$.

2.15 Show that, if $x \neq n\cdot 2\pi$ $(n \in \mathbf{Z})$,

$$\tfrac{1}{2} + \sum_{k=1}^{\infty} \cos kx = 0 \qquad (C,1).$$

2.16 Prove that

$$\sum_{n=0}^{\infty} z^n = \frac{1}{1-z} \quad (C,1) \qquad \text{for } |z| \leq 1,\ z \neq 1.$$

2.4 Positive summation kernels

In this section we prove a theorem that is useful in many situations for recovering the values of a function from various kinds of transforms. The main idea is summarized in the following formulation.

Theorem 2.1 *Let $I = (-a,a)$ be an interval (finite or infinite). Suppose that $\{K_n\}_{n=1}^{\infty}$ is a sequence of real-valued, Riemann-integrable functions defined on I, with the following properties:*

$$(1)\quad K_n(s) \geq 0.$$

$$(2)\quad \int_{-a}^{a} K_n(s)\,ds = 1.$$

$$(3)\quad \text{If } \delta > 0, \text{ then } \lim_{n\to\infty} \int_{\delta<|s|<a} K_n(s)\,ds = 0.$$

If $f : I \to \mathbf{C}$ is integrable and bounded on I and continuous for $s = 0$, we then have

$$\lim_{n\to\infty} \int_{-a}^{a} K_n(s)\,f(s)\,ds = f(0).$$

Proof. Let ε be a positive number. Since f is continous at the origin there exists a number $\delta > 0$ such that

$$|s| \leq \delta \quad \Rightarrow \quad |f(s) - f(0)| < \varepsilon.$$

Furthermore, f is bounded on I, i.e., there exists a number M such that $|f(s)| \leq M$ for all s. Because of the property 2 we have

$$\Delta := \int_{-a}^{a} K_n(s)\,f(s)\,ds - f(0) = \int_{-a}^{a} K_n(s)\,f(s)\,ds - f(0)\int_{-a}^{a} K_n(s)\,ds$$
$$= \int_{-a}^{a} K_n(s)\big(f(s) - f(0)\big)ds.$$

We want to prove that $\Delta \to 0$ as $n \to \infty$. Let us estimate the absolute value of Δ, assuming that $|s| \leq \delta$:

$$|\Delta| = \left| \int_{-a}^{a} K_n(s)\big(f(s) - f(0)\big)ds \right| \leq \int_{-a}^{a} K_n(s)\,|f(s) - f(0)|\,ds$$

$$= \int_{-\delta}^{\delta} K_n(s)\,|f(s) - f(0)|\,ds + \int_{\delta<|s|<a} K_n(s)\,|f(s) - f(0)|\,ds$$

$$\leq \varepsilon \int_{-\delta}^{\delta} K_n(s)\,ds + \int_{\delta<|s|<a} K_n(s)\,2M\,ds \leq \varepsilon + 2M \int_{\delta<|s|<a} K_n(s)\,ds.$$

The last integral tends to zero, by the assumptions, and so the second term of the last member is also less than ε if n is large enough. This means that for large n we have $|\Delta| < 2\varepsilon$, which proves the theorem. □

A sequence $\{K_n\}_{n=1}^{\infty}$ having the properties 1–3 is called a *positive summation kernel*. We illustrate with a few simple examples.

Example 2.7. Define $K_n : \mathbf{R} \to \mathbf{R}$ by

$$K_n(s) = \begin{cases} n, & |s| < 1/(2n), \\ 0, & |s| > 1/(2n) \end{cases}$$

(see Figure 2.1a). It is obvious that the conditions 1–3 are fullfilled. See also Exercise 2.17. □

Example 2.8. Let $\varphi(s) = e^{-s^2/2}/\sqrt{2\pi}$, the density function of the normal probability distribution (Figure 2.1b). Define $K_n(s) = n\varphi(ns)$. Then $\{K_n\}$ is a positive summation kernel on $\mathbf{R}$ (check it!). □

Example 2.9. The preceding example can be generalized in the following way: Let $\psi : \mathbf{R} \to \mathbf{R}$ be some function satisfying $\psi(s) \geq 0$ and $\int_{\mathbf{R}} \psi(s)\,ds = 1$. Putting $K_n(s) = n\psi(ns)$, we have a positive summation kernel. □

The examples should help the reader to understand what is going on: a positive summation kernel creates a weighted mean value of the function f, with the weight being successively concentrated towards the point $s = 0$. If f is continuous at that point, the limit will yield precisely the value of f at $s = 0$.

A corollary of Theorem 2.1 is the following, where we move the concentration of mass to some other point than the origin:

Corollary 2.1 *If $\{K_n\}_{n=1}^{\infty}$ is a positive summation kernel on the interval I, s_0 is an interior point of I, and f is continuous at $s = s_0$, then*

$$\lim_{n\to\infty} \int_I K_n(s)\,f(s_0 - s)\,ds = f(s_0).$$

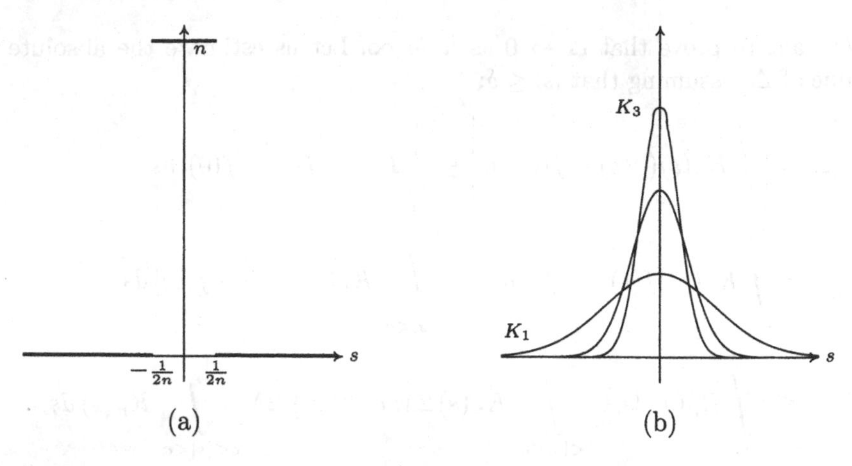

FIGURE 2.1.

The proof is left as an exercise (do the change of variable $s_0 - s = u$).

Remark. The choice of the interval I is often rather unimportant. It is also easy to see that the condition 2 can be weakened, e.g., it suffices that the integrals of K_n over the interval *tend to* 1 as $n \to \infty$. In consequence, kernels on all of $\mathbf{R}$ can also be used on any subinterval $\mathbf{R}$ having the origin in its interior. □

Remark. The reader who is familiar with the notion of *uniform continuity*, can appreciate a sharper formulation of the corollary: if f is continuous on a compact interval K, the functions

$$F_n(t) = \int_I K_n(s) f(t-s)\, ds$$

will converge to $f(t)$ *uniformly on* K. The proof is practically unchanged, with the only addition that the number δ occuring in the proof of Theorem 2.1 can be chosen so that it can be used simultaneously for all values of t that are involved. □

Exercises

2.17 Prove directly, without using the theorem, that if K_n is as in Example 2.7 and f is continuous at the origin, then $\lim_{n\to\infty} \int_{\mathbf{R}} K_n(s) f(s)\, ds = f(0)$.

2.18 Prove that the "roof functions" g_n, defined by $g_n(t) = n - n^2 t$ for $0 \le t \le 1/n$, $g_n(t) = 0$ for $t > 1/n$ and $g_n(-t) = g_n(t)$, make up a positive summation kernel. Draw pictures!

2.19 (a) Show that $K_n(t) = \frac{1}{2} n e^{-n|t|}$ describes a positive summation kernel.
(b) Suppose that f is bounded and piecewise continuous on $\mathbf{R}$, and $\lim_{t \nearrow 0} f(t) = 1$, $\lim_{t \searrow 0} f(t) = 3$. Show that

$$\lim_{n\to\infty} \frac{n}{2} \int_{\mathbf{R}} e^{-n|t|} f(t)\, dt = 2.$$

2.20 Show that if f is bounded on $\mathbf{R}$ and has a derivative f' that is also bounded on $\mathbf{R}$ and continuous at the origin, then

$$\lim_{n\to\infty} \frac{n^3}{\sqrt{2\pi}} \int_{\mathbf{R}} s\, e^{-n^2 s^2/2} f(s)\, ds = f'(0).$$

2.21 Let φ be defined by $\varphi(x) = \frac{15}{16}(x^2-1)^2$ for $|x| < 1$ and $\varphi(x) = 0$ otherwise. Let f be a function with a continuous derivative. Find the limit

$$\lim_{n\to\infty} \int_{-1}^{1} n^2 \varphi'(nx)\, f(x)\, dx.$$

2.5 The Riemann–Lebesgue lemma

The following theorem plays a central role in Fourier Analysis. It takes its name from the fact that it holds even for functions that are integrable according to the definition of Lebesgue. We prove it for functions that are absolutely integrable in the Riemann sense. First, let us very briefly recall what this means.

A bounded function f on a finite interval $[a, b]$ is integrable if it can be approximated by Riemann sums from above and below in such a way that the difference of the integrals of these sums can be made as small as we wish. This definition is then extended to unbounded functions and infinite intervals by taking limits; these cases are often called improper integrals. If I is any interval and f is a function on I such that the (possibly improper) integral

$$\int_I |f(u)|\, du$$

has a finite value, then f is said to be absolutely integrable on I.

Theorem 2.2 (Riemann–Lebesgue lemma) *Let f be absolutely integrable in the Riemann sense on a finite or infinite interval I. Then*

$$\lim_{\lambda\to\infty} \int_I f(u) \sin \lambda u\, du = 0.$$

Proof. We do it in four steps. First, assume that the interval is compact, $I = [a, b]$, and that f is constant and equal to 1 on the entire interval. Then

$$\int_a^b f(u) \sin \lambda u\, du = \int_a^b \sin \lambda u\, du = \left[-\frac{\cos \lambda u}{\lambda}\right]_{u=a}^{u=b} = \frac{1}{\lambda}(\cos \lambda a - \cos \lambda b),$$

which gives

$$\left| \int_a^b f(u) \sin \lambda u\, du \right| \le \frac{2}{\lambda} \longrightarrow 0 \quad \text{as } \lambda \to \infty.$$

The assertion is thus true for this f.

Now assume that f is *piecewise constant*, which means that I (still assumed to be compact) is subdivided into a finite number of subintervals $I_k = (a_{k-1}, a_k)$, $k = 1, 2, \ldots, N$ ($a_0 = a$, $a_N = b$), and that $f(u)$ has a certain constant value c_k for $u \in I_k$. This means that we can write

$$f(u) = \sum_{k=1}^{N} c_k \, g_k(u),$$

where $g_k(u) = 1$ on I_k and $g_k(u) = 0$ outside of I_k. We get

$$\int_a^b f(u) \sin \lambda u \, du = \sum_{k=1}^{N} \int_a^b c_k \, g_k(u) \sin \lambda u \, du = \sum_{k=1}^{N} c_k \int_{a_{k-1}}^{a_k} \sin \lambda u \, du.$$

This is a sum of finitely many terms, and by the preceding case each of these terms tends to zero as $\lambda \to \infty$. Thus the assertion is true also for this f.

Let now f be an arbitrary function that is Riemann integrable on $I = [a, b]$. Let ε be an arbitrary positive number. By the definition of the Riemann integral, there exists a piecewise constant function g such that

$$\int_a^b |f(u) - g(u)| \, du < \frac{\varepsilon}{2}.$$

(Let g be a function whose integral is a Riemann sum of f.) Then,

$$\begin{aligned}
\left| \int_a^b f(u) \sin \lambda u \, du \right| &= \left| \int_a^b (f(u) - g(u)) \sin \lambda u \, du + \int_a^b g(u) \sin \lambda u \, du \right| \\
&\leq \int_a^b |f(u) - g(u)| |\sin \lambda u| \, du + \left| \int_a^b g(u) \sin \lambda u \, du \right| \\
&\leq \int_a^b |f(u) - g(u)| \, du + \left| \int_a^b g(u) \sin \lambda u \, du \right|.
\end{aligned}$$

The last integral tends to zero as $\lambda \to \infty$, by the preceding case. Thus there is a value λ_0 such that this integral is less that $\varepsilon/2$ for all $\lambda > \lambda_0$. For these λ, the left-hand member is thus less than ε, which proves the assertion.

Finally, we no longer require that I is compact. Let $\varepsilon > 0$ be prescribed. Since f is absolutely integrable, there is a compact subinterval $J \subset I$ such that $\int_{I \setminus J} |f(u)| \, du < \varepsilon$. We can write

$$\left| \int_I f(u) \sin \lambda u \, du \right| \leq \left| \int_J f(u) \sin \lambda u \, du \right| + \int_{I \setminus J} |f(u)| \, du,$$

where the first term tends to zero by the preceding case, and thus it is less than ε if λ is large enough; the second term is always less than ε. This completes the proof. □

The intuitive content of the theorem is not hard to understand: For large values of $|\lambda|$, the integrated function $f(u)\sin\lambda u$ is an amplitude-modulated sine function with a high frequency; its mean value over a fixed interval should reasonably approach zero as the frequency increases. Of course, the factor $\sin\lambda u$ in the integral can be replaced by $\cos\lambda u$ or the complex function $e^{i\lambda u}$, with the same result. And, of course, we can just as well let λ tend to $-\infty$.

2.6 *Some simple distributions

In this section, we introduce, in an informal way, a sort of generalization of the notion of a function. (A more coherent and systematic way of defining these objects is given in Chapter 8.) As a motivation for this generalization, we begin with a few "examples."

Example 2.10. In Sec. 1.3 (on the wave equation) we saw difficulties in the usual requirement that solutions of a differential equation of order n shall actually have (maybe even continuous) derivatives of order n. Quite natural solutions are disqualified for reasons that seem more of a "bureaucratic" nature than physically motivated. This indicates that it would be a good thing to widen the notion of differentiability in one way or another. □

Example 2.11. Ever since the days of NEWTON, physicists have been dealing with situations where some physical entity assumes a very large magnitude during a very short period of time; often this is idealized so that the value is infinite at one point in time. A simple example is an elastic collision of two bodies, where the forces are thought of as infinite at the moment of impact. Nevertheless, a *finite* and well-defined amount of impulse is transferred in the collision. How is this to be treated mathematically? □

Example 2.12. A situation that is mathematically analogous to the previous one is found in the theory of electricity. An electron is considered (at least in classical quantum theory) to be a *point charge*. This means that there is a certain finite amount of electric charge localized at one point in space. The charge density is infinite at this point, but the charge itself has an exact, finite value. What mathematical object describes this? □

Example 2.13. In Sec. 2.4 we studied positive summation kernels. These consisted of sequences of nonnegative functions with integral equal to 1, that concentrate toward a fixed point as a parameter, say, N, tends to infinity, for example. Can we invent a mathematical object that can be interpreted as the limit of such a sequence? □

The problems in Examples 2.11 and 2.12 above have been addressed by many physicists ever since the later years of the nineteenth century by using the following trick. Let us assume that the independent variable is t. Introduce a "function" $\delta(t)$ with the following properties:

$$\begin{aligned}
&(1) \quad \delta(t) \geq 0 \text{ for } -\infty < t < \infty, \\
&(2) \quad \delta(t) = 0 \text{ for } t \neq 0, \\
&(3) \quad \int_{-\infty}^{\infty} \delta(t)\, dt = 1.
\end{aligned}$$

Regrettably, there is no ordinary real-(or complex)-valued function that satisfies these conditions. Condition 2 irrevocably implies that the integral in condition 3 must be zero. Nevertheless, using formal calculations involving the "function" δ, it was possible to arrive at results that were both physically meaningful and "correct." A name that is commonly associated with this is P. DIRAC, but he was not the only person (nor even the first one) to reason in this way. He has, however, given his name to the object δ: it is often called the *Dirac delta function* (or the Dirac measure, or the Dirac distribution).

One way of making legitimate the formal δ calculus is to follow an idea that is indicated in Example 2.13. If δ occurs in a formula, it is at first replaced by a positive summation kernel K_N; upon this we then do our calculations, and finally we pass to the limit. In a certain sense (which will be made precise in Chapter 8), it is true that $\delta = \lim_{N\to\infty} K_N$.

In this section, and in certain star-marked sections in the following chapters, we shall accept the delta function and some of its relatives in an intuitive way. Thus, $\delta(t)$ stands for an object that acts on a continuous function φ according to the formula

$$\int \delta(t)\varphi(t)\, dt = \varphi(0),$$

where the integral is taken over some interval that contains the origin in its interior. We also introduce the *translates* δ_a, which can be described by either $\delta_a(t) = \delta(t-a)$ or $\int \delta_a(t)\varphi(t)\, dt = \varphi(a)$. It is essential that the point a, where the "pulse" appears, is located in the interior of the interval of integration. If the point coincides with an endpoint of the interval, the integral is not considered to be well-defined.

Example 2.14. The LAPLACE *transform* of a function f is defined to be another function $\widetilde{f}$, given by

$$\widetilde{f}(s) = \int_0^{\infty} f(t)e^{-st}\, dt$$

for all s such that the integral is convergent (see Chapter 3). The Laplace transform of δ cannot be defined in this way. We can, however, modify the definition so as to include the origin. It is indeed customary to write

$$\widetilde{f}(s) = \int_{0-}^{\infty} f(t)e^{-st}\,dt = \lim_{k\nearrow 0} \int_{k}^{\infty} f(t)e^{-st}\,dt.$$

With this definition one finds that $\widetilde{\delta}(s) = 1$ for all s. Similarly, $\widetilde{\delta}_a(s) = e^{-as}$, if $a > 0$. □

The HEAVISIDE function, or *unit step* function, H is defined by

$$H(t) = \begin{cases} 0 \text{ for } t < 0, \\ 1 \text{ for } t > 0. \end{cases}$$

The value of $H(0)$ is mostly left undefined, because it is normally of no importance. The Heaviside function is useful in many contexts. One of these is when we are dealing with functions that are given by different formulae in different intervals.

If $a < b$, the expression $H(t-a) - H(t-b)$ is equal to 1 for $a < t < b$ and equal to 0 outside the interval $[a, b]$. It might be called a "window" that lights up the interval (a, b) (we do not in these situations care much about whether an interval is open or closed). For unbounded intervals we can also find "windows": the function $H(t-a)$ lights up the interval (a, ∞), and the expression $1 - H(t-b)$ the interval $(-\infty, b)$.

Example 2.15. Consider the function $f : \mathbf{R} \to \mathbf{R}$ that is given by

$$f(t) = \begin{cases} 1 - t^2 \text{ for } t < -2, \\ t + 2 \text{ for } -2 < t < 1, \\ 1 - t \text{ for } t > 1. \end{cases}$$

This can now be compressed into one formula:

$$\begin{aligned} &f(t) \\ &= (1-t^2)(1 - H(t+2)) + (t+2)(H(t+2) - H(t-1)) + (1-t)H(t-1) \\ &= (1-t^2) + (-1 + t^2 + t + 2)H(t+2) + (-t - 2 + 1 - t)H(t-1) \\ &= 1 - t^2 + (t^2 + t + 1)H(t+2) - (2t+1)H(t-1). \end{aligned}$$

□

Heaviside's function is connected with the δ function via the formula

$$H(t) = \int_{-\infty}^{t} \delta(u)\,du.$$

A very bold differentiation of this formula would give the result

$$H'(t) = \delta(t). \tag{2.5}$$

Since H is constant on the intervals $]-\infty, 0[$ and $]0, \infty[$, and $\delta(t)$ is considered to be zero on these intervals, the formula (2.5) is reasonable for $t \neq 0$. What is new is that the "derivative" of the jump discontinuity of H should be considered to be the "pulse" of δ. In fact, this assertion can be given a completely coherent background; this will be done in Chapter 8.

If φ is a function in the class C^1, i.e., it has a continuous derivative, and if in addition φ is zero outside some finite interval, the following calculation is clear:

$$\int_{-\infty}^{\infty} \varphi'(t)H(t)\,dt = \int_{0}^{\infty} \varphi'(t)\,dt = \big[\varphi(t)\big]_{t=0}^{\infty} = 0 - \varphi(0) = -\varphi(0).$$

The same result can also be obtained by the following formal integration by parts:

$$\begin{aligned}\int_{-\infty}^{\infty} \varphi'(t)H(t)\,dt &= \Big[\varphi(t)H(t)\Big]_{-\infty}^{\infty} - \int_{-\infty}^{\infty} \varphi(t)H'(t)\,dt \\ &= (0-0) - \int_{-\infty}^{\infty} \varphi(t)\delta(t)\,dt = -\varphi(0).\end{aligned}$$

This is characteristic of the way in which these generalized functions can be treated: if they occur in an integral together with an "ordinary" function of sufficient regularity, this integral can be treated formally, and the results will be true facts.

One can go further and introduce derivatives of the δ functions. What would be, for example, the first derivative of $\delta = \delta_0$? One way of finding out is by operating formally as in the preceding situation. Let φ be a function in C^1, and let it be understood that all integrals are taken over an interval that contains 0 in its interior. Since $\delta(t) = 0$ if $t \neq 0$, it is reasonable that also $\delta'(t) = 0$ for $t \neq 0$. Integration by parts gives

$$\int_a^b \delta'(t)\varphi(t)\,dt = \Big[\delta(t)\varphi(t)\Big]_a^b - \int_a^b \delta(t)\varphi'(t)\,dt = (0-0) - \varphi'(0) = -\varphi'(0).$$

If δ itself serves to pick out the value of a function at the origin, the derivative of δ can thus be used to find the value at the same place of the derivative of a function.

Another way of seeing δ' is to consider δ to be the limit of a differentiable positive summation kernel, and taking the derivative of the kernel. An example is actually given in Exercise 2.20. As in Example 2.8 on page 23, we study the summation kernel

$$K_n(t) = \frac{n}{\sqrt{2\pi}}\, e^{-n^2t^2/2},$$

(which consists in rescaling the normal probability density function). The derivatives are

$$K_n{}'(t) = -\frac{n^3 t}{\sqrt{2\pi}}\, e^{-n^2t^2/2}.$$

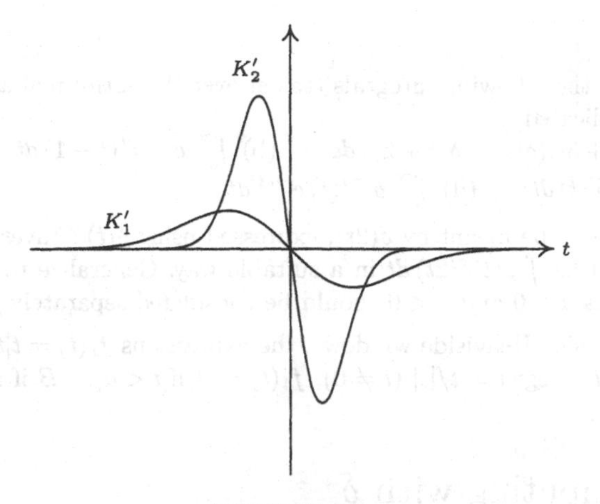

FIGURE 2.2.

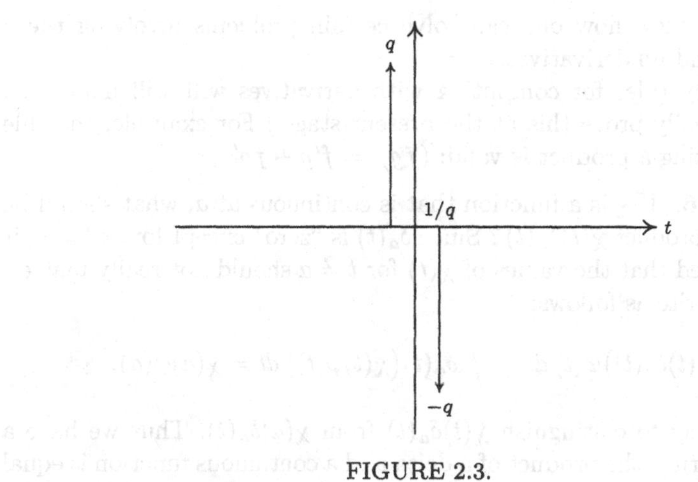

FIGURE 2.3.

These are illustrated in Figure 2.2. The fact that they approach $-\delta'(t)$ is proved by integration by parts (which is what Exercise 2.20 is all about).

In the theory of electricity, there occurs a phenomenon known as an *electric dipole*. This consists of two equal but opposite charges $\pm q$ at a small distance from each other (see Figure 2.3). If the distance is made smaller and charges increase in proportion to the inverse of the distance, the "limit object" is an idealized dipole. A mathematical model of this object consists of δ', just as a a point charge can be represented by δ.

Higher derivatives of δ can also be defined. Using integration by parts one finds that the nth derivative $\delta^{(n)}$ should act according to the formula

$$\int \delta^{(n)}(t)\varphi(t)\,dt = (-1)^n \,\varphi^{(n)}(0),$$

provided the function φ has an nth derivative that is continuous at the origin.

Exercises

2.22 Compute the following integrals (taken over the entire real axis if nothing else is indicated):
(a) $\int (t^2+3t)(\delta(t)-\delta(t+2))\,dt$ (b) $\int_0^\infty e^{-st}\delta'(t-1)\,dt$
(c) $\int e^{2t}\delta'(t)\,dt$ (d) $\int_{0-}^\infty \delta^{(n)}(t)\,e^{-st}\,dt$

2.23 What should be meant by $\delta(2t)$, expressed using $\delta(t)$? Investigate this by manipulating $\int \varphi(t)\delta(2t)\,dt$ in a suitable way. Generalize to $\delta(at)$, $a \neq 0$. (The cases $a > 0$ and $a < 0$ should be considered separately.)

2.24 Rewrite, using Heaviside windows, the expressions $f_1(t) = t|t+1|$, $f_2(t) = e^{-|t|}$, $f_3(t) = \operatorname{sgn} t = t/|t|$ $(t \neq 0)$, $f_4(t) = A$ if $t < a$, $= B$ if $t > a$.

2.7 *Computing with δ

We shall now show how one can solve certain problems involving the δ distribution and its derivatives.

The ordinary rules for computing with derivatives will still hold true. (We cannot really prove this at the present stage.) For example, the rule for differentiating a product is valid: $(fg)' = f'g + fg'$.

Example 2.16. If χ is a function that is continuous at a, what should be meant by the product $\chi(t)\delta_a(t)$? Since $\delta_a(t)$ is "zero" except for at $t = a$, it can be expected that the values of $\chi(t)$ for $t \neq a$ should not really matter. And we can write as follows:

$$\int \bigl(\chi(t)\delta_a(t)\bigr)\varphi(t)\,dt = \int \delta_a(t)\bigl(\chi(t)\varphi(t)\bigr)\,dt = \chi(a)\varphi(a).$$

There is no way to distinguish $\chi(t)\delta_a(t)$ from $\chi(a)\delta_a(t)$. Thus we have a simplification rule: the product of a delta and a continuous function is equal to a scalar multiple of the delta, with coefficient equal to the value of the function at the point where the pulse sits:

$$\chi(t)\delta_a(t) = \chi(a)\delta_a(t). \tag{2.6}$$

If we encounter derivatives of δ, the matter is more complicated. What happens is this: start from (2.6) and differentiate:

$$\chi'(t)\delta_a(t) + \chi(t)\delta_a'(t) = \chi(a)\delta_a'(t).$$

In the first term we can replace $\chi'(t)$ by $\chi'(a)$ and then move this term to the other side. We get

$$\chi(t)\delta_a'(t) = \chi(a)\delta_a'(t) - \chi'(a)\delta_a(t).$$

(On second thought, it should not be surprising that the product of a function and a δ' somehow takes into account the value of the derivative of the function as well.)

What happens when the second derivative is multiplied by a function is left to the reader to find out (in Exercise 2.25). □

Example 2.17. Find the first two derivatives of $f(t) = |t|$.
Solution. Rewrite the function without modulus signs, using Heaviside windows:

$$f(t) = |t| = -t(1 - H(t)) + tH(t) = 2tH(t) - t.$$

Differentiation then gives

$$f'(t) = 2H(t) + 2t\delta(t) - 1 = 2H(t) - 1.$$

In the last step we used the formula (2.6). In plain language, the derivative of $|t|$ is plus one for positive t and minus one for negative t, just as we know from elementary calculus; at the origin, the value of the derivative is undecided. We proceed to the second derivative:

$$f''(t) = 2\delta(t) - 0 = 2\delta(t).$$

This formula reflects the fact that f' has derivative zero everywhere outside the origin; whereas at the origin, the delta term indicates that f' has a positive jump of two units. This is characteristic of the derivative of a function with jumps. □

Example 2.18. Another example of the same type, though more complicated. The function $f(x) = |x^2 - 1|$ can be rewritten as

$$\begin{aligned} f(x) &= (x^2 - 1)H(x - 1) + (1 - x^2)(H(x + 1) - H(x - 1)) \\ &\qquad + (x^2 - 1)(1 - H(x + 1)) \\ &= (x^2 - 1)\big(2H(x - 1) - 2H(x + 1) + 1\big). \end{aligned}$$

This formula can be differentiated, using the rule for differentiating a product:

$$\begin{aligned} f'(x) &= 2x\big(2H(x - 1) - 2H(x + 1) + 1\big) + (x^2 - 1)\big(2\delta(x - 1) - 2\delta(x + 1)\big) \\ &= 2x\big(2H(x - 1) - 2H(x + 1) + 1\big). \end{aligned}$$

In the last step, we used (2.6). One more differentiation gives

$$\begin{aligned} f''(x) &= 2\big(2H(x - 1) - 2H(x + 1) + 1\big) + 2x\big(2\delta(x - 1) - 2\delta(x + 1)\big) \\ &= 2\big(2H(x - 1) - 2H(x + 1) + 1\big) + 4\delta(x - 1) + 4\delta(x + 1). \end{aligned}$$

The first term contains the classical second derivative of $|x^2 - 1|$, which exists for $x \neq \pm 1$; the two δ terms demonstrate that f' has upward jumps of size 4 for $x = \pm 1$. The reader should draw pictures of f, f', and f''. □

In the two last examples, the first derivative at the "corners" of f is considered to be *undecided*. The classical point of view is to say that f does not have a derivative at such a point; when working with distributions, the derivative is thought of as more of a global notion, that always exists, but may lack a precise value at certain points.

Example 2.19. Solve the differential equation $y' + 2y = \delta(t-1)$ for $t > 0$ with the initial value $y(0) = 1$.

Solution. The method of integrating factor can be used. An integrating factor is e^{2t}:

$$e^{2t}y' + 2e^{2t}y = e^{2t}\delta(t-1) \quad \Longleftrightarrow \quad \frac{d}{dt}(e^{2t}y) = e^2\delta(t-1).$$

In rewriting the right-hand side we used (2.6). Now we can integrate:

$$e^{2t}y = e^2 H(t-1) + C,$$

where C is a constant. To satisfy the initial condition, we must take $C = 1$. Thus the solution is

$$y = e^{2-2t}H(t-1) + e^{-2t}, \quad t > 0.$$

(The reader is recommended to check the solution by differentiation and substitution into the original equation.) □

Example 2.20. Find all solutions of the differential equation $y'' + 4y = \delta$.

Solution. The classical method for this sort of problem amounts to first finding the general solution of the corresponding homogeneous equation, which is $y_H = C_1 \cos 2t + C_2 \sin 2t$, where C_1 and C_2 are arbitrary constants. Then we should find some particular solution of the inhomogeneous equation. What kind of expression y_P could possibly, after differentiation and substitution into the left-hand side of the equation, yield the result δ? Apparently, something drastic happens at $t = 0$. Since $\delta(t) = 0$ for $t < 0$, the equation can be said to be homogeneous during this period of time. Let us then guess that there is a particular solution of the form $y_P(t) = u(t)H(t)$, where $u(t)$ is to be determined. Differentiation gives

$$\begin{aligned} y_P'(t) &= u'(t)H(t) + u(t)H'(t) = u'(t)H(t) + u(0)\delta(t), \\ y_P''(t) &= u''(t)H(t) + u'(t)H'(t) + u(0)\delta'(t) \\ &= u''(t)H(t) + u'(0)\delta(t) + u(0)\delta'(t). \end{aligned}$$

Substitution into the equation gives

$$\big(u''(t)H(t) + u'(0)\delta(t) + u(0)\delta'(t)\big) + 4u(t)H(t) = \delta(t)$$

or

$$\big(u''(t) + 4u(t)\big)H(t) + u'(0)\delta(t) + u(0)\delta'(t) = \delta(t).$$

The function u should be chosen so that $u'' + 4u = 0$, $u'(0) = 1$ and $u(0) = 0$. This means that $u(t) = a\cos 2t + b\sin 2t$, where $0 = u(0) = a$ and $1 = u'(0) = 2b$. Thus, $u = \frac{1}{2}\sin 2t$, and $y_P = \frac{1}{2}\sin 2t\, H(t)$. The solutions of the problem are thus

$$y = C_1\cos 2t + (C_2 + \tfrac{1}{2}H(t))\sin 2t.$$

Again, the reader is recommended to check the solution. □

Example 2.21. In Sec. 1.3, on the wave equation, the final example turned out to have a solution that was not really a differentiable function. Now we can put this right, by allowing the generalized derivatives introduced in this section. The solution involved the function φ, defined by

$$\varphi(s) = \tfrac{1}{2}s - \tfrac{1}{2}C \text{ for } s > 0, \quad \varphi(s) = -\tfrac{1}{2}s - \tfrac{1}{2}C \text{ for } s < 0.$$

We can rewrite this definition, using Heaviside windows:

$$\varphi(s) = (-\tfrac{1}{2}s - \tfrac{1}{2}C)(1 - H(s)) + (\tfrac{1}{2}s - \tfrac{1}{2}C)H(s) = -\tfrac{1}{2}s - \tfrac{1}{2}C + sH(s).$$

The two first derivatives are

$$\varphi'(s) = -\tfrac{1}{2} + H(s) + s\delta(s) = -\tfrac{1}{2} + H(s), \quad \varphi''(s) = \delta(s).$$

The complete solution of the problem in Sec. 1.3 can be written

$$u(x,t) = \varphi(x-t) + \psi(x+t) = \varphi(x-t) + \tfrac{3}{2}(x+t) + \tfrac{1}{2}C.$$

Differentiating, and trusting that the chain rule holds as usual (which it does, as will be proved in Chapter 8), we find

$$u_x = \varphi'(x-t) + \tfrac{3}{2} = 1 + H(x-t), \quad u_{xx} = \delta(x-t)$$

and

$$u_t = -\varphi'(x-t) + \tfrac{3}{2} = 2 - H(x-t), \quad u_{tt} = \delta(x-t).$$

Thus, $u_{xx} = u_{tt}$ as distributions, and u can be considered as a worthy solution of the wave equation. □

Exercises

2.25 Find a simpler expression for $\chi(t)\delta_a''(t)$, where χ is a C^2 function.

2.26 Determine the derivatives of order ≤ 2 of the functions $f(t) = e^{-|t|}$, $g(t) = |t|e^{-|t|}$ and $h(t) = |\sin t|$. Draw pictures!

2.27 Let $f : \mathbf{R} \to \mathbf{R}$ be given by $f(x) = 1 - x^2$ if $-1 < x < 1$ and $f(x) = 0$ otherwise. Find f'', and then simplify the expression $(x^2 - 1)f''(x)$ as far as possible.

2.28 Find the derivatives f' and f'', if $f(t) = |t^3 - t|$. Sketch the graphs of f, f' of f'' in separate pictures.

2.29 Find the general solution of the differential equation $\frac{dy}{dt} + 2ty = \delta(t-a)$.

2.30 Solve the problems (a) $y''-y = tH(t+1)$, (b) $y''+3y'+2y = tH(t)+\delta'(t)$.

2.31 Find $y = y(x)$ that satisfies $(1+x^2)y' - 2xy = \delta(x-1)$ and $y(0) = 1$.

2.32 Establish the following formula for an antiderivative (F being an antiderivative of f):

$$\int f(t)H(t-a)\,dt = (F(t) - F(a))H(t-a) + C.$$

2.33 Find a function $y = y(x)$ such that $y' + 2xy = 2xH(x) - \delta(x-1)$ and $y(2) = 1$. (Hint: the result of the preceding exercise may be useful.)

Historical notes

Complex numbers began to pop up as early as the Renaissance era, when scholars such as CARDANO began solving equations of third and fourth degrees. But not until LEONHARD EULER (1707–83) did they begin to be accepted as just as natural as the real numbers. The study of complex-valued functions was intensified in the nineteenth century; some famous names are AUGUSTIN CAUCHY (1789–1857), BERNHARD RIEMANN (1826–66), and KARL WEIERSTRASS (1815–97).

The idea of "summing" certain divergent series was made precise by mathematicians such as the young Norwegian NIELS HENRIK ABEL (1802–29) and CARL FRIEDRICH GAUSS (1777–1855). The method presented in Sec. 2.3 is due to the Italian mathematician ERNESTO CESÀRO (1859–1906), but the German OTTO HÖLDER (1859–1937) had the same idea at about the same time.

RIEMANN is the originator of an integral definition which is even today in universal use for elementary education. His definition has certain disadvantages, that were remedied by HENRI LEBESGUE (1875–1941) in his 1900 thesis. The Lebesgue integral is, however, even after one century considered to be too complicated to be included in elementary courses.

The theory of distributions is chronicled after Chapter 8.

Problems for Chapter 2

2.34 Show that the function $f(z) = e^{4z}$ has period $\pi i/2$.

2.35 Let f be a continuous function on $\mathbf{R}$. Assume that we know that it has period 2π and that it satisfies the equation

$$f(t) = \tfrac{1}{2}\left(f\left(t - \tfrac{1}{2}\pi\right) + f\left(t + \tfrac{1}{2}\pi\right)\right) \quad \text{for all } t \in \mathbf{R}.$$

Show that f in fact has a period shorter than 2π, and determine this period.

2.36 Let φ be a C^1 function such that φ and φ' are bounded on the real axis. Compute the limit

$$\lim_{n\to\infty} \frac{2n^3}{\pi} \int_{-\infty}^{\infty} \frac{x}{(1+n^2x^2)^2}\,\varphi(x)\,dx.$$

2.37 Let $F(x) = (1 - x^2)(H(x+1) - H(x-1))$. Let g be continuous on the interval $[-1, 1]$. Find the limit

$$\lim_{n\to\infty} \frac{3}{4} \int_{-1}^{1} nF(nx)\, g(x)\, dx.$$

2.38 Find the derivatives of order ≤ 4 of $f(t) = t^2 H(t)$.

2.39 Find $y(x)$ that solves the differential equation $y' + \dfrac{x^2+1}{x} y = \delta(x-2)$ and satisfies $y(1) = 1$.

2.40 Let $f : \mathbf{R} \to \mathbf{R}$ be described by $f(x) = (x^2 - 1)^2 (H(x+1) - H(x-1))$. Show that f belongs to the class C^1 but not to C^2. Also compute its third derivative.

3
Laplace and Z transforms

3.1 The Laplace transform

Let f be a function defined on the interval $\mathbf{R}_+ = [0, \infty[$. Alternatively, we can think of $f(t)$ as being defined for all real t, but satisfying $f(t) = 0$ for all $t < 0$. This can be expressed by writing

$$f(t) = f(t)H(t),$$

where H is the Heaviside function. Now let s be a real (or complex, if you like) number. If the integral

$$\widetilde{f}(s) = \int_0^\infty f(t)\, e^{-st}\, dt \tag{3.1}$$

exists (with a finite value), we say that it is the *Laplace transform of f, evaluated at the point s*. We shall write, interchangeably, $\widetilde{f}(s)$ or $\mathcal{L}[f](s)$. In applications, one also often uses the notation $F(s)$ (capital letter for the transform of the corresponding lower-case letter).

Example 3.1. Let $f(t) = e^{at}$, $t \geq 0$. Then,

$$\int_0^\infty f(t)\, e^{-st}\, dt = \int_0^\infty e^{at-st}\, dt = \left[\frac{e^{(a-s)t}}{a-s}\right]_{t=0}^{\infty} = \frac{1}{s-a},$$

provided that $a - s < 0$ so that the evaluation at infinity yields zero. Thus we have $\widetilde{f}(s) = 1/(s-a)$ for $s > a$, or

$$\mathcal{L}[e^{at}](s) = \frac{1}{s-a}, \quad s > a.$$

In particular, if $a = 0$, we have the Laplace transform of the constant function 1: it is equal to $1/s$ for $s > 0$. □

Example 3.2. Let $f(t) = t$, $t > 0$. Then, integrating by parts, we get

$$\begin{aligned}\widetilde{f}(s) &= \int_0^\infty t\,e^{-st}\,dt = \left[t\cdot\frac{e^{-st}}{-s}\right]_{t=0}^\infty + \frac{1}{s}\int_0^\infty 1\cdot e^{-st}\,dt \\ &= 0 + \frac{1}{s}\mathcal{L}[1](s) = \frac{1}{s^2}.\end{aligned}$$

This works for $s > 0$. □

It may happen that the Laplace transform does not exist for any real value of s. Examples of this are given by $f(t) = 1/t$, $f(t) = e^{t^2}$.

A profound understanding of the workings of the Laplace transform requires considering it to be a so-called analytic function of a complex variable, but in most of this book we shall assume that the variable s is real. We shall, however, permit the function f to take complex values: it is practical to be allowed to work with functions such as $f(t) = e^{i\alpha t}$.

Furthermore, we shall assume that the integral (3.1) is not merely convergent, but that it actually converges *absolutely*. This enables us to estimate integrals, using the inequality $|\int f| \le \int |f|$.

Example 3.3. Let $f(t) = e^{ibt}$. Then we can imitate Example 3.1 above and write

$$\begin{aligned}\int_0^\infty f(t)\,e^{-st}\,dt &= \int_0^\infty e^{(ib-s)t}\,dt = \left[\frac{e^{(ib-s)t}}{ib-s}\right]_{t=0}^\infty \\ &= \frac{1}{ib-s}\left[e^{-st}(\cos bt + i\sin bt)\right]_{t=0}^\infty.\end{aligned}$$

For $s > 0$ the substitution as $t \to \infty$ will tend to zero, because the factor e^{-st} tends to zero and the rest of the expression is bounded. The result is thus that $\mathcal{L}[e^{ibt}](s) = 1/(s - ib)$, which means that the formula that we proved in Example 3.1 holds true also when a is purely imaginary. It is left to the reader to check that the same formula holds if a is an arbitrary complex number and $s > \operatorname{Re} a$. □

It would be convenient to have some simple set of conditions on a function f that ensure that the Laplace transform is absolutely convergent for some value of s. Such a set of conditions is given in the following definition.

Definition 3.1 *Let k be a positive number. Assume that f has the following properties:*

(i) *f is continuous on $[0, \infty[$ except possibly for a finite number of jump discontinuities in every finite subinterval;*

(ii) *there is a positive number M such that $|f(t)| \le Me^{kt}$ for all $t \ge 0$.*

Then we say that f belongs to the class $\mathcal{E}_k$. If $f \in \mathcal{E}_k$ for some *value of k, we say that $f \in \mathcal{E}$.*

Using set notation we can say that $\mathcal{E} = \bigcup_{k>0} \mathcal{E}_k$. Condition (ii) means that f grows at most *exponentially;* this word lies behind the use of the letter $\mathcal{E}$. If $f \in \mathcal{E}_k$ for one value of k, then also $f \in \mathcal{E}_k$ for all *larger* k.

Theorem 3.1 *If $f \in \mathcal{E}_k$, then $\widetilde{f}(s)$ exists for all $s > k$.*

Proof. We begin by observing that condition (i) for the class $\mathcal{E}_k$ implies that the integral

$$\int_0^T f(t)\, e^{-st}\, dt$$

exists finitely for all s and all $T > 0$. Now assume $s > k$. Thus there exists a number M and a number t_0 so that $f(t)e^{-kt} \le M$ for $t > t_0$. Then we can estimate as follows:

$$\begin{aligned}\int_{t_0}^T |f(t)|\, e^{-st}\, dt &= \int_{t_0}^T |f(t)|\, e^{-kt}\, e^{-(s-k)t}\, dt \le \int_{t_0}^T M e^{-(s-k)t}\, dt \\ &\le M \int_{t_0}^{\infty} e^{-(s-k)t}\, dt \le M \int_0^{\infty} e^{-(s-k)t}\, dt = \frac{M}{s-k} < \infty.\end{aligned}$$

This means that the generalized integral over $[t_0, \infty[$ converges absolutely, and then this is equally true for the integral over $[0, \infty[$. □

The result of the theorem can be "bootstrapped" in the following way. If $\sigma_0 = \inf\{k : f \in \mathcal{E}_k\}$, then the Laplace transform exists for all $s > \sigma_0$. Indeed, let $k = (s + \sigma_0)/2$, so that $\sigma_0 < k < s$; then $f \in \mathcal{E}_k$ (why?), and the theorem can be applied. The number σ_0 is a reasonably exact measure of the rate of growth of the function f. In what follows we shall sometimes use the notation σ_0 or $\sigma_0(f)$ for this measure.

As a consequence of the theorem we now know that a large set of common functions do have Laplace transforms. Among them are, e.g., polynomials, trigonometric functions such as sin and cos and ordinary exponential functions; also sums and products of such functions. If you have studied simple differential equations you may recall that these functions are precisely the possible solutions of homogeneous linear differential equations with constant coefficients, such as, for example,

$$y^{(\mathrm{v})} + 4y^{(\mathrm{iv})} - 8y''' + 15y'' - 24y' = 0.$$

We shall soon see that Laplace transforms give us a new technique for solving these equations. We shall also be able to solve more general problems, like integral equations of this kind:

$$\int_0^t f(t-x)\, f(x)\, dx + 3 \int_0^t f(x)\, dx + 2t = 0, \quad t > 0. \tag{3.2}$$

Another consequence of the theorem is worth emphasizing: if a Laplace transform exists for *one* value of s, then it is also defined for all *larger* values of s. If we are dealing with several different transforms having various domains, we can always be sure that they are all defined at least in one common semi-infinite interval. It is customary to be rather sloppy about specifying the domains of definition for Laplace transforms: we make a tacit agreement that s is large enough so that all transforms occuring in a given situation are defined.

Exercises

3.1 Let $f(t) = e^{t^2}$, $g(t) = e^{-t^2}$. Show that $f \notin \mathcal{E}$, whereas $g \in \mathcal{E}_k$ for all k.

3.2 Compute the Laplace transform of $f(t) = e^{at}$, where $a = \alpha + i\beta$ is a complex constant.

3.3 Let $f(t) = \sin t$ for $0 \le t \le \pi$, $f(t) = 0$ otherwise. Find $\widetilde{f}(s)$.

3.2 Operations

The Laplace transformation obeys some simple rules of computation and also some less simple rules. The simplest ones are collected in the following table. Everywhere we assume that s takes sufficiently large values, as discussed at the end of the preceding section.

1. $\mathcal{L}[\alpha f + \beta g](s) = \alpha\widetilde{f}(s) + \beta\widetilde{g}(s)$, if α and β are constants.
2. $\mathcal{L}[e^{at} f(t)](s) = \widetilde{f}(s-a)$, if a is a constant (damping rule).
3. If we define $f(t) = 0$ for $t < 0$ and if $a > 0$, then
$$\mathcal{L}[f(t-a)](s) = e^{-as}\widetilde{f}(s) \quad \text{(delaying rule)}.$$
4. $\mathcal{L}[f(at)](s) = \dfrac{1}{a}\widetilde{f}(s/a)$, if $a > 0$.

The proofs of these rules are easy. As an example we give the computations that yield rules 3 and 4:

$$\begin{aligned}
\mathcal{L}[f(t-a)](s) &= \int_0^\infty f(t-a)\, e^{-st}\, dt \left\{ \begin{array}{c} u = t - a \\ du = dt \\ t = 0 \Leftrightarrow u = -a \end{array} \right\} \\
&= \int_{-a}^\infty f(u)\, e^{-s(u+a)}\, du = e^{-as} \int_{-a}^\infty f(u)\, e^{-su}\, du \\
&= e^{-as} \int_0^\infty f(u)\, e^{-su}\, du = e^{-as}\widetilde{f}(s);
\end{aligned}$$

$$\begin{aligned}
\mathcal{L}[f(at)](s) &= \int_0^\infty f(at)\, e^{-st}\, dt \left\{ \begin{array}{c} u = at \\ du = a\, dt \end{array} \right\} = \int_0^\infty f(u)\, e^{-s\cdot u/a} \frac{du}{a} \\
&= \frac{1}{a} \int_0^\infty f(u) \exp\!\left(-\frac{s}{a}\cdot u\right) du = \frac{1}{a}\widetilde{f}\!\left(\frac{s}{a}\right).
\end{aligned}$$

Example 3.4. Using rule 1 and the result of Example 3.3 in the preceding section, we can find the Laplace transforms of cos and sin:

$$\mathcal{L}[\cos bt](s) = \tfrac{1}{2}\mathcal{L}[e^{ibt} + e^{-ibt}](s) = \tfrac{1}{2}\left(\frac{1}{s-ib} + \frac{1}{s+ib}\right) = \frac{s}{s^2+b^2},$$
$$\mathcal{L}[\sin bt](s) = \frac{1}{2i}\mathcal{L}[e^{ibt} - e^{-ibt}](s) = \frac{1}{2i}\left(\frac{1}{s-ib} - \frac{1}{s+ib}\right) = \frac{b}{s^2+b^2}.$$

□

Example 3.5. Applying rule 2 to the result of Example 3.4 we get

$$\mathcal{L}[e^{at}\cos bt](s) = \frac{s-a}{(s-a)^2+b^2}, \qquad \mathcal{L}[e^{at}\sin bt](s) = \frac{b}{(s-a)^2+b^2}.$$

□

A couple of deeper rules are given in the following theorems.

Theorem 3.2 *If $f \in \mathcal{E}_{k_0}$, then $(t \mapsto tf(t)) \in \mathcal{E}_{k_1}$ for $k_1 > k_0$ and*

$$\mathcal{L}[tf(t)](s) = -\frac{d}{ds}\widetilde{f}(s).$$

Proof. We shall use a theorem on differentiation of integrals. In order to keep it lucid, we assume that f is continuous on the whole of $\mathbf{R}_+$; otherwise we would have to split into integrals over subintervals where f *is* continuous, and this introduces certain purely technical complications. Since $f \in \mathcal{E}_{k_0}$, we know that $|f(t)| \le Me^{k_0 t}$ for some number M and all sufficiently large t, say $t > t_1$. Let $\delta > 0$. Then there is a t_2 such that $|t| < e^{\delta t}$ for $t > t_2$. If $t > t_0 = \max(t_1, t_2)$ we have

$$|tf(t)| \le e^{\delta t} \cdot Me^{k_0 t} = Me^{(k_0+\delta)t} = Me^{k_1 t},$$

which means that $tf(t)$ belongs to $\mathcal{E}_{k_1}$ and has a Laplace transform.

If we differentiate the formula $\widetilde{f}(s) = \int_0^\infty f(t)\,e^{-st}\,dt$ formally with respect to s, we get $(\widetilde{f})'(s) = \int_0^\infty (-t)f(t)\,e^{-st}\,dt$. According to the theorem concerning differentiation of integrals, this maneuver is permitted if we can find a "dominating" function g (that may depend on t but not on s) such that the integrand in the differentiated formula can be estimated by g for all $t \ge 0$ and all values of s that we consider, and which is such that $\int_0^\infty g$ is convergent. Let a be a number greater than the constant k_1 and put $g(t) = |tf(t)\,e^{-at}|$. For all $s \ge a$ we have then $|(-t)\,f(t)\,e^{-st}| \le g(t)$, and

$$\int_0^\infty g(t)\,dt = \int_0^\infty |tf(t)|e^{-at}\,dt \le M\int_0^\infty e^{k_1 t}\cdot e^{-at}\,dt$$
$$= M\int_0^\infty e^{-(a-k_1)t}\,dt = \frac{M}{a-k_1} < \infty.$$

This shows that the conditions for differentiating formally are fulfilled, and the theorem is proved. □

Example 3.6. We know that $\mathcal{L}[1](s) = 1/s$ for $s > 0$. Then we can say that

$$\mathcal{L}[t](s) = \mathcal{L}[t \cdot 1](s) = -\frac{d}{ds}\frac{1}{s} = -\left(-\frac{1}{s^2}\right) = \frac{1}{s^2}, \quad s > 0.$$

Repeating this argument (do it!) we find that

$$\mathcal{L}[t^n](s) = \frac{n!}{s^{n+1}}, \quad s > 0.$$

□

Example 3.7. Also, rule 2 allows us to conclude that

$$\mathcal{L}[t^n\, e^{at}](s) = \frac{n!}{(s-a)^{n+1}}, \quad s > 0.$$

□

A sort of reverse of Theorem 3.2 is the following. The notation $f(0+)$ stands for the right-hand limit $\lim_{t \to 0+} f(t) = \lim_{t \searrow 0} f(t)$.

Theorem 3.3 *Assume that $f \in \mathcal{E}$ is continuous on $\mathbf{R}_+$. Also assume that the derivative $f'(t)$ exists for all $t \geq 0$ (with $f'(0)$ interpreted as the right-hand derivative) and that $f' \in \mathcal{E}$. Then*

$$\mathcal{L}[f'](s) = s\,\widetilde{f}(s) - f(0+).$$

Proof. Suppose that $f \in \mathcal{E}_{k_0}$ and $f' \in \mathcal{E}_{k_1}$, and take s to be larger than both k_0 and k_1. Let T be a positive number. Integration by parts gives

$$\int_0^T f'(t)\, e^{-st}\, dt = f(T)\, e^{-sT} - f(0+)\, e^0 + s \int_0^T f(t)\, e^{-st}\, dt.$$

When $T \to \infty$, the first term in the right-hand member tends to zero, and the result is the desired formula. □

Theorem 3.3 will be used for solving differential equations.

The following theorem states a few additional properties of the Laplace transform.

Theorem 3.4 (a) *If $f \in \mathcal{E}$, then*

$$\lim_{s \to \infty} \widetilde{f}(s) = 0. \tag{3.3}$$

(b) The initial value rule: *If $f(0+)$ exists, then*

$$\lim_{s \to \infty} s\widetilde{f}(s) = f(0+). \tag{3.4}$$

(c) The final value rule: *If $f(t)$ has a limit as $t \to +\infty$, then*

$$\lim_{s \searrow 0+} s\widetilde{f}(s) = f(+\infty) = \lim_{t \to \infty} f(t). \tag{3.5}$$

In applications, the rule (3.5) is useful for deciding the ultimate or "steady-state" behavior of a function or a signal.

Proof. (a) Let $\varepsilon > 0$ be given and choose $\delta > 0$ so small that

$$\int_0^\delta |f(t)|\,dt < \varepsilon.$$

Let $k > 0$ be such that $f \in \mathcal{E}_k$ and let $s_0 > k$. Then for $s > s_0$ we get

$$\begin{aligned}
|\widetilde{f}(s)| &\le \int_0^\delta |f(t)|\,e^{-st}\,dt + \int_\delta^\infty |f(t)|\,e^{-st}\,dt \\
&\le \int_0^\delta |f(t)|\,dt + \int_\delta^\infty |f(t)|\,e^{-s_0t}e^{-(s-s_0)t}\,dt \\
&\le \varepsilon + e^{-(s-s_0)\delta}\int_\delta^\infty |f(t)|e^{-s_0t}\,dt \le \varepsilon + Ce^{-(s-s_0)\delta} = \varepsilon + Ce^{\delta s_0}\cdot e^{-\delta s}.
\end{aligned}$$

The last term tends to zero as $s \to \infty$ and thus it is less than ε if s is large enough. This proves that $|\widetilde{f}(s)| < 2\varepsilon$ for all sufficiently large s, and since ε can be arbitrarily small, we have proved (3.3).

(b) The idea of proof is similar to the preceding. $\varepsilon > 0$ is arbitrary, but now we choose $\delta > 0$ so small that $|f(t) - f(0+)| < \varepsilon$ for $0 < t < \delta$. With s_0 as above we get, for $s > s_0$,

$$\begin{aligned}
&s\widetilde{f}(s) \\
&= s\int_0^\delta (f(t) - f(0+))\,e^{-st}\,dt + sf(0+)\int_0^\delta e^{-st}\,dt + s\int_\delta^\infty f(t)\,e^{-st}\,dt.
\end{aligned}$$

The modulus of the first term is

$$\le s\varepsilon\int_0^\delta e^{-st}\,dt \le s\varepsilon\int_0^\infty e^{-st}\,dt = s\varepsilon\cdot\frac{1}{s} = \varepsilon, \quad \text{if } s > 0.$$

The second term can be computed:

$$= sf(0+)\frac{1-e^{-s\delta}}{s} = f(0+)(1-e^{-s\delta}) \to f(0+) \quad \text{as } s \to \infty.$$

Finally, the modulus of the third term can be estimated:

$$\le s\int_\delta^\infty |f(t)|e^{-s_0t}e^{-(s-s_0)t}\,dt \le se^{-s\delta}\cdot e^{s_0\delta}\int_\delta^\infty |f(t)|\,e^{-s_0t}\,dt = Cse^{-\delta s},$$

which tends to zero as $s \to \infty$. Just as in the proof of (3.3) we can draw the conclusion (3.4).

(c) This proof also runs along similar paths. We begin by writing

$$s\tilde{f}(s) = s\int_0^T f(t)\,e^{-st}\,dt + s\int_T^\infty (f(t) - f(\infty))e^{-st}\,dt + f(\infty)e^{-sT}.$$

Choose T so large that $|f(t) - f(\infty)| < \varepsilon$ for $t \geq T$. The modulus of the first term can be estimated by $s\int_0^T |f| \to 0$ as $s \to 0+$, and the modulus of the second one is

$$\leq s\int_T^\infty \varepsilon \cdot e^{-st}\,dt = \varepsilon\, e^{-sT} \leq \varepsilon.$$

The proof is finished in an analogous way to the others. □

We round off this section by a generalization of the rule for Laplace transformation of a power of t (cf. Example 3.6). To this end we need a generalization of factorials to non-integers. This is provided by EULER's *Gamma function*, whis is defined by

$$\Gamma(x) = \int_0^\infty u^{x-1}e^{-u}\,du, \quad x > 0.$$

It is easy to see that this integral converges for positive x. It is also easy to see that $\Gamma(1) = 1$. Integrating by parts we find

$$\Gamma(x+1) = \int_0^\infty u^x e^{-u}\,du = \Big[-u^x e^{-u}\Big]_0^\infty + x\int_0^\infty u^{x-1}e^{-u}\,du = x\Gamma(x).$$

From this we deduce that $\Gamma(2) = 1 \cdot \Gamma(1) = 1$, $\Gamma(3) = 2$, and, by induction, $\Gamma(n+1) = n!$ for integral n. Thus, this function can be viewed as an interpolation of the factorial.

Now we let $f(t) = t^a$, where $a > -1$. It is then clear that f has a Laplace transform, and we find, for $s > 0$,

$$\begin{aligned}\tilde{f}(s) &= \int_0^\infty t^a e^{-st}\,dt \left\{\begin{matrix} st = u \\ dt = du/s \end{matrix}\right\} = \int_0^\infty \left(\frac{u}{s}\right)^a e^{-u}\,\frac{du}{s} \\ &= \frac{1}{s^{a+1}}\int_0^\infty u^a e^{-u}\,du = \frac{\Gamma(a+1)}{s^{a+1}}.\end{aligned}$$

If a is an integer, this reduces to the formula of Example 3.6.

Exercises

3.4 Find the Laplace transforms of (a) $2t^2 - e^{-t}$
(b) $(t^2+1)^2$ (c) $(\sin t - \cos t)^2$ (d) $\cosh^2 4t$ (e) $e^{2t}\sin 3t$ (f) $t^3 \sin 3t$.

3.5 Compute the Laplace transform of $f(t) = \begin{cases} 1/\varepsilon & \text{for } 0 < t < \varepsilon, \\ 0 & \text{otherwise.} \end{cases}$

3.6 Find the transform of $f(t) = \begin{cases} (t-1)^2 & \text{for } t > 1, \\ 0 & \text{otherwise.} \end{cases}$

3.7 Solve the same problem for $f(t) = \displaystyle\int_0^t \frac{1-e^{-u}}{u}\,du$.

3.8 Compute $\displaystyle\int_0^\infty t\,e^{-3t}\sin t\,dt$. (Hint: $\widetilde{f}(3)$!)

3.9 Find the Laplace transform of f, if we define $f(t) = t\sin t$ for $0 \le t \le \pi$, $f(t) = 0$ otherwise. (Hint: use the result of Exercise 3.3, p. 42.)

3.10 Find the Laplace transform of the function f defined by

$$f(t) = na \quad \text{for} \quad n-1 \le x < n, \qquad n = 1, 2, 3, \ldots.$$

3.11 Compute $\mathcal{L}[te^{-t}\sin t](s)$.

3.12 Explain why the function $\dfrac{s^2}{s^2+1}$ cannot be the Laplace transform of any $f \in \mathcal{E}$.

3.13 Show that if f is periodic with period a, then

$$\widetilde{f}(s) = \frac{1}{1-e^{-as}} \int_0^a f(t)\,e^{-st}\,dt.$$

(Hint: $\int_0^\infty = \sum_0^\infty \int_{ak}^{a(k+1)}$. Let $u = t - ak$, use the formula for the sum of a geometric series.)

3.14 Find the Laplace transform of the function with period 1 that is described by $f(t) = t$ for $0 < t < 1$.

3.15 Verify the final value rule (3.5) for $\widetilde{f}(s) = 1/(s(s+1))$ by comparing $f(t)$ and $\displaystyle\lim_{s\to 0+} s\widetilde{f}(s)$.

3.16 Prove that $\Gamma\left(\frac{1}{2}\right) = \sqrt{\pi}$. What are the values of $\Gamma\left(\frac{3}{2}\right)$ and $\Gamma\left(\frac{5}{2}\right)$?

3.3 Applications to differential equations

Example 3.8. Let us try to solve the initial value problem

$$y'' - 4y' + 3y = t, \quad t > 0; \qquad y(0) = 3, \quad y'(0) = 2. \qquad (3.6)$$

We assume that $y = y(t)$ is a solution such that y, as well as y' and y'', has a Laplace transform. By Theorem 3.3 we have then

$$\begin{aligned} \mathcal{L}[y'](s) &= s\widetilde{y} - y(0) = s\widetilde{y} - 3, \\ \mathcal{L}[y''](s) &= s\mathcal{L}[y'](s) - y'(0) = s(s\widetilde{y} - 3) - 2 = s^2\widetilde{y} - 3s - 2. \end{aligned}$$

Due to linearity, we can transform the left-hand side of the equation to get

$$(s^2\widetilde{y} - 3s - 2) - 4(s\widetilde{y} - 3) + 3\widetilde{y} = (s^2 - 4s + 3)\widetilde{y} - 3s + 10,$$

and this must be equal to the transform of the right-hand side, which is $1/s^2$. The result is an algebraic equation, which we can solve for $\widetilde{y}$:

$$(s^2-4s+3)\widetilde{y}-3s+10=\frac{1}{s^2} \iff \widetilde{y}=\frac{3s^3-10s^2+1}{s^2(s^2-4s+3)}=\frac{3s^3-10s^2+1}{s^2(s-1)(s-3)}.$$

The last expression can be expanded into partial fractions. Assume that

$$\frac{3s^3-10s^2+1}{s^2(s-1)(s-3)}=\frac{A}{s^2}+\frac{B}{s}+\frac{C}{s-1}+\frac{D}{s-3}.$$

Multiplying by the common denominator and identifying coefficients we find that $A=\frac{1}{3}$, $B=\frac{4}{9}$, $C=3$, and $D=-\frac{4}{9}$. Thus we have

$$\widetilde{y}=\tfrac{1}{3}\cdot\frac{1}{s^2}+\tfrac{4}{9}\cdot\frac{1}{s}+3\cdot\frac{1}{s-1}-\tfrac{4}{9}\cdot\frac{1}{s-3}.$$

It so happens that there exists a function with precisely this Laplace transform, namely, the function

$$z=\tfrac{1}{3}t+\tfrac{4}{9}+3e^t-\tfrac{4}{9}e^{3t}.$$

Could it be the case that $y=z$? One way of finding this out is by differentiating and investigating if indeed z does satisfy the equation and initial conditions. And it does (check for yourself)! By the general theory of differential equations, the problem (3.6) has a unique solution, and it follows that z must be the solution we are looking for. □

The example demonstrates a very useful method for treating linear initial value problems. There is one difficulty that is revealed at the end of the example: could it be possible that two *different* functions might have the same Laplace transform? This question is answered by the following theorem.

Theorem 3.5 (Uniqueness for Laplace transforms) *If f and g both belong to $\mathcal{E}$, and $\widetilde{f}(s)=\widetilde{g}(s)$ for all (sufficiently) large values of s, then $f(t)=g(t)$ for all values of t where f and g are continuous.*

We omit the proof of this at this point. It is given in Sec. 7.10. In that section we also prove a formula for the reconstruction of $f(t)$ when $\widetilde{f}(s)$ is known — a so-called *inversion formula* for the Laplace transform. The present theorem, however, gives us the possibility to invert Laplace transforms by *recognizing* functions, just as we did in the example.

This requires that we have access to a table of Laplace transforms of such functions that can be expected to occur. Such a table is found at the end of the book (p. 247 ff), and similar tables are included in all decent handbooks on the subject. Several of the entries in such tables have already been proved in the examples of this chapter; others can be done as exercises by the interested student.

We point out that the uniqueness result as such does not rule out the possibility that a differential equation (or other problem) may have solutions that have no Laplace transforms, e.g., solutions that grow faster than exponentially. To preclude such solutions one must look into the theory of differential equations. For linear equations there is a result on unique solutions for initial value problems, which may serve the purpose. If the coefficients are constants and the equation is homogeneous, one actually knows that all solutions have at most exponential growth.

The Laplace transform method is ideally adapted to solving initial value problems. Strictly speaking, the method takes into consideration only what goes on for $t \geq 0$. Very often, however, the expressions obtained for the solutions are also valid for $t < 0$.

We include some examples on using a table of Laplace transforms in a few more complicated situations. The technique may remind the reader of the integration of rational functions.

Example 3.9. Find $f(t)$, when $\widetilde{f}(s) = \dfrac{2s+3}{s^2+4s+13}$.

Solution. Complete the square in the denominator: $s^2+4s+13 = (s+2)^2+9$. Then split the numerator to enable us to recognize transforms of cosines and sines:

$$\frac{2s+3}{s^2+4s+13} = \frac{2(s+2)-1}{(s+2)^2+3^2} = 2 \cdot \frac{s+2}{(s+2)^2+3^2} - \frac{1}{3} \cdot \frac{3}{(s+2)^2+3^2},$$

and now we can see that this is the transform of $f(t) = 2e^{-2t}\cos 3t - \frac{1}{3}e^{-2t}\sin 3t$. □

Example 3.10. Find $g(t)$, if $\widetilde{g}(s) = \dfrac{2s}{(s^2+1)^2}$.

Solution. We recognize the transform as a derivative:

$$\widetilde{g}(s) = -\frac{d}{ds}\frac{1}{s^2+1}.$$

By Theorem 3.2 and the known transform of the sine we get $g(t) = t\sin t$. □

Example 3.11. Solve the initial value problem

$$y'' + 4y' + 13y = 13, \quad y(0) = y'(0) = 0.$$

Solution. Transformation gives

$$(s^2+4s+13)\widetilde{y} = \frac{13}{s} \iff \widetilde{y} = \frac{13}{s\big((s+2)^2+9\big)}.$$

Expand into partial fractions:

$$\widetilde{y} = \frac{1}{s} - \frac{s+4}{(s+2)^2+9} = \frac{1}{s} - \frac{s+2}{(s+2)^2+9} - \frac{2}{3} \cdot \frac{3}{(s+2)^2+9}.$$

The solution is found to be

$$y(t) = \big(1 - e^{-2t}(\cos 3t + \tfrac{2}{3}\sin 3t)\big)H(t).$$

(Here we have multiplied the result by a Heaviside factor, to indicate that we are considering the solution only for $t \geq 0$. This factor is often omitted. Whether or not it should be there is often a matter of dispute among users of the transform.) □

We can also treat *systems* of differential equations.

Example 3.12. Solve the initial value problem

$$\begin{cases} x' = x + 3y, \\ y' = 3x + y; \end{cases} \qquad x(0) = 5, \quad y(0) = 1.$$

Solution. Laplace transformation gives

$$\begin{cases} s\widetilde{x} - 5 = \widetilde{x} + 3\widetilde{y} \\ s\widetilde{y} - 1 = 3\widetilde{x} + \widetilde{y} \end{cases} \iff \begin{cases} (1-s)\widetilde{x} + 3\widetilde{y} = -5 \\ 3\widetilde{x} + (1-s)\widetilde{y} = -1 \end{cases}$$

We can, for example, solve the second equation for $\widetilde{x} = \frac{1}{3}(s-1)\widetilde{y} - \frac{1}{3}$ and substitute this into the first, whereupon simplification yields $(s^2-2s-8)\widetilde{y} = s+14$ and

$$\widetilde{y} = \frac{s+14}{(s-4)(s+2)} = \frac{3}{s-4} - \frac{2}{s+2}.$$

We see that $y = 3e^{4t} - 2e^{-2t}$, and then we deduce, in one way or another, that $x = 3e^{4t} + 2e^{-2t}$. (Think of at least three different ways of performing this last step!) □

Finally, we demonstrate how even a *partial* differential equation can be treated by Laplace transforms. The trick is to transform with respect to one of the independent variables and let the others stand. Using this technique often involves taking rather bold chances in the hope that rules of computation be valid. One way of regarding this is to view it precisely as taking chances – if we arrive at a tentative solution, it can always be checked by substitution in the original problem.

Example 3.13. Find a solution of the problem

$$\frac{\partial^2 u}{\partial x^2} = \frac{\partial u}{\partial t}, \quad 0 < x < 1,\ t > 0; \qquad \begin{aligned} &u(0,t) = 1,\ u(1,t) = 1, \quad t > 0; \\ &u(x,0) = 1 + \sin \pi x, \quad 0 < x < 1. \end{aligned}$$

Solution. We introduce the Laplace transform $U(x, s)$ of $u(x, t)$, i.e.,

$$U(x, s) = \mathcal{L}[t \mapsto u(x, t)](s) = \int_0^\infty u(x, t)\, e^{-st}\, dt.$$

Here, x is thought of as a constant. Then we change our attitude and assume that this integral can be differentiated with respect to x, indeed twice, so that

$$\frac{\partial^2 U}{\partial x^2} = \frac{\partial^2}{\partial x^2} \int_0^\infty u(x, t)\, e^{-st}\, dt = \int_0^\infty \frac{\partial^2}{\partial x^2} u(x, t)\, e^{-st}\, dt.$$

The differential equation is then transformed into

$$\frac{\partial^2 U}{\partial x^2} = sU - (1 + \sin \pi x), \qquad 0 < x < 1,$$

and the boundary conditions into

$$U(0, s) = \frac{1}{s}, \qquad U(1, s) = \frac{1}{s}.$$

Now we switch attitudes again: think of s as a constant and solve the boundary value problem. Just to feel comfortable we could write the equation as

$$U'' - sU = -1 - \sin \pi x. \tag{3.7}$$

The homogeneous equation has a characteristic equation $r^2 - s = 0$ and its solution is $U_H = Ae^{x\sqrt{s}} + Be^{-x\sqrt{s}}$. (Here, the "constants" A and B are in general functions of s.) A particular solution to the inhomogeneous equation could have the form $U_P = a + b \sin \pi x + c \cos \pi x$, and insertion and identification gives $a = 1/s$, $b = 1/(s + \pi^2)$, $c = 0$. Thus the general solution of (3.7) is

$$U(x, s) = A(s)e^{x\sqrt{s}} + B(s)e^{-x\sqrt{s}} + \frac{1}{s} + \frac{\sin \pi x}{s + \pi^2}.$$

The boundary conditions force us to take $A(s) = B(s) = 0$, so we are left with $U(x, s) = \dfrac{1}{s} + \dfrac{\sin \pi x}{s + \pi^2}$. Now we again consider x as a constant and recognize that U is the Laplace transform of $u(x, t) = 1 + e^{-\pi^2 t} \sin \pi x$. The fact that this function really does solve the original problem must be checked directly (since we have made an assumption on differentiability of an integral, which might have been too bold). □

Remark. This problem can also be attacked by other methods developed in later parts of the book (Chapter 6). □

Exercises

3.17 Invert the following Laplace transforms: (a) $\dfrac{1}{s(s+1)}$ (b) $\dfrac{3}{(s-1)^2}$ (c) $\dfrac{1}{s(s+2)^2}$ (d) $\dfrac{5}{s^2(s-5)^2}$ (e) $\dfrac{1}{(s-a)(s-b)}$ (f) $\dfrac{1}{s^2+4s+29}$.

3.18 Use partial fractions to find f when $\widetilde{f}(s)$ is given by (a) $s^{-2}(s+1)^{-1}$, (b) $b^2 s^{-1}(s^2+b^2)^{-1}$, (c) $s(s-3)^{-5}$, (d) $(s^2+2)s^{-1}(s+1)^{-1}(s+2)^{-1}$.

3.19 Invert the following Laplace transforms: (a) $\dfrac{1+e^{-s}}{s}$ (b) $\dfrac{e^{-s}}{(s-1)(s-2)}$ (c) $\ln\dfrac{s+3}{s+2}$ (d) $\ln\dfrac{s^2+1}{s(s+3)}$ (e) $\dfrac{s+1}{s^{4/3}}$ (f) $\dfrac{\sqrt{s}-1}{s}$.

3.20 Solve the initial value problem $y''+y=2e^t$, $t>0$, $y(0)=y'(0)=2$.

3.21 Solve the initial value problem $\begin{cases} y''(t)-2y'(t)+y(t)=t\,e^t\sin t, \\ y(0)=0,\ y'(0)=0. \end{cases}$

3.22 Solve $\begin{cases} y^{(3)}(t)-y''(t)+4y'(t)-4y(t)=-3e^t+4e^{2t}, \\ y(0)=0,\ y'(0)=5,\ y''(0)=3. \end{cases}$

3.23 Solve the system $\begin{cases} x'(t)+y'(t)=t, \\ x''(t)-y(t)=e^{-t}, \\ x(0)=3,\ x'(0)=-2,\ y(0)=0. \end{cases}$

3.24 Solve the system $\begin{cases} x'(t)-y'(t)-2x(t)+2y(t)=\sin t, \\ x''(t)+2y'(t)+x(t)=0, \\ x(0)=x'(0)=y(0)=0. \end{cases}$

3.25 Solve the problem

$$y''(t)-3y'(t)+2y(t)=\begin{cases} 1, & t>2 \\ 0, & t<2 \end{cases}; \quad y(0)=1,\ y'(0)=0.$$

3.26 Solve the system

$$\begin{cases} \dfrac{dy}{dt}=2z-2y+e^{-t} \\ \dfrac{dz}{dt}=y-3z \end{cases} \qquad t>0; \qquad y(0)=1, \quad z(0)=2.$$

3.27 Solve the differential equation

$$2y^{(\mathrm{iv})}+y'''-y''-y'-y=t+2, \quad t>0,$$

with initial conditions $y(0)=y'(0)=0$, $y''(0)=y'''(0)=1$.

3.28 Solve the differential equation

$$y''+3y'+2y=e^{-t}\sin t, \quad t>0; \qquad y(0)=1, \quad y'(0)=-3.$$

3.4 Convolution

In control theory, for example, one studies the effect on an incoming signal by a "black box" that transforms it into an "outsignal":

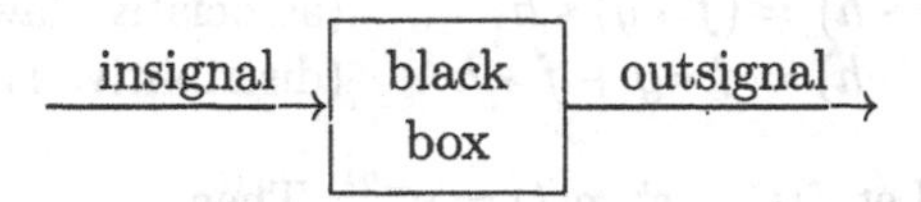

Let the insignal be represented by the function $t \mapsto x(t)$, $t \geq 0$, and the outsignal by $t \mapsto y(t)$, $t \geq 0$. We assume that the system has four important properties:

(a) it is *linear*, which means that a linear combination of inputs results in the corresponding linear combination of outputs;

(b) it is *translation invariant*, which means, loosely, that the black box operates in the same way at all points in time;

(c) it is *continuous* in the sense that "small" changes in the input generate "small" changes in the output (which should be formulated more precisely when necessary);

(d) it is *causal*, i.e., the outsignal at a certain moment t does not depend on the insignal at moments *later* than t.

It can then be shown (see Appendix A) that there exists a function $t \mapsto g(t)$, $t \geq 0$, such that

$$y(t) = \int_0^t x(u)g(t-u)\,du = \int_0^t x(t-u)g(u)\,du. \tag{3.8}$$

The function g can be said to contain all information about the system.

The formula (3.8) is an example of a notion called the *convolution* of the two functions x and g. (We shall encounter other versions of convolution in other parts of this book.) We shall now study this notion from a mathematical point of view.

Thus, we assume that f and g are two functions, both belonging to $\mathcal{E}$. The *convolution* $f * g$ is a new function defined by the formula

$$(f * g)(t) = f * g(t) = \int_0^t f(u)\,g(t-u)\,du, \qquad t \geq 0.$$

It is not hard to see that this function is continuous on $[0, \infty[$, and it might possibly belong to $\mathcal{E}$. Indeed, it is not very difficult to show directly that if $f \in \mathcal{E}_{k_1}$ and $g \in \mathcal{E}_{k_2}$, then $f * g \in \mathcal{E}_k$ for all $k > \max(k_1, k_2)$. (See Exercise 3.38.) Using the notation $\sigma_0(f)$, introduced after Theorem 3.1, we could express this as $\sigma_0(f * g) \leq \max(\sigma_0(f), \sigma_0(g))$.)

Convolution can be regarded as an operation for functions, a sort of "multiplication." For this operation a few simple rules hold; the reader is invited to check them out:

$$\begin{aligned} f*g &= g*f && \text{(commutative law)} \\ f*(g*h) &= (f*g)*h && \text{(associative law)} \\ f*(g+h) &= f*g+f*h && \text{(distributive law)} \end{aligned}$$

Example 3.14. Let $f(t) = e^t$, $g(t) = e^{-2t}$. Then

$$\begin{aligned} f*g(t) &= \int_0^t e^u\, e^{-2(t-u)}\, du = \int_0^t e^{u-2t+2u}\, du = e^{-2t}\int_0^t e^{3u}\, du \\ &= e^{-2t}\left[\tfrac{1}{3}\, e^{3u}\right]_{u=0}^{u=t} = \tfrac{1}{3}e^{-2t}\left(e^{3t}-1\right) = \frac{e^t - e^{-2t}}{3}. \end{aligned}$$

□

Example 3.15. If $g(t) = 1$, then $f*g(t) = \int_0^t f(u)\, du$. Thus, "integration" can be considered to be convolution with the function 1. □

When dealing with convolutions, the Laplace transform is useful because of the following theorem.

Theorem 3.6 *The Laplace transform of a convolution is the product of the Laplace transforms of the two convolution factors:*

$$\mathcal{L}[f*g](s) = \widetilde{f}(s)\,\widetilde{g}(s).$$

Proof. Let s be so large that both $\widetilde{f}(s)$ and $\widetilde{g}(s)$ exist. We have agreed in section 3.1 that this means that the corresponding integrals converge absolutely. Now consider the improper double integral

$$\iint_Q |f(u)g(v)|e^{-s(u+v)}\, du\, dv,$$

where Q is the first quadrant in the uv plane. The integrated function being positive, the integral can be calculated just as we choose. For example, we can write

$$\begin{aligned} \iint_Q |f(u)g(v)|e^{-s(u+v)}\, du\, dv &= \int_0^\infty du \int_0^\infty |f(u)||g(v)|e^{-su}e^{-sv}\, dv \\ &= \int_0^\infty |f(u)|e^{-su}\, du \int_0^\infty |g(v)|e^{-sv}\, dv. \end{aligned}$$

The two one-dimensional integrals here are assumed to be convergent, which means that the double integral also converges. But this in turn means that the improper double integral *without* modulus signs,

$$\Phi(s) = \iint_Q f(u)g(v)e^{-s(u+v)}\, du\, dv$$

is absolutely convergent. It can then also be computed in any manner, and we do it in two ways. One way is imitating the previous calculation:

$$\begin{aligned}\Phi(s) &= \int_0^\infty du \int_0^\infty f(u)g(v)e^{-su}e^{-sv}\,dv \\ &= \int_0^\infty f(u)e^{-su}\,du \int_0^\infty g(v)e^{-sv}\,dv = \widetilde{f}(s)\,\widetilde{g}(s).\end{aligned}$$

Another way is integrating on triangles $D_T : u \geq 0,\ v \geq 0,\ u + v \leq T$. But

$$\begin{aligned}\int_0^T f * g(t)e^{-st}\,dt &= \int_0^T \left(\int_0^t f(u)g(t-u)\,du \right) e^{-su}\,dt \\ &= \int_0^T dt \int_0^t f(u)e^{-su}\,g(t-u)e^{-s(t-u)}\,du \\ &= \int_0^T f(u)e^{-su}\,du \int_u^T g(t-u)e^{-s(t-u)}\,dt = \left\{ \begin{matrix} t-u=v \\ dt=dv \end{matrix} \right\} \\ &= \int_0^T f(u)e^{-su}\,du \int_0^{T-u} g(v)e^{-sv}\,dv \\ &= \iint_{D_T} f(u)g(v)e^{-su}e^{-sv}\,du\,dv \to \Phi(s)\end{aligned}$$

as $T \to \infty$. This proves the formula in the theorem. □

Example 3.16. As an illustration of the theorem we can take the situation in Example 3.14. There we have

$$\widetilde{f}(s) = \frac{1}{s-1}, \qquad \widetilde{g}(s) = \frac{1}{s+2},$$

$$\widetilde{f}(s)\widetilde{g}(s) = \frac{1}{(s-1)(s+2)} = \frac{\frac{1}{3}}{s-1} - \frac{\frac{1}{3}}{s+2} = \mathcal{L}[f * g](s).$$

□

Example 3.17. Find a function f that satisfies the integral equation

$$f(t) = 1 + \int_0^\infty f(t-u)\sin u\,du, \qquad t \geq 0.$$

Solution. Suppose that $f \in \mathcal{E}$. Then we can transform the equation to get

$$\widetilde{f}(s) = \frac{1}{s} + \widetilde{f}(s) \cdot \frac{1}{s^2+1},$$

from which we solve

$$\widetilde{f}(s) = \frac{s^2+1}{s^2} \cdot \frac{1}{s} = \frac{s^2+1}{s^3} = \frac{1}{s} + \frac{1}{s^3},$$

and we see that $f(t) = 1 + \frac{1}{2}t^2$ ought to be a solution. Indeed it is, because this function belongs to $\mathcal{E}$, and then our successive steps make up a sequence of equivalent statements. (It is also possible to check the solution by substitution in the given integral equation. This should be done, if time permits.) □

Exercises

3.29 Calculate directly the convolution of e^{at} and e^{bt} (consider separately the cases $a \neq b$ and $a = b$). Check the result by taking Laplace transforms.

3.30 Use the convolution formula to determine f if $\widetilde{f}(s)$ is given by (a) $s^{-1}(s+1)^{-1}$, (b) $s^{-1}(s^2+a^2)^{-1}$.

3.31 Find a function with the Laplace transform $\dfrac{s^2}{(s^2+1)^2}$.

3.32 Find a function f such that

$$\int_0^x e^{-y} \cos y \, f(x-y) \, dy = x^2 \, e^{-x}, \quad x \geq 0.$$

3.33 Find a solution of the integral equation

$$\int_0^t (t-u)^2 f(u) \, du = t^3, \quad t \geq 0.$$

3.34 Find two solutions of the integral equation (3.2) on page 41.

3.35 Find a function $y(t)$ that satisfies $y(0) = 0$ and

$$2 \int_0^t (t-u)^2 \, y(u) \, du + y'(t) = (t-1)^2 \quad \text{for } t > 0.$$

3.36 Find a function $f(t)$ for $t \geq 0$, that satisfies

$$f(0) = 1, \qquad f'(t) + 3f(t) + \int_0^t f(u) e^{u-t} \, du = \begin{cases} 0, & 0 \leq t < 2, \\ 1, & t > 2 \end{cases}.$$

3.37 Find a solution f of the integral-differential equation

$$5e^{-t} \int_0^t e^y \cos 2(t-y) \, f(y) \, dy = f'(t) + f(t) - e^{-t}, \quad f(0) = 0.$$

3.38 Prove the following result: if $f \in \mathcal{E}_{k_1}$ and $g \in \mathcal{E}_{k_2}$, then $f * g \in \mathcal{E}_k$ for all $k > \max\{k_1, k_2\}$.

3.5 *Laplace transforms of distributions

Laplace transforms can be used in the study of physical phenomena that take place in a time interval that starts at a certain moment, at which the clock is set to $t = 0$. It is possible to allow the functions to include instantaneous pulses and even more far-reaching generalizations of the classical notion of a function – i.e., to allow so-called distributions into the game. When we do so, it will normally be a good thing to allow such things to happen also at the very moment $t = 0$, so we modify slightly the definition of the Laplace transform into the following formula:

$$\widetilde{f}(s) = \int_{0-}^{\infty} f(t)e^{-st}\,dt = \lim_{\varepsilon \searrow 0} \int_{-\varepsilon}^{\infty} f(t)e^{-st}\,dt.$$

If f is an ordinary function, the modified definition agrees with the former one. But if f is a distribution, something new may occur.

As an example, let $\delta_a(t)$ be the Dirac pulse at the point a, where $a \geq 0$. Then

$$\widetilde{\delta_a}(s) = \int_{0-}^{\infty} \delta_a(t)e^{-st}\,dt = e^{-as}.$$

In particular, if $a = 0$, we get $\widetilde{\delta}(s) = 1$. We see that the rule that a Laplace transform must tend to zero as $s \to \infty$ no longer need hold for transforms of distributions.

The formula for the transform of a derivative must also be slightly modified. Indeed, integration by parts gives

$$\widetilde{f'}(s) = \int_{0-}^{\infty} f'(t)e^{-st}\,dt = \Big[f(t)e^{-st}\Big]_{0-}^{\infty} + s\int_{0-}^{\infty} f(t)e^{-st}\,dt = s\widetilde{f}(s) - f(0-),$$

where $f(0-)$ is the left-hand limit of $f(t)$ at 0. This may cause some confusion when dealing with functions that are considered to be zero for negative t but nonzero for positive t. In this case it may now happen that f' includes a multiple of δ, which explains the different appearance of the formula. In this situation, it is preferable to be very explicit in supplying the factor $H(t)$ in the description of functions.

Example 3.18. Solve the initial value problem

$$y'' + 4y' + 13y = \delta'(t), \quad y(0-) = y'(0-) = 0.$$

Solution. Transformation gives

$$(s^2+4s+13)\widetilde{y} = s \iff \widetilde{y} = \frac{s}{(s+2)^2+9} = \frac{s+2}{(s+2)^2+9} - \frac{2}{3}\cdot\frac{3}{(s+2)^2+9}.$$

The solution is found to be

$$y(t) = e^{-2t}(\cos 3t - \tfrac{2}{3}\sin 3t)H(t).$$

We check it by differentiating:

$$\begin{aligned} y'(t) &= e^{-2t}(-2\cos 3t + \tfrac{4}{3}\sin 3t - 3\sin 3t - 2\cos 3t)H(t) + \delta(t) \\ &= e^{-2t}(-4\cos 3t - \tfrac{5}{3}\sin 3t)H(t) + \delta(t), \\ y''(t) &= e^{-2t}(8\cos 3t + \tfrac{10}{3}\sin 3t + 12\sin 3t - 5\cos 3t)H(t) - 4\delta(t) + \delta'(t) \\ &= e^{-2t}(3\cos 3t + \tfrac{46}{3}\sin 3t)H(t) - 4\delta(t) + \delta'(t). \end{aligned}$$

Substituting this into the left-hand member of the equation, one sees that it indeed solves the problem. □

Example 3.19. Find the *general* solution of the differential equation $y'' + 3y' + 2y = \delta$.

Solution. It should be wellknown that the solution can be written as the sum of the general solution y_H of the corresponding *homogeneous* equation $y'' + 3y' + 2y = 0$, and one particular solution y_P of the given equation. We easily find $y_H = C_1 e^{-t} + C_2 e^{-2t}$, and proceed to look for y_P. In doing this we assume that $y_P(0-) = y_P'(0-) = 0$, which gives the simplest Laplace transforms. Indeed, $\widetilde{y_P'} = s\widetilde{y_P}$ and $\widetilde{y_P''} = s^2\widetilde{y_P}$, so that

$$s^2\widetilde{y_P} + 3s\widetilde{y_P} + 2\widetilde{y_P} = 1 \iff \widetilde{y_P} = \frac{1}{(s+1)(s+2)} = \frac{1}{s+1} - \frac{1}{s+2}.$$

Thus it turns out that

$$y_P = (e^{-t} - e^{-2t})H(t).$$

This means that the solution of the given problem is

$$\begin{aligned} y &= C_1 e^{-t} + C_2 e^{-2t} + (e^{-t} - e^{-2t})H(t) \\ &= (C_1 + H(t))e^{-t} + (C_2 - H(t))e^{-2t} \\ &= \begin{cases} C_1 e^{-t} + C_2 e^{-2t}, & t < 0, \\ (C_1 + 1)e^{-t} + (C_2 - 1)e^{-2t}, & t > 0. \end{cases} \end{aligned}$$

We can see that in each of the intervals $t < 0$ and $t > 0$ these expressions are solutions of the homogeneous equation, which is in accordance with the fact that $\delta = 0$ in the intervals. What happens at $t = 0$ is that the constants change value in such a way that the first derivative has a jump discontinuity and the second derivative contains a δ pulse (draw pictures!). □

The particular solution y_P found in the preceding problem is called a *fundamental solution* of the equation. Let us now denote it by E; thus,

$$E(t) = (e^{-t} - e^{-2t})H(t).$$

It is useful in the following situation. Let f be any function, continuous for $t \geq 0$. We want to find a solution of the problem $y'' + 3y' + 2y = f$. If we assume $y(0-) = y'(0-) = 0$, we get

$$\widetilde{y} = \frac{\widetilde{f}(s)}{s^2 + 3s + 2} = \widetilde{f}(s) \cdot \frac{1}{s^2 + 3s + 2} = \widetilde{f}(s)\widetilde{E}(s).$$

This means that y can be found as the convolution of f and E:

$$y(t) = f * E(t) = \int_0^t f(t-u)\big(e^{-u} - e^{-2u}\big)\,du.$$

The fundamental solution thus provides a means for finding a particular solution for any inhomogeneuous equation with the given left-hand side.

This idea can be applied to any linear differential equation with constant coefficients. The left-hand member of such an equation can be written in the form $P(D)y$, where D is the differentiation operator and $P(\cdot)$ is a polynomial. For example, if $P(r) = r^2 + 3r + 2$, then

$$P(D)y = (D^2 + 3D + 2)y = y'' + 3y' + 2y.$$

The fundamental solution E is, in the general case, the function such that

$$\widetilde{E}(s) = \frac{1}{P(s)}, \qquad E(t) = 0 \text{ for } t < 0.$$

Exercises

3.39 Find a solution of the differential equation $y''' + 3y'' + 3y' + y = H(t-1) + \delta(t-2)$, that satisfies $y(0) = y'(0) = y''(0) = 0$.

3.40 Solve the differential equation $y'' + 4y' + 5y = \delta(t)$, $y(t) = 0$ for $t < 0$. Then deduce a formula for a particular solution of the equation $y'' + 4y' + 5y = f(t)$, where f is any continuous function such that $f(t) = 0$ for $t < 0$.

3.41 Find fundamental solutions for the following equations: (a) $y'' + 4y = \delta$, (b) $y'' + 4y' + 8y = \delta$, (c) $y''' + 3y'' + 3y' + y = \delta$.

3.42 Find a function y such that $y(t) = 0$ for $t \leq 0$ and

$$y'(t) + 3y(t) + 2\int_0^t y(u)\,du = 2\big(H(t-1) - H(t-2)\big) \text{ for } t > 0.$$

3.43 Find a function $f(t)$ such that $f(t) = 0$ for $t < 0$ and

$$e^{-t}\int_{0-}^{t+} f(p)\,e^p\,dp - f(t) + f'(t) = \delta(t) - t\,e^{-t}\,H(t), \qquad -\infty < t < \infty.$$

3.6 The Z transform

In this section we sketch the theory of a *discrete* analogue of the Laplace transform. We have so far been considering functions $t \mapsto f(t)$, where t is a real variable (mostly thought of as representing time). Now, we shall think of t as a variable that only assumes the values $0, 1, 2, \ldots$, i.e., non-negative integer values. In applications, this is sometimes more realistic than considering a continuous variable; it corresponds to taking measurements at equidistant points in time.

A function of an integer variable is mostly written as a sequence of numbers. This will be the way we do it, at least at the beginning of the section.

Let $\{a_n\}_{n=0}^{\infty}$ be a sequence of numbers. We form the infinite series

$$A(z) = \sum_{n=0}^{\infty} \frac{a_n}{z^n} = \sum_{n=0}^{\infty} a_n z^{-n}.$$

If the series is convergent for some z, then it converges absolutely outside of some circle in the complex plane. More precisely, the domain of convergence is a set of the type $|z| > \sigma$, where $0 \le \sigma \le \infty$. (It may also happen that the series converges at certain points on the circle $|z| = \sigma$, but this is rarely of any importance.) Power series of this kind, that may encompass both positive and negative powers of z, are called LAURENT *series.* (A particular case is Taylor series that do not contain any negative powers of z; in the present situation we are considering a reversed case, with no positive powers.) A necessary and sufficient condition for the series to converge at all is that there exist constants M and R such that $|a_n| \le MR^n$ for all n. This condition is analogous to the condition of exponential growth for functions to have a Laplace transform.

The function $A(z)$ is called the Z *transform* of the sequence $\{a_n\}_{n=0}^{\infty}$. It can be employed to solve certain problems concerning sequences, in a manner that is largely analogous to the way that Laplace transforms can be used for solving problems for ordinary functions. Important applications occur in the theory of electronics, systems engineering, and automatic control.

When working with the Z transformation, one should be familiar with the geometric series. Recall that this is the series

$$\sum_{n=0}^{\infty} w^n,$$

where w is a real or complex number. It is convergent precisely if $|w| < 1$, and its sum is then $1/(1-w)$. This fact is used "in both directions," as the following example shows.

Example 3.20. If $a_n = 1$ for all $n \geq 0$, the Z transform is

$$\sum_{n=0}^{\infty} \frac{1}{z^n} = \sum_{n=0}^{\infty} \left(\frac{1}{z}\right)^n = \frac{1}{1 - \dfrac{1}{z}} = \frac{z}{z-1},$$

which is convergent for all z such that $|z| > 1$. On the other hand, if λ is a nonzero complex number, we can rewrite the function $B(z) = z/(z - \lambda)$ in this way:

$$B(z) = \frac{z}{z-\lambda} = \frac{1}{1 - \dfrac{\lambda}{z}} = \sum_{n=0}^{\infty} \left(\frac{\lambda}{z}\right)^n = \sum_{n=0}^{\infty} \frac{\lambda^n}{z^n}, \quad |z| > |\lambda|,$$

which shows that $B(z)$ is the transform of the sequence $b_n = \lambda^n$ $(n \geq 0)$. (Here we actually use the fact that Laurent expansions are unique, which implies that two different sequences cannot have the same transform.) □

We next present a simple, but typical, problem where the transform can be used.

Example 3.21. If we know that $a_0 = 1$, $a_1 = 2$ and

$$a_{n+2} = 3a_{n+1} - 2a_n, \qquad n = 0, 1, 2, \ldots, \tag{3.9}$$

find a formula for a_n.

An equation of the type (3.9) is often called a *difference equation*. In many respects, it is analogous to a differential equation: if differential equations are used for the description of processes taking place in "continuous time," difference equations can do the corresponding thing in "discrete time."

To solve the problem in Example 3.21, we multiply the formula (3.9) by z^{-n} and add up for $n = 0, 1, 2, \ldots$:

$$\sum_{n=0}^{\infty} a_{n+2} z^{-n} = 3 \sum_{n=0}^{\infty} a_{n+1} z^{-n} - 2 \sum_{n=0}^{\infty} a_n z^{-n}. \tag{3.10}$$

Now we introduce the Z transform of the sequence $\{a_n\}_{n=0}^{\infty}$:

$$A(z) = \sum_{n=0}^{\infty} a_n z^{-n} = 1 + \frac{2}{z} + \frac{a_2}{z^2} + \frac{a_3}{z^3} + \cdots. \tag{3.11}$$

We notice that, firstly,

$$\sum_{n=0}^{\infty} a_{n+1} z^{-n} = \sum_{k=1}^{\infty} a_k z^{-(k-1)} = z \left(\sum_{k=1}^{\infty} a_k z^{-k} \right) = z \left(\sum_{n=0}^{\infty} a_n z^{-n} - a_0 \right)$$
$$= z(A(z) - 1),$$

and, secondly,

$$\sum_{n=0}^{\infty} a_{n+2} z^{-n} = \sum_{k=2}^{\infty} a_k z^{-(k-2)} = z^2 \left(\sum_{k=2}^{\infty} a_k z^{-k} \right)$$
$$= z^2 \left(\sum_{n=0}^{\infty} a_n z^{-n} - a_0 - \frac{a_1}{z} \right) = z^2 \left(A(z) - 1 - \frac{2}{z} \right).$$

Thus, the equation (3.10) can be written as

$$z^2 \left(A(z) - 1 - \frac{2}{z} \right) = 3z(A(z) - 1) - 2A(z),$$

from which $A(z)$ can be solved. After simplification we have

$$A(z) = \frac{z}{z-2}.$$

We saw in the preceding example that this is the Z transform of the sequence

$$a_n = 2^n, \quad n = 0, 1, 2, \ldots.$$

We can check the result by returning to the statement of the problem: $a_0 = 1$ and $a_1 = 2$ are all right; and if $a_n = 2^n$ and $a_{n+1} = 2^{n+1}$, then

$$3a_{n+1} - 2a_n = 3 \cdot 2^{n+1} - 2 \cdot 2^n = 3 \cdot 2^{n+1} - 2^{n+1} = 2 \cdot 2^{n+1} = 2^{n+2},$$

which is also right. □

In the example, it is obvious from the beginning that the solution is unique. If a_0 and a_1 are given, the formula (3.9) produces the subsequent values of the a_n in an unequivocal way. In general, problems about number sequences are often uniquely determined in the same manner. However, just as for the Laplace transform, the Z transform cannot be expected to give solutions if these are very fast-growing sequences.

We take a closer look at the correspondence between sequences $\{a_n\}_{n=0}^{\infty}$ and their Z transforms $A(z)$. In order to have an efficient notation we write $a = \{a_n\}_{n=0}^{\infty}$ and $A = \mathcal{Z}[a]$. Thus, $\mathcal{Z}$ denotes a mapping from (a subset of) the set of number sequences to the set of Laurent series convergent outside of some circle.

Example 3.22. We have already seen that if $a = \{\lambda^n\}_0^{\infty}$, then

$$\mathcal{Z}[a](z) = \sum_{n=0}^{\infty} \lambda^n z^{-n} = \frac{z}{z-\lambda}, \quad |z| > |\lambda|.$$

□

Example 3.23. If $a = \{1/n!\}_0^{\infty}$, then $\mathcal{Z}[a](z) = \sum_{n=0}^{\infty} \frac{z^{-n}}{n!} = e^{1/z}$, $|z| > 0$.

□

Example 3.24. The sequence $a = \{n!\}_0^\infty$ has no Z transform, because the series $\sum_{n=0}^{\infty} n!\, z^{-n}$ diverges for all z. □

As stated at the beginning of this section, a sufficient (and actually necessary) condition for $A(z)$ to exist is that the numbers a_n grow at most exponentially: $|a_n| \le MR^n$ for some numbers M and R. It is easy to see that this condition implies the convergence of the series for all z with $|z| > R$.

Some computational rules for the transformation $\mathcal{Z}$ have been collected in the following theorem. In the interest of brevity we introduce some notation for operations on number sequences (which can be viewed as functions $\mathbf{N} \to \mathbf{C}$). If we let $a = \{a_n\}_{n=0}^\infty$ and $b = \{b_n\}_{n=0}^\infty$, we write $a+b = \{a_n + b_n\}_{n=0}^\infty$; and if furthermore λ is a complex number, we put $\lambda a = \{\lambda a_n\}_{n=0}^\infty$. We also agree to write

$$A = \mathcal{Z}[a], \quad B = \mathcal{Z}[b].$$

The "radius of convergence" of the Z transform of a is denoted by σ_a: this means that the series is convergent for $|z| > \sigma_a$ (and divergent for $|z| < \sigma_a$).

Theorem 3.7 (i) *The transformation $\mathcal{Z}$ is linear, i.e.,*

$$\begin{aligned} \mathcal{Z}[\lambda a](z) &= \lambda \mathcal{Z}[a](z), && |z| > \sigma_a, \\ \mathcal{Z}[a+b](z) &= \mathcal{Z}[a](z) + \mathcal{Z}[b](z), && |z| > \max(\sigma_a, \sigma_b). \end{aligned}$$

(ii) *If λ is a complex number and $b_n = \lambda^n a_n$, $n = 0, 1, 2, \ldots$, then*

$$B(z) = A(z/\lambda), \quad |z| > \lambda \sigma_a.$$

(iii) *If k is a fixed integer > 0 and $b_n = a_{n+k}$, $n = 0, 1, 2, \ldots$, then*

$$\begin{aligned} B(z) &= z^k \left(A(z) - a_0 - \frac{a_1}{z} - \cdots - \frac{a_{k-1}}{z^{k-1}} \right) \\ &= z^k A(z) - a_0 z^k - a_1 z^{k-1} - \cdots - a_{k-1} z, \qquad |z| > \sigma_a. \end{aligned}$$

(iv) *Conversely, if k is a positive integer and $b_n = a_{n-k}$ for $n \ge k$ and $b_n = 0$ for $n < k$, then $B(z) = z^{-k} A(z)$.*

(v) *If $b_n = n a_n$, $n = 0, 1, 2, \ldots$, then*

$$B(z) = -z\, A'(z), \quad |z| > \sigma_a.$$

Proof. The assertions follow rather immediately from the definitions. We saw a couple of cases of (iii) in Example 3.21 above. We content ourselves by sketching the proofs of (ii) and (v). For (ii) we find

$$B(z) = \sum_{n=0}^{\infty} b_n z^{-n} = \sum_{n=0}^{\infty} \lambda^n a_n z^{-n} = \sum_{n=0}^{\infty} a_n \left(\frac{z}{\lambda} \right)^{-n} = A(z/\lambda).$$

And as for (v), the right-hand side is

$$-z \cdot \frac{d}{dz} \sum_{n=0}^{\infty} a_n z^{-n} = -z \sum_{n=0}^{\infty} (-n) a_n z^{-n-1} = \sum_{n=0}^{\infty} n a_n z^{-n} = \text{left-hand side.}$$

□

Example 3.25. Example 3.23 and rule (ii) give us the transform of the sequence $\{\lambda^n/n!\}_0^\infty$, viz.,

$$\Lambda(z) = e^{1/(z/\lambda)} = e^{\lambda/z}.$$

□

When solving problems concerning the Z transform, you should have a table at hand, containing rules of computation as well as actual transforms. Such a table is included at the end of this book (p. 250).

Example 3.26. Find a formula for the so-called FIBONACCI numbers, which are defined by $f_0 = f_1 = 1$, $f_{n+2} = f_{n+1} + f_n$ for $n \geq 0$.

Solution. Let $F = \mathcal{Z}[f]$. If we Z-transform the recursion formula, using (iii) from the theorem, we get

$$z^2 F(z) - z^2 - z = (zF(z) - z) + F(z),$$

whence $(z^2 - z - 1)F(z) = z^2$ and

$$F(z) = \frac{z^2}{z^2 - z - 1} = z \cdot \frac{z}{z^2 - z - 1}.$$

In order to recover f_n, a good idea would be to expand into partial fractions, in the hope that simple expressions could be looked up in the table on page 250. A closer look at this table reveals, however, that it would be a good thing to have a z in the numerator of the partial fractions, instead of just a constant. Thus, here we have peeled off a factor z from $F(z)$ and proceed to expand the remaining expression:

$$\frac{F(z)}{z} = \frac{z}{z^2 - z - 1} = \frac{A}{z - \alpha} + \frac{B}{z - \beta},$$

where

$$\alpha = \frac{1 + \sqrt{5}}{2}, \quad \beta = \frac{1 - \sqrt{5}}{2}, \quad A = \frac{\sqrt{5} + 1}{2\sqrt{5}}, \quad B = \frac{\sqrt{5} - 1}{2\sqrt{5}}.$$

This gives

$$F(z) = \frac{Az}{z - \alpha} + \frac{Bz}{z - \beta}$$

and from the table we conclude that

$$f_n = A\alpha^n + B\beta^n = \frac{\sqrt{5}+1}{2\sqrt{5}}\left(\frac{1+\sqrt{5}}{2}\right)^n + \frac{\sqrt{5}-1}{2\sqrt{5}}\left(\frac{1-\sqrt{5}}{2}\right)^n.$$

This can be rewritten as

$$f_n = \frac{1}{\sqrt{5}}\left[\left(\frac{1+\sqrt{5}}{2}\right)^{n+1} - \left(\frac{1-\sqrt{5}}{2}\right)^{n+1}\right], \quad n = 0, 1, 2, \ldots.$$

(In spite of all the appearances of $\sqrt{5}$ in the expression, it is an integer for all $n \geq 0$.) □

As you can see in this example, the method of expanding rational functions into partial fractions can be useful in dealing with Z transforms, provided one starts out by securing an extra factor z to be reintroduced in the numerators after the expansion.

If a and b are two number sequences, we can form a third sequence, c, called the *convolution* of a and b, by writing

$$c_n = \sum_{k=0}^{n} a_{n-k}b_k = \sum_{k=0}^{n} a_k b_{n-k}, \quad n = 0, 1, 2, \ldots.$$

One writes $c = a * b$, and we also permit ourselves to write things like $c_n = (a * b)_n$. We determine the Z transform $C = \mathcal{Z}[c]$:

$$\begin{aligned} C(z) &= \sum_{n=0}^{\infty}\sum_{k=0}^{n} a_{n-k}b_k\, z^{-n} = \sum_{k=0}^{\infty}\sum_{n=k}^{\infty} a_{n-k}b_k\, z^{-n} \\ &= \sum_{k=0}^{\infty}\sum_{n=k}^{\infty} a_{n-k}z^{-(n-k)}\, b_k z^{-k} = \sum_{k=0}^{\infty} b_k z^{-k} \sum_{n=k}^{\infty} a_{n-k}z^{-(n-k)} \\ &= \sum_{k=0}^{\infty} b_k z^{-k} \sum_{m=0}^{\infty} a_m z^{-m} = A(z)B(z). \end{aligned}$$

The manipulations of the double series are permitted for $|z| > \max(\sigma_a, \sigma_b)$, because in that region everything converges absolutely.

This notion of convolution appears in, e.g., control theory, if a system is considered in discrete time (see Appendix A).

Example 3.27. Find $x(t)$, $t = 0, 1, 2, \ldots$, from the equation

$$\sum_{k=0}^{t} 3^{-k}\, x(t-k) = 2^{-t}, \quad t = 0, 1, 2, \ldots.$$

Solution. The left-hand side is the convolution of x and the function $t \mapsto (1/3)^t$, so that taking Z transforms of both members gives

$$\frac{z}{z-\frac{1}{3}} \cdot X(z) = \frac{z}{z-\frac{1}{2}}.$$

(We have used the result of Example 3.22.) We get

$$z) = \frac{z-\frac{1}{3}}{z-\frac{1}{2}} = \frac{z}{z-\frac{1}{2}} - \frac{1}{3} \cdot \frac{1}{z-\frac{1}{2}},$$

and, using Example 3.22 and rule (iv) of Theorem 3.7, we see that

$$x(t) = \begin{cases} 1 & \text{for } t = 0, \\ \left(\frac{1}{2}\right)^t - \frac{1}{3} \cdot \left(\frac{1}{2}\right)^{t-1} & \text{for } t \geq 1. \end{cases}$$

The final expression can be rewritten as

$$x(t) = \left(\tfrac{1}{2} - \tfrac{1}{3}\right) \cdot \left(\tfrac{1}{2}\right)^{t-1} = \tfrac{1}{6} \cdot 2^{1-t} = \tfrac{1}{3} \cdot 2^{-t}, \quad t \geq 1.$$

□

In a final example, we indicate a way of viewing the Z transform as a particular case of the Laplace transform. Here we use translates of the Dirac delta "function," as in Sec. 3.5.

Example 3.28. Let $\{a_n\}_{n=0}^{\infty}$ be a sequence having a Z transform $A(z)$, and define a function f by

$$f(t) = \sum_{n=0}^{\infty} a_n \delta_n(t) = \sum_{n=0}^{\infty} a_n \delta(t-n).$$

The convergence of this series is no problem, because for any particular t at most one of the terms is different from zero. Its Laplace transform must be

$$\widetilde{f}(s) = \sum_{n=0}^{\infty} \int_{0-}^{\infty} e^{-st} a_n \delta(t-n)\, dt = \sum_{n=0}^{\infty} a_n e^{-ns} = \sum_{n=0}^{\infty} a_n \left(e^s\right)^{-n} = A(e^s).$$

Thus, via a change of variable $z = e^s$, the two transforms are more or less the same thing. □

Exercises

3.44 Determine the Z transforms of the following sequences $\{a_n\}_{n=0}^{\infty}$:

(a) $a_n = \dfrac{1}{2^n}$ (b) $a_n = n \cdot 3^n$ (c) $a_n = n^2 \cdot 2^n$

(d) $a_n = \dbinom{n}{p} = \dfrac{n(n-1)\cdots(n-p+1)}{p!}$ for $n \geq p$, $= 0$ for $0 \leq n \leq p$ (p is a fixed integer).

3.45 Determine the sequence $a = \{a_n\}_{n=0}^{\infty}$, if its Z transform is (a) $A(z) = \dfrac{z}{3z-2}$, (b) $A(z) = \dfrac{1}{z}$.

3.46 Determine the numbers a_n and b_n, $n = 0, 1, 2, \ldots$, if $a_0 = 0$, $b_0 = 1$ and

$$\begin{cases} a_{n+1} + b_n = -2n, \\ a_n + b_{n+1} = 1, \end{cases} \qquad n = 0, 1, 2, \ldots.$$

3.47 Find the numbers a_n and b_n, $n = 0, 1, 2, \ldots$, if $a_0 = 0$, $b_0 = 1$ and

$$\begin{cases} a_{n+1} + b_n = 2, \\ a_n - b_{n+1} = 0, \end{cases} \qquad n = 0, 1, 2, \ldots.$$

3.48 Find a_n, $n = 0, 1, 2, \ldots$, such that $a_0 = a_1 = 0$ and $a_{n+2} - 3a_{n+1} + 2a_n = 1 - 2n$ for $n = 0, 1, 2, \ldots$.

3.49 Find a_n, $n = 0, 1, 2, \ldots$, if $a_0 = a_1 = 0$ and

$$a_{n+2} + 2a_{n+1} + a_n = (-1)^n n, \quad n = 0, 1, 2, \ldots.$$

3.50 Find a_n, $n = 0, 1, 2, \ldots$, if $a_0 = 1$, $a_1 = 3$ and $a_{n+2} + a_n = 2n + 4$ when $n \geq 0$.

3.51 Determine the numbers $y(t)$ for $t = 0, 1, 2, \ldots$, so that

$$\sum_{k=0}^{t} (t-k)\, 3^{t-k} y(k) = \begin{cases} 0, & t = 0, \\ 1, & t = 1, 2, 3, \ldots. \end{cases}$$

3.52 Find a_n for $n \geq 0$, if $a_0 = 0$ and $\displaystyle\sum_{k=0}^{n} k a_{n-k} - a_{n+1} = 2^n$ for $n \geq 0$.

3.53 Determine $x(n)$ for $n = 0, 1, 2, \ldots$, so that

$$x(n) + 2\sum_{k=0}^{n} (n-k)\, x(k) = 2^n, \quad n = 0, 1, 2, \ldots.$$

3.7 Applications in control theory

We return to the "black box" of Sec. 3.4 (p. 53). Such a box can often be described by a differential equation of the type $P(D)y(t) = x(t)$, where x is the input and y the output. If $x(t)$ is taken to be a unit pulse, $x(t) = \delta(t)$, the solution $y(t)$ with $y(t) = 0$ for $t < 0$ is called the *pulse response*, or *impulse response*, of the black box. The pulse response is the same thing as the fundamental solution. In the general case, Laplace transformation will give $P(s)\widetilde{y}(s) = 1$ and thus $\widetilde{y}(s) = 1/P(s)$. The function

$$G(s) = \frac{1}{P(s)}$$

is called the *transfer function* of the box. When solving the general problem

$$P(D)y(t) = x(t), \quad y(t) = 0 \text{ for } t < 0,$$

Laplace transformation will now result in

$$P(s)\widetilde{y}(s) = \widetilde{x}(s)$$

or

$$\widetilde{y}(s) = G(s)\widetilde{x}(s).$$

This formula is actually the Laplace transform of the convolution formula (3.8) of page 53. It provides a quick way of finding the outsignal y to any insignal x. The function g in the convolution is actually the impulse response.

In control theory, great importance is attached to the notion of *stability*. A black box is *stable*, if its impulse response is *transient*, i.e., $g(t)$ tends to zero as time goes by. This means that disturbances in the input will affect the output only for a short time and will not accumulate. If $P(s)$ is a polynomial, the impulse response will be transient if and only if all its zeroes have negative real parts.

Example 3.29. The polynomial $P_1(s) = s^2 + 2s + 2$ has zeroes $s = -1 \pm i$. Both have real part -1, so that the device described by the equation $y'' + 2y' + 2y = x(t)$ is stable. In contrast, the polynomial $P_2(s) = s^2 + 2s - 1$ has zeroes $s = -1 \pm \sqrt{2}$. One of these is positive, which implies that the corresponding black box is unstable. Finally, the polynomial $P_3(s) = s^2 + 1$ has zeroes $s = \pm i$. These have real part zero; the impulse reponse is $g(t) = \sin t$, which is not transient. The situation is considered as unstable. (It is unstable also inasmuch as a small disturbance of the coefficients of $P_3(s)$ can cause the zeroes to move into the right half-plane, which gives rise to exponentially growing solutions.) □

So far, we have assumed that the black box is described in continuous time. In the real world, it is often more realistic to assume that time is discrete, i.e., that input and output are *sampled* at equidistant points in time. For simplicity, we assume that the sampling is done at $t = 0, 1, 2, \ldots$, and that the input signal $x(t)$ and the output $y(t)$ are both zero for $t < 0$. Then, of course, the Z transform is the adequate tool.

A black box is often described by a difference equation of the type

$$y(t+k) + a_{k-1}y(t+k-1) + \cdots + a_2 y(t+2) + a_1 y(t+1) + a_0 y(t) = x(t), \; t \in \mathbf{Z}. \tag{3.12}$$

We introduce the *characteristic polynomial*

$$P(z) = z^k + a_{k-1}z^{k-1} + \cdots + a_2 z^2 + a_1 z + a_0.$$

We assumed that $x(t)$ and $y(t)$ were both zero for negative t. Putting $t = -k$ in (3.12), we find that

$$y(0) = x(-k) - a_{k-1}y(-1) - \cdots - a_1 y(-k+1) - a_0 y(-k),$$

which implies that also $y(0) = 0$. Consequently, putting $t = -k+1$, also $y(1) = 0$, and so on. Not until we have an $x(t)$ that is different from zero do we find a $y(t+k)$ different from zero. Thus we have initial values $y(0) = \cdots = y(k-1) = 0$. By the rules for the Z transform, we can then easily transform the equation (3.12). With obvious notation we get

$$P(z)Y(z) = X(z).$$

Thus,

$$Y(z) = \frac{X(z)}{P(z)} = G(z)X(z),$$

where $G(z) = 1/P(z)$ is the *transfer function.* Just as in the previous situation, it is also the *impulse response,* because it is the output resulting from inputting the signal

$$\delta(t) = 1 \text{ for } t = 0, \quad \delta(t) = 0 \text{ otherwise.}$$

The *stability* of equation (3.12) hinges on the localization of the zeroes of the polynomial $P(z)$. As can be seen from a table of Z transforms, a zero a of $P(z)$ implies that the solution contains terms involving a^t. Thus we have stability precisely if *all the zeroes of* $P(z)$ *are in the interior of the unit disc* $|z| < 1$.

Example 3.30. The difference equation $y(t+2)+\frac{1}{2}y(t+1)+\frac{1}{4}y(t) = x(t)$ has $P(z) = z^2+\frac{1}{2}z+\frac{1}{4}$ with zeroes $z = -\frac{1}{4}\pm\frac{\sqrt{3}}{4}i$. These satisfy $|z| = \frac{1}{2} < 1$, so that the equation is stable. The equation

$$y(t+3) + 2y(t+2) - y(t+1) + 2y(t) = x(t)$$

is unstable. This can be seen from the constant term ($= 2$) of the characteristic polynomial; as is well known, this term is (plus or minus) the product of the zeroes, which implies that these cannot all be of modulus less than one. □

More sophisticated methods for localizing the zeroes of polynomials can be found in the literature on complex analysis and in books dealing with these applications.

Exercises

3.54 Investigate the stability of the following equations:
(a) $y''+2y'+3y = x(t)$, (b) $y'''+3y''+3y'+y = x(t)$, (c) $y''+4y = x(t)$.

3.55 Are these difference equations stable or unstable?
(a) $2y(t+2) - 2y(t+1) + y(t) = x(t)$,
(b) $y(t+2) - y(t+1) + y(t) = x(t)$,
(c) $2y(t+3) - y(t+2) + 3y(t+1) + 3y(t) = x(t)$.

Summary of Chapter 3

To provide an overview of the results of this chapter, we collect the main definitions and theorems here. The precise details of the conditions for the validity of the results are sometimes indicated rather sketchily. Thus, this summary should serve as a memory refresher. Details should be looked up in the core of the text. Facts that rather belong in a table of transforms, such as rules of computation, are not included here, but can be found at the end of the book (p. 247 ff).

Definition
If $f(t)$ is defined for $t \in \mathbf{R}$ and $f(t) = 0$ for $t < 0$, its *Laplace transform* is defined by

$$\widetilde{f}(s) = \int_0^\infty f(t)e^{-st}\,dt,$$

provided the integral is abolutely convergent for some value of s.

Theorem
For $\widetilde{f}$ to exist it is sufficient that f grows at most exponentially, i.e., that $|f(t)| \le Me^{kt}$ for some constants M and k.

Theorem
If $\widetilde{f}(s) = \widetilde{g}(s)$ for all (sufficiently large) s, then $f(t) = g(t)$ for all t where both f and g are continuous.

Theorem
If we define the *convolution* $h = f * g$ by

$$h(t) = f * g(t) = \int_0^t f(t-u)g(u)\,du = \int_0^t f(u)g(t-u)\,du,$$

then its Laplace transform is $\widetilde{h} = \widetilde{f}\widetilde{g}$.

Definition
If $\{a_n\}_{n=0}^\infty$ is a sequence of numbers, its *zeta transform* is defined by

$$A(z) = \sum_{n=0}^\infty a_n z^{-n},$$

provided the series is convergent for some value of z. This holds if $|a_n| \le MR^n$ for some constants M and R.

Historical notes

The Laplace transform is, not surprisingly, found in the works of Pierre Simon de Laplace, notably his *Théorie analytique des probabilités* of 1812. In this book, he made free use of Laplace transforms and also generating functions (which are related to the Z transform) in a way that baffled his contemporaries. During the nineteenth century, the technique was developed further, and also influenced by similar ideas such as the "operational calculus" of OLIVER HEAVISIDE (British physicist and applied mathematician, 1850–1925). With the development of modern technology in computing and control theory, the importance of these methods has grown enormously.

Problems for Chapter 3

3.56 Solve the system $y' - 2z = (1-t)e^{-t}$, $z' + 2y = 2te^{-t}$, $t > 0$, with initial conditions $y(0) = 0$, $z(0) = 1$.

3.57 Solve the problem $y'' + 2y' + 2y = 5e^t$, $t > 0$; $y(0) = 1$, $y'(0) = 0$.

3.58 Solve the problem $y''' + y'' + y' - 3y = 1$, $t > 0$, when $y(0) = y'(0) = 0$, $y''(0) = 1$.

3.59 Solve the problem $y'' + 4y = f(t)$, $t > 0$; $y(0) = 0$, $y'(0) = 1$, where

$$f(t) = \begin{cases} (t-1)^2, & t \geq 1 \\ 0, & 0 < t < 1. \end{cases}$$

3.60 Find $y = y(t)$ for $t > 0$ that solves $y'' - 4y' + 5y = \varphi(t)$, $y(0) = 2$, $y'(0) = 0$, where $\varphi(t) = 0$ for $t < 2$, $\varphi(t) = 5$ for $t > 2$.

3.61 Find $f(t)$ for $t \geq 0$, such that $f(0) = 1$ and

$$8\int_0^t f(t-u)\,e^{-u}\,du + f'(t) - 3f(t) + 2e^{-t} = 0, \quad t > 0.$$

3.62 Let f be the function described by

$$f(t) = 0,\ t \leq 0; \quad f(t) = t,\ 0 < t \leq 1; \quad f(t) = 1,\ t > 1.$$

Solve the differential equation $y''(t) + y(t) = f(t)$ with initial values $y(0) = 0$, $y'(0) = 1$.

3.63 Solve $y''' + y' = t - 1$, $y(0) = 2$, $y'(0) = y''(0) = 0$.

3.64 Solve $y''' + 3y'' + 3y' + y = t + 3$, $t > 0$; $y(0) = 0$, $y'(0) = 1$, $y''(0) = 2$.

3.65 Solve the initial value problem

$$\begin{cases} z'' - y' = e^{-t}, \\ y'' + y' + z' + z = 0, \end{cases} \quad t > 0; \quad \begin{array}{ll} y(0) = 0, & y'(0) = 1; \\ z(0) = 0, & z'(0) = -1. \end{array}$$

3.66 Find f such that $f(0) = 1$ and

$$2e^{-t}\int_0^t (t-u)\,e^{u}\,f(u)\,du + f'(t) + 2t^2e^{-t} = 0.$$

3.67 Solve the problem

$$\begin{cases} y''(t) + 2z'(t) - y(t) = 4e^t, \\ z''(t) - 2y'(t) - z(t) = 0, \end{cases} \quad t > 0; \quad \begin{array}{r} y(0) = 0, \quad y'(0) = 2, \\ z(0) = z'(0) = 0. \end{array}$$

3.68 Solve $y''' + y'' + y' + y = 4e^{-t}$, $t > 0$; $y(0) = 0$, $y'(0) = 3$, $y''(0) = -6$.

3.69 Find f that solves

$$\int_0^t f(u)(t-u)\sin(t-u)\,du - 2f'(t) = 12e^{-t}, \quad t > 0; \quad f(0) = 6.$$

3.70 Solve $y''' + 3y'' + y' - 5y = 0$, $t > 0$; $y(0) = 1$, $y'(0) = -2$, $y''(0) = 3$.

3.71 Solve $y'''(t) + y''(t) + 4y'(t) + 4y(t) = 8t + 4$, $t > 0$, with initial values $y(0) = -1$, $y'(0) = 4$, $y''(0) = 0$.

3.72 Solve $y''' + y'' + y' + y = 2e^{-t}$, $t > 0$; $y(0) = 0$, $y'(0) = 2$, $y''(0) = -2$.

3.73 Find a solution $y = y(t)$ for $t > 0$ to the initial value problem $y'' + 2ty' - 4y = 1$, $y(0) = y'(0) = 0$.

3.74 Find a solution of the partial differential equation $u_{tt} + 2u_t + xu_x + u = xt$ for $x > 0$, $t > 0$, such that $u(x,0) = u_t(x,0) = 0$ for $x > 0$ and $u(0,t) = 0$ for $t > 0$.

3.75 Use Laplace transformation to find a solution of

$$y''(t) - ty'(t) + y(t) = 5, \quad t > 0; \qquad y(0) = 5, \quad y'(0) = 3.$$

3.76 Find f such that $f(t) = 0$ for $t < 0$ and

$$5e^{-t}\int_0^t e^y \cos 2(t-y)\, f(y)\,dy = f'(t) + f(t) - e^{-t}, \quad t > 0.$$

3.77 Solve the integral equation $y(t) + \int_0^t (t-u)\,y(u)\,du = 3\sin 2t$.

3.78 Solve the difference equation $a_{n+2} - 2a_{n+1} + a_n = b_n$ for $n \geq 0$ with initial values $a_0 = a_1 = 0$ and right-hand member (a) $b_n = 1$, (b) $b_n = e^n$, (c) $b_0 = 1$, $b_n = 0$ for $n > 0$.

4
Fourier series

4.1 Definitions

We are going to solve, as far as we can, the approximation problem that was presented in Sec. 1.4. The strategy will perhaps appear somewhat surprising: starting from a function f, we shall define a certain series, and in due time we shall find that the function can be recovered from this series in various ways.

All functions that we consider will have period 2π. The whole theory could just as well be carried through for functions having some other period. This is equivalent to the standard case that we treat, via a simple linear transformation of the independent variable. The formulae that hold in the general case are collected in Sec. 4.5.

A function defined on $\mathbf{R}$ with period 2π can alternatively be thought of as defined on the unit circle $\mathbf{T}$, the variable being the polar coordinate. We shall frequently take this point of view. For example, the integral of f over an interval of one period can be written $\int_{\mathbf{T}} f(t)\,dt$. When we want to compute this integral, we can choose any convenient period interval for the actual calculations:

$$\int_{\mathbf{T}} = \int_{-\pi}^{\pi} = \int_{0}^{2\pi} = \int_{a}^{a+2\pi}, \qquad a \in \mathbf{R}.$$

(If $\mathbf{T}$ is viewed as a circle, the integral $\int_{\mathbf{T}} f(t)\,dt$ is *not* to be considered as a line integral of the sort used to calculate amounts of work in mechanics, or that appears in complex analysis. Instead, it is a line integral *with respect to arc length.*)

One must be careful when working on $\mathbf{T}$ and speaking of notions such as *continuity.* The statement that $f \in C(\mathbf{T})$ must mean that f is continuous at all points of the circle. If we switch to viewing f as a 2π-periodic function, this function must also be continuous. The formula $f(t) = t$ for $-\pi < t < \pi$, for instance, defines a function that *cannot* be made continuous on $\mathbf{T}$: at the point on $\mathbf{T}$ that corresponds to $t = \pm\pi$, the limits of $f(t)$ from different directions are different.

Similar care must be taken when speaking of functions belonging to $C^k(\mathbf{T})$, i.e., having continuous derivatives of orders up to and including k. As an example, the definition $g(t) = t^2$, $|t| \leq \pi$, describes a function that is in $C(\mathbf{T})$, but not in $C^1(\mathbf{T})$. The first derivative does not exist at $t = \pm\pi$. This can be seen graphically by drawing the periodic continuation, which has corners at these points (sketch a picture!).

Let us now do a preparatory maneuver. Suppose that a function f is the sum of a series

$$f(t) = \sum_{n=-\infty}^{\infty} c_n\, e^{int} = \sum_{n\in\mathbf{Z}} c_n\, e^{int}. \tag{4.1}$$

We assume that the coefficients c_n are complex numbers such that

$$\sum_{n\in\mathbf{Z}} |c_n| < \infty.$$

By the Weierstrass M-test, the series actually converges absolutely and uniformly, since $|e^{int}|$ is always equal to 1. Each term of the series is continuous and has period 2π, and the sum function f inherits both these properties.

Now let m be any integer (positive, negative, or zero), and multiply the series by e^{-imt}. It will still converge uniformly, and it can be integrated term by term over a period, such as the interval $(-\pi, \pi)$:

$$\int_{-\pi}^{\pi} f(t)\, e^{-imt}\, dt = \int_{-\pi}^{\pi} \sum_{n\in\mathbf{Z}} c_n\, e^{i(n-m)t}\, dt = \sum_{n\in\mathbf{Z}} c_n \int_{-\pi}^{\pi} e^{i(n-m)t}\, dt.$$

But it is readily seen that

$$\int_{-\pi}^{\pi} e^{ikt}\, dt = \begin{cases} 2\pi, & k = 0, \\ 0, & k \neq 0. \end{cases}$$

It follows that all the terms in the sum vanish, except the one where $n-m = 0$, which is the same thing as $n = m$, and the result is that

$$\int_{-\pi}^{\pi} f(t)\, e^{-imt}\, dt = 2\pi c_m.$$

Thus, for an absolutely convergent series of the form (4.1), the coefficients can be computed from the sum function using this formula. This fact can be taken as a motivation for the following definition.

Definition 4.1 *Let f be a function with period 2π that is absolutely Riemann-integrable over a period. Define the numbers c_n, $n \in \mathbf{Z}$, by*

$$c_n = \frac{1}{2\pi} \int_{\mathbf{T}} f(t)\, e^{-int}\, dt = \frac{1}{2\pi} \int_{-\pi}^{\pi} f(t)\, e^{-int}\, dt.$$

These numbers are called the Fourier coefficients *of f, and the* Fourier series *of f is the series*

$$\sum_{n \in \mathbf{Z}} c_n\, e^{int}.$$

Notice that the definition does not state anything about the convergence of the series, even less what its sum might be if it happens to converge. It is the main task of this chapter to investigate these questions.

When dealing simultaneously with several functions and their Fourier coefficients it is convenient to indicate to what function the coefficients belong by writing things like $c_n(f)$. Another commonly used way of denoting the Fourier coefficients of f is $\widehat{f}(n)$.

When we want to state, as a formula, that f has a certain Fourier series, we write

$$f(t) \sim \sum_{n \in \mathbf{Z}} c_n\, e^{int}.$$

This means nothing more or less than the fact that the numbers c_n are computable from f using certain integrals.

There are a number of alternative ways of writing the terms in a Fourier series. For instance, when dealing with real-valued functions, the complex-valued functions e^{int} are often felt to be rather "unnatural." One can then write $e^{int} = \cos nt + i \sin nt$ and reshape the two terms corresponding to $\pm n$ like this:

$$\begin{aligned} c_n e^{int} + c_{-n} e^{-int} &= c_n(\cos nt + i \sin nt) + c_{-n}(\cos nt - i \sin nt) \\ &= (c_n + c_{-n}) \cos nt + i(c_n - c_{-n}) \sin nt = a_n \cos nt + b_n \sin nt, \\ &\qquad\qquad n = 1, 2, \ldots. \end{aligned}$$

In the special case $n = 0$ we have only one term, c_0. This gives a series of the form

$$c_0 + \sum_{n=1}^{\infty} (a_n \cos nt + b_n \sin nt).$$

The coefficients in this series are given by new integral formulae:

$$a_n = c_n + c_{-n} = \frac{1}{2\pi} \int_{\mathbf{T}} f(t) e^{-int}\, dt + \frac{1}{2\pi} \int_{\mathbf{T}} f(t) e^{int}\, dt$$

$$= \frac{1}{\pi}\int_{\mathbf{T}} f(t)\,\tfrac{1}{2}(e^{int}+e^{-int})\,dt = \frac{1}{\pi}\int_{\mathbf{T}} f(t)\cos nt\,dt, \quad n = 1,2,3,\ldots,$$

and similarly one shows that

$$b_n = \frac{1}{\pi}\int_{\mathbf{T}} f(t)\sin nt\,dt, \quad n = 1,2,3,\ldots.$$

If we extend the validity of the formula for a_n to $n = 0$, we find that $a_0 = 2c_0$. For this reason the Fourier series is commonly written

$$f(t) \sim \tfrac{1}{2}a_0 + \sum_{n=1}^{\infty}(a_n \cos nt + b_n \sin nt). \tag{4.2}$$

This is sometimes called the "real" or *trigonometric* version of the Fourier series for f. It should be stressed that this is nothing but a different way of writing the series — it is really the same series as in the definition.

The terms in the series (4.2) can be interpreted as vibrations of different frequencies. The constant term $\frac{1}{2}a_0$ is a "DC component," the term $a_1 \cos t + b_1 \sin t$ has period 2π, the term with $n = 2$ has half the period length, for $n = 3$ the period is one-third of 2π, etc. These terms can be written in yet another way, that emphasizes this physical interpretation. The reader should be familiar with the fact that the sum of a cosine and a sine with the same period can always be rewritten as a single cosine (or sine) function with a phase angle:

$$a\cos nt + b\sin nt = \sqrt{a^2+b^2}\left(\frac{a}{\sqrt{a^2+b^2}}\cos nt + \frac{b}{\sqrt{a^2+b^2}}\sin nt\right)$$
$$= \sqrt{a^2+b^2}(\cos\alpha\cos nt + \sin\alpha\sin nt) = \sqrt{a^2+b^2}\cos(nt-\alpha),$$

where the *phase angle* α is a number such that $\cos\alpha = a/\sqrt{a^2+b^2}$, $\sin\alpha = b/\sqrt{a^2+b^2}$. This means that (4.2) can be written in the form

$$\sum_{n=0}^{\infty} A_n \cos(nt - \alpha_n). \tag{4.3}$$

This is sometimes called the *physical* version of the Fourier series. In this formula one can immediately see the *amplitude* A_n of each partial frequency. In this text, however, we shall not work with this form of the series, since it is slightly unwieldy from a mathematical point of view.

When asked to compute the Fourier series of a specific function, it is normally up to the reader to choose what version to work with. This is illustrated by the following examples.

Example 4.1. Define f by saying that $f(t) = e^t$ for $-\pi < t < \pi$ and $f(t + 2\pi) = f(t)$ for all t. (This leaves $f(t)$ undefined for $t = (2n+1)\pi$,

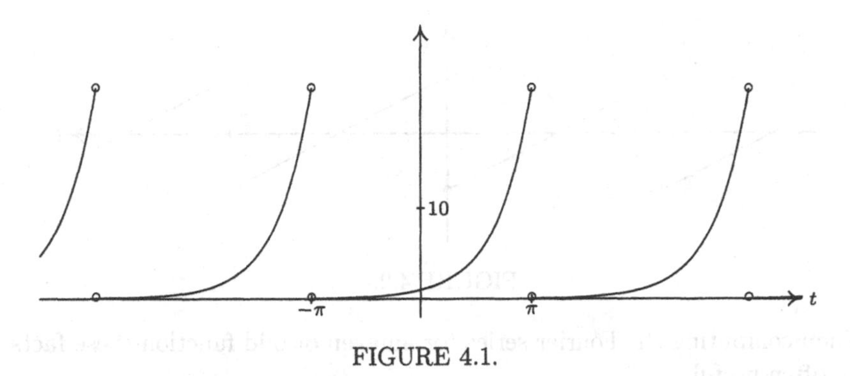

FIGURE 4.1.

but this does not matter. The value of a function at one point or another does not affect the values of its Fourier coefficients!) We get a function with period 2π (see Figure 4.1). Its Fourier coefficients are

$$\begin{aligned} c_n &= \frac{1}{2\pi}\int_{-\pi}^{\pi} e^t\, e^{-int}\,dt = \frac{1}{2\pi}\int_{-\pi}^{\pi} e^{(1-in)t}\,dt = \frac{1}{2\pi}\left[\frac{e^{(1-in)t}}{1-in}\right]_{t=-\pi}^{\pi} \\ &= \frac{e^{\pi-in\pi}-e^{-\pi+in\pi}}{2\pi(1-in)} = \frac{(-1)^n(e^\pi-e^{-\pi})}{2\pi(1-in)} = \frac{(-1)^n\sinh\pi}{\pi(1-in)}. \end{aligned}$$

Here we used the fact that $e^{\pm in\pi} = (-1)^n$. Now we can write

$$f(t) \sim \frac{1}{\pi}\sum_{n\in\mathbf{Z}}\frac{(-1)^n\sinh\pi}{1-in}\,e^{int} = \frac{\sinh\pi}{\pi}\sum_{n\in\mathbf{Z}}\frac{(-1)^n}{1-in}\,e^{int}.$$

□

We remind the reader of a couple of notions of symmetry that turn out to be useful in connection with Fourier series. A function f defined on $\mathbf{R}$ is said to be *even*, if $f(-t) = f(t)$ for all $t \in \mathbf{R}$. A function f is *odd*, if $f(-t) = -f(t)$. (The terms should bring to mind the special function $f(t) = t^n$, which is even if n is an even integer, odd if n is an odd integer.) An odd function f on a symmetric interval $(-a, a)$ has the property that the integral over $(-a, a)$ is equal to zero. This has useful consequences for the so-called real Fourier coefficients a_n and b_n. If f is even and has period 2π, the sine coefficients b_n will be zero, and furthermore the cosine coefficients will be given by the formula

$$f \text{ even} \quad \Rightarrow \quad a_n = \frac{2}{\pi}\int_0^{\pi} f(t)\cos nt\,dt.$$

In an analogous way, an odd function has all cosine coefficients equal to zero, and its sine coefficients are given by

$$f \text{ odd} \quad \Rightarrow \quad b_n = \frac{2}{\pi}\int_0^{\pi} f(t)\sin nt\,dt.$$

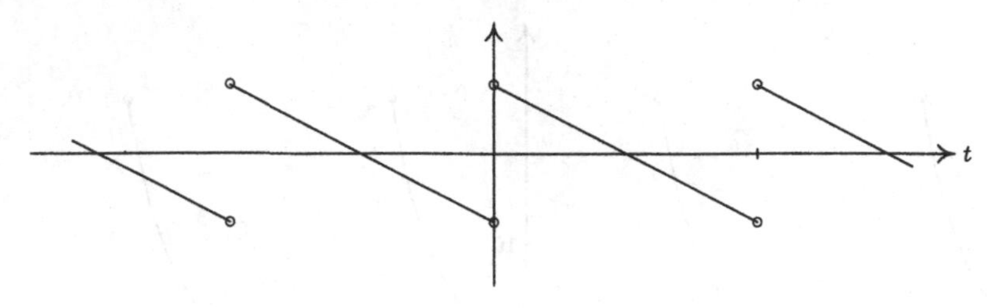

FIGURE 4.2.

When computing the Fourier series for an even or odd function these facts are often useful.

Example 4.2. Let f be an odd function with period 2π, that satisfies $f(t) = (\pi - t)/2$ for $0 < t < \pi$. Find its Fourier series! (See Figure 4.2.)

Solution. Notice that the description as given actually determines the function completely (except for its value at one point in each period, which does not matter). Because the function is odd we have $a_n = 0$ and

$$\begin{aligned} b_n &= \frac{2}{\pi}\int_0^\pi \frac{\pi - t}{2}\sin nt\,dt \\ &= \frac{1}{\pi}\left[(\pi - t)\frac{-\cos nt}{n}\right]_{t=0}^{\pi} + \frac{1}{n\pi}\int_0^\pi (-1)\cos nt\,dt \\ &= \frac{1}{n} - \frac{1}{n^2\pi}\Big[\sin nt\Big]_{t=0}^{\pi} = \frac{1}{n}. \end{aligned}$$

Thus,

$$f(t) \sim \sum_{n=1}^{\infty} \frac{\sin nt}{n}.$$

□

Example 4.3. Let $f(t) = t^2$ for $|t| \le \pi$ and define f outside of this interval by proclaiming it to have period 2π (draw a picture!). Find the Fourier series of this function.

Solution. Now the function is even, and so $b_n = 0$ and

$$a_n = \frac{2}{\pi}\int_0^\pi t^2\cos nt\,dt \overset{\substack{n\neq 0\\ \downarrow}}{=} \frac{2}{\pi}\left[t^2\frac{\sin nt}{n}\right]_0^\pi - \frac{2}{n\pi}\int_0^\pi 2t\sin nt\,dt$$

$$= -\frac{4}{n\pi}\left[t\,\frac{-\cos nt}{n}\right]_0^\pi - \frac{4}{n^2\pi}\int_0^\pi 1\cdot\cos nt\,dt = \frac{4\pi\cos n\pi}{n^2\pi} - 0 = \frac{4(-1)^n}{n^2}.$$

For $n = 0$ we must do a separate calculation:

$$a_0 = \frac{2}{\pi}\int_0^\pi t^2\,dt = \frac{2}{\pi}\cdot\frac{\pi^3}{3} = \frac{2\pi^2}{3}.$$

Collecting the results we get

$$f(t) \sim \frac{\pi^2}{3} + 4\sum_{n=1}^{\infty} \frac{(-1)^n}{n^2} \cos nt.$$

□

The series obtained in Example 4.3 is clearly convergent; indeed it even converges uniformly, by Weierstrass. At this stage we cannot tell what its sum is. The goal of the next few sections is to investigate this. For the moment, we can notice two facts about Fourier coefficients:

Lemma 4.1 *Suppose that f is as in the definition of Fourier series. Then*

1. *The sequence of Fourier coefficients is bounded; more precisely,*

$$|c_n| \le \frac{1}{2\pi} \int_{\mathbf{T}} |f(t)|\, dt \quad \textit{for all } n.$$

2. *The Fourier coefficients tend to zero as $|n| \to \infty$.*

Proof. For the c_n we have

$$|c_n| = \frac{1}{2\pi} \left| \int_{\mathbf{T}} f(t)\, e^{-int}\, dt \right| \le \frac{1}{2\pi} \int_{\mathbf{T}} |f(t)||e^{-int}|\, dt = \frac{1}{2\pi} \int_{\mathbf{T}} |f(t)|\, dt = M,$$

where M is a fixed number that does not depend on n. (In just the same way one can estimate a_n and b_n.) The second assertion of the lemma is just a case of Riemann–Lebesgue's lemma. □

The constant term in a Fourier series is of particular interest:

$$c_0 = \frac{a_0}{2} = \frac{1}{2\pi} \int_{-\pi}^{\pi} f(t)\, dt.$$

This can be interpreted as the *mean value* of the function f over one period (or over $\mathbf{T}$). This can often be useful in problem-solving. It is also intuitively reasonable in that all the other terms of the series have mean value 0 over any period (think of the graph of, say, $\sin nt$).

Exercises

4.1 Prove the formulae $c_n = \frac{1}{2}(a_n - ib_n)$ and $c_{-n} = \frac{1}{2}(a_n + ib_n)$ for $n \ge 0$ (where $b_0 = 0$).

4.2 Assume that f and g are odd functions and h is even. Find out which of the following functions are odd or even: $f + g$, fg, fh, f^2, $f + h$.

4.3 Show that an arbitrary function f on a symmetric interval $(-a, a)$ can be decomposed as $f_E + f_O$, where f_E is even and f_O is odd. Also show that this decomposition is unique. Hint: put $f_E(t) = (f(t) + f(-t))/2$.

4.4 Determine the Fourier series of the 2π-periodic function described by $f(t) = t+1$ for $|t| < \pi$.

4.5 Prove the following relations for a (continuous) function f and its "complex" Fourier coefficients c_n:
(a) If f is even, then $c_n = c_{-n}$ for all n.
(b) If f is odd, then $c_n = -c_{-n}$ for all n.
(c) If f is real-valued, then $\overline{c_n} = c_{-n}$ for all n (where $\overline{}$ denotes complex conjugation).

4.6 Find the Fourier series (in the "real" version) of the functions (a) $f(t) = \cos 2t$, (b) $g(t) = \cos^2 t$, (c) $h(t) = \sin^3 t$. *Sens moral?*

4.7 Let f have the Fourier coefficients $\{c_n\}$. Prove the following rules for Fourier coefficients (F.c.'s):
(a) Let $a \in \mathbf{Z}$. Then the function $t \mapsto e^{iat} f(t)$ has F.c.'s $\{c_{n-a}\}$.
(b) Let $b \in \mathbf{R}$. Then the function $t \mapsto f(t-b)$ has F.c.'s $\{e^{-inb} c_n\}$.

4.8 Find the Fourier series of $h(t) = e^{3it} f(t-4)$, when f has period 2π and satisfies $f(t) = 1$ for $|t| < 2$, $f(t) = 0$ for $2 < |t| < \pi$.

4.9 Compute the Fourier series of f, where $f(t) = e^{-|t|}$, $|t| < \pi$, $f(t+2\pi) = f(t)$, $t \in \mathbf{R}$.

4.10 Let f and g be defined on $\mathbf{T}$ with Fourier coefficients $c_n(f)$ resp. $c_n(g)$. Define the function h by

$$h(t) = \frac{1}{2\pi} \int_{\mathbf{T}} f(t-u)\, g(u)\, du.$$

Show that h is welldefined on $\mathbf{T}$ (i.e., h has also period 2π), and prove that $c_n(h) = c_n(f)\, c_n(g)$. (The function h is called the *convolution* of f and g.)

4.2 Dirichlet's and Fejér's kernels; uniqueness

It is a regrettable fact that a Fourier series need not be convergent. For example, it is possible to construct a continuous function such that its Fourier series diverges at a specified point (see, for example, the book by Thomas Körner mentioned in the bibliography). We shall see, in due time, that if we impose somewhat harder requirements on the function, such as differentiability, the results are more positive.

It is, however, true that the Fourier series of a continuous function is Cesàro summable to the values of the function, and this is the main result of this section.

We start by establishing a closed formula for the partial sums of a Fourier series. To this end we shall use the following formula:

Lemma 4.2

$$D_N(u) := \frac{1}{\pi}\left(\tfrac{1}{2} + \sum_{n=1}^{N} \cos nu\right) = \frac{1}{2\pi} \sum_{n=-N}^{N} e^{inu} = \frac{\sin(N+\frac{1}{2})u}{2\pi \sin \frac{1}{2}u}.$$

Proof. The equality of the two sums follows easily from Euler's formulae. Let us then start from the "complex" version of the sum and compute it as a finite geometric sum:

$$\begin{aligned}
2\pi D_N(u) &= \sum_{n=-N}^{N} e^{inu} = e^{-iNu} \sum_{n=0}^{2N} e^{inu} = e^{-iNu} \cdot \frac{1 - e^{i(2N+1)u}}{1 - e^{iu}} \\
&= e^{-iNu} \cdot \frac{e^{i(N+\frac{1}{2})u}\left(e^{-i(N+\frac{1}{2})u} - e^{i(N+\frac{1}{2})u}\right)}{e^{iu/2}\left(e^{-iu/2} - e^{iu/2}\right)} \\
&= \frac{e^{-iNu+i(N+\frac{1}{2})u}}{e^{iu/2}} \cdot \frac{-2i\sin(N+\frac{1}{2})u}{-2i\sin\frac{1}{2}u} = \frac{\sin(N+\frac{1}{2})u}{\sin\frac{1}{2}u}.
\end{aligned}$$

□

The function D_N is called the DIRICHLET kernel. Its graph is shown in Figure 4.3 on page 87.

When discussing the convergence of Fourier series, the natural partial sums are those containing all frequencies up to a certain value. Thus we *define* the partial sum $s_N(t)$ to be

$$s_N(t) := \tfrac{1}{2}a_0 + \sum_{n=1}^{N}(a_n \cos nt + b_n \sin nt) = \sum_{n=-N}^{N} c_n e^{int}.$$

Using the Dirichlet kernel we can obtain an integral formula for this sum, assuming the c_n to be the Fourier coefficients of a function f:

$$\begin{aligned}
s_N(t) &= \sum_{n=-N}^{N} c_n e^{int} = \sum_{n=-N}^{N} \frac{1}{2\pi}\int_{-\pi}^{\pi} f(u)\, e^{-inu}\, du \cdot e^{int} \\
&= \frac{1}{\pi}\int_{-\pi}^{\pi} f(u) \cdot \tfrac{1}{2}\sum_{n=-N}^{N} e^{in(t-u)}\, du = \int_{-\pi}^{\pi} f(u)\, D_N(t-u)\, du \\
&= \frac{1}{2\pi}\int_{-\pi}^{\pi} f(t-u)\, \frac{\sin(N+\frac{1}{2})u}{\sin\frac{1}{2}u}\, du.
\end{aligned}$$

In the last step we change the variable ($t - u$ is replaced by u) and make use of the periodicity of the integrand. We shall presently take another step and form the arithmetic means of the $N + 1$ first partial sums. To achieve this we need a formula for the mean of the corresponding Dirichlet kernels:

Lemma 4.3

$$F_N(u) := \frac{1}{N+1}\sum_{n=0}^{N} D_n(u) = \frac{1}{2\pi(N+1)}\left(\frac{\sin\frac{1}{2}(N+1)u}{\sin\frac{1}{2}u}\right)^2.$$

The proof can be done in a way similar to Lemma 4.2 (or in some other way). It is left as an exercise. The function $F_N(t)$ is called the FEJÉR kernel.

Now we can form the mean of the partial sums:

$$\sigma_N(t) = \frac{s_0(t)+s_1(t)+\cdots+s_N(t)}{N+1} = \frac{1}{N+1}\sum_{n=0}^{N}\int_{-\pi}^{\pi} f(t-u)\,D_n(u)\,du$$
$$= \int_{-\pi}^{\pi} f(t-u)\cdot\frac{1}{N+1}\sum_{n=0}^{N} D_n(u)\,du = \int_{-\pi}^{\pi} f(t-u)\,F_N(u)\,du.$$

Lemma 4.4 *The Fejér kernel $F_N(u)$ has the following properties:*

1. *F_N is an even function, and $F_N(u) \geq 0$.*
2. *$\int_{-\pi}^{\pi} F_N(u)\,du = 1$.*
3. *If $\delta > 0$, then $\lim_{N\to\infty} \int_{\delta}^{\pi} F_N(u)\,du = 0$.*

Proof. Property 1 is obvious. Number 2 follows from

$$\int_{-\pi}^{\pi} D_n(u)\,du = \frac{1}{\pi}\int_{-\pi}^{\pi} \left(\tfrac{1}{2}+\cos u+\cdots+\cos nu\right)du = 1, \quad n = 0,1,2,\ldots,N,$$

and the fact that F_N is the mean of these Dirichlet kernels. Finally, property 3 can be proved thus:

$$0 \leq \int_{\delta}^{\pi} F_N(u)\,du = \frac{1}{2\pi(N+1)}\int_{\delta}^{\pi} \frac{\sin^2 \frac{1}{2}(N+1)u}{\sin^2 \frac{1}{2}u}\,du$$
$$\leq \frac{1}{2\pi(N+1)}\int_{\delta}^{\pi} \frac{1}{\sin^2 \frac{1}{2}\delta}\,du = \frac{1}{2\pi(N+1)}\,\frac{\pi-\delta}{\sin^2 \frac{1}{2}\delta} = \frac{C_\delta}{N+1} \to 0$$

as $N \to \infty$. □

The lemma implies that $\{F_N\}_{N=1}^{\infty}$ is a positive summation kernel such as the ones studied in Sec. 2.4. Applying Corollary 2.1 we then have the result on Cesàro sums of Fourier series.

Theorem 4.1 (Fejér's theorem) *If f is piecewise continuous on* **T** *and continuous at the point t, then $\lim_{N\to\infty} \sigma_N(t) = f(t)$.*

Remark. Using the remark following Corollary 2.1, we can sharpen the result of the theorem a bit. If f is continuous in an interval $I_0 =]a_0, b_0[$, and $I = [a, b]$ is a compact subinterval of I_0, then $\sigma_N(t)$ will converge to $f(t)$ *uniformly on I*. □

If a series is convergent in the traditional sense, then its sum coincides with the Cesàro limit. This means that if a *continuous* function happens to have a Fourier series, which is *seen* to be convergent, in one way or another, then it actually converges to the function it comes from. In particular we have the following theorem.

Theorem 4.2 *If f is continuous on* $\mathbf{T}$ *and its Fourier coefficients c_n are such that $\sum |c_n|$ is convergent, then the Fourier series is convergent with sum $f(t)$ for all $t \in \mathbf{T}$, and the convergence is even uniform on* $\mathbf{T}$.

The uniform convergence follows using the Weierstrass M-test just as at the beginning of this chapter.

This result can be applied to Example 4.3 of the previous section, where we computed the Fourier series of $f(t) = t^2$ ($|t| \leq \pi$). Applying the usual comparison test, the series obtained is easily seen to be convergent, and now we know that its sum is also equal to $f(t)$. We now have this formula:

$$t^2 = \frac{\pi^2}{3} + 4 \sum_{n=1}^{\infty} \frac{(-1)^n}{n^2} \cos nt, \quad -\pi \leq t \leq \pi. \tag{4.4}$$

(Why does this formula hold even for $t = \pm\pi$?) In particular, we can amuse ourselves by inserting various values of t just to see what we get. For $t = 0$ the result is

$$0 = \frac{\pi^2}{3} + 4 \sum_{n=1}^{\infty} \frac{(-1)^n}{n^2}.$$

From this we can conclude that

$$\sum_{n=1}^{\infty} \frac{(-1)^n}{n^2} = -\frac{\pi^2}{12}.$$

If $t = \pi$ is substituted into (4.4), we have

$$\pi^2 = \frac{\pi^2}{3} + 4 \sum_{n=1}^{\infty} \frac{(-1)^n}{n^2} (-1)^n = \frac{\pi^2}{3} + 4 \sum_{n=1}^{\infty} \frac{1}{n^2},$$

which enables us to state that

$$\sum_{n=1}^{\infty} \frac{1}{n^2} = \frac{\pi^2}{6}.$$

Thus, Fourier series provide a means of computing the sums of numerical series. Regrettably, it can hardly be called a "method": if one faces a more-or-less randomly chosen series, there is no general method to find a function whose Fourier expansion will help us to sum it. As an illustration we mention that it is rather easy to find nice expressions for the values of

$$\zeta(s) = \sum_{n=1}^{\infty} \frac{1}{n^s}$$

for $s = 2, 4, 6, \ldots$, but no one has so far found such an expression for, say, $\zeta(3)$.

The following uniqueness result is also a consequence of Theorem 4.2.

Theorem 4.3 *Suppose that f is piecewise continuous and that all its Fourier coefficients are 0. Then $f(t) = 0$ at all points where f is continuous.*

In fact, all the partial sums are zero and the series is trivially convergent, and by Theorem 4.2 it must then converge to the function from which it is formed.

Corollary 4.1 *If two continuous functions f and g have the same Fourier coefficients, then $f = g$.*

Proof. Apply Theorem 4.3 to the function $h = f - g$. □

Exercises

4.11 Prove the formula for the Fejér kernel (i.e., Lemma 4.3).

4.12 Study the function $f(t) = t^4 - 2\pi^2 t^2$, $|t| < \pi$, and compute the value of $\zeta(4)$.

4.13 Determine the Fourier series of $f(t) = |\cos t|$. Prove that the series converges uniformly to f and find the value of

$$s = \sum_{n=1}^{\infty} \frac{(-1)^n}{4n^2 - 1}.$$

4.14 Prove converse statements to the assertions in Exercise 4.5; i.e., show that if f is continuous (say), we can say that
(a) if $c_n = c_{-n}$ for all n, then f is even;
(b) If $c_n = -c_{-n}$ for all n, then f is odd;
(c) If $\overline{c_n} = c_{-n}$ for all n, then f is real-valued.

4.3 Differentiable functions

Suppose that $f \in C^1(\mathbf{T})$, which means that both f and its derivative f' are continuous on $\mathbf{T}$. We compute the Fourier coefficients of the derivative:

$$\begin{aligned} c_n(f') &= \frac{1}{2\pi} \int_{-\pi}^{\pi} f'(t)\, e^{-int}\, dt = \frac{1}{2\pi} \left[f(t)\, e^{-int} \right]_{-\pi}^{\pi} - \frac{1}{2\pi} \int_{-\pi}^{\pi} f(t)(-in) e^{-int}\, dt \\ &= \frac{1}{2\pi} \left(f(\pi)(-1)^n - f(-\pi)(-1)^n \right) + in\, c_n(f) = in\, c_n(f). \end{aligned}$$

(The fact that f is continuous on $\mathbf{T}$ implies that $f(-\pi) = f(\pi)$.) This means that if f has the Fourier series $\sum c_n\, e^{int}$, then f' has the series $\sum in\, c_n\, e^{int}$. This indeed means that the Fourier series can be differentiated termwise (even if we have no information at all concerning the convergence of either of the two series).

If $f \in C^2(\mathbf{T})$, the argument can be repeated, and we find that the Fourier series of the second derivative is $\sum(-n^2)c_n\, e^{int}$. Since the Fourier coefficients of f'' are bounded, by Lemma 4.1, we conclude that $|-n^2 c_n| \le M$ for some constant M, which implies that $|c_n| \le M/n^2$ for $n \ne 0$. But then we can use Theorem 4.2 to conclude that the Fourier series of f converges to $f(t)$ for all t. Here we have a first, simple, sufficient condition on the function f itself that ensures a nice behavior of its Fourier series.

In the next section, we shall see that C^2 can be improved to C^1 and indeed even less demanding conditions.

By iteration of the argument above, the following general result follows.

Theorem 4.4 *If $f \in C^k(\mathbf{T})$, then $|c_n| \le M/|n|^k$ for some constant M.*

The smoother the function, the smaller the Fourier coefficients: a function with high differentiability contains small high-frequency components.

The assertion of the theorem is really rather weak. Indeed, one can say more, which is exemplified in Exercises 4.15 and 4.17.

The situation concerning integration of Fourier series is extremely favorable. It turns out that termwise integration is *always* possible, both when talking about antiderivatives and integrals over an interval. There is one complication: if the constant term in the series is not zero, the formally integrated series is no longer a Fourier series. However, we postpone the treatment of these matters until later on, when it will be easier to carry through. (Sec. 5.4, Theorem 5.9 on p. 122.)

The fact that termwise differentiation is possible can be used when looking for periodic solutions of differential equations and similar problems. We give an example of this.

Example 4.4. Find a solution $y(t)$ with period 2π of the differential-difference equation $y'(t) + 2y(t-\pi) = \sin t$, $-\infty < t < \infty$.

Solution. Assume the solution to be the sum of a "complex" Fourier series (a "real" series could also be used):

$$y(t) = \sum_{n \in \mathbf{Z}} c_n\, e^{int}.$$

If we differentiate termwise and substitute into the given equation, we get

$$y'(t)+2y(t-\pi) = \sum inc_n\, e^{int} + 2\sum c_n\, e^{int-in\pi} = \sum (in+2(-1)^n)c_n\, e^{int}. \tag{4.5}$$

This should be equal to $\sin t = (e^{it} - e^{-it})/(2i) = \frac{1}{2}\,i\,e^{-it} - \frac{1}{2}\,i\,e^{it}$. The equality must imply that the coefficients in the last series of (4.5) are zeroes for all $n \ne \pm 1$, and furthermore

$$(i-2)c_1 = -\frac{i}{2}, \qquad (-i-2)c_{-1} = \frac{i}{2}.$$

From this we solve $c_1 = \frac{1}{10}(2i-1)$, $c_{-1} = \frac{1}{10}(-2i-1)$ (and $c_n = 0$ for all other n), which gives

$$\begin{aligned} y(t) &= c_1 e^{it} + c_{-1} e^{-it} = -\tfrac{1}{10}(e^{it} + e^{-it}) + \tfrac{2}{10}\, i(e^{it} - e^{-it}) \\ &= -\tfrac{1}{5}\cos t + \tfrac{1}{5}\, i \cdot 2i \sin t = \tfrac{1}{5}(-\cos t - 2\sin t). \end{aligned}$$

Check the solution by substituting into the original equation! □

Exercises

4.15 Prove the following improvement on Theorem 4.4: If $f \in C^k(\mathbf{T})$, then $\lim_{n\to\pm\infty} n^k c_n = 0$.

4.16 Find the values of the constant a for which the problem $y''(t) + ay(t) = y(t+\pi)$, $t \in \mathbf{R}$, has a solution with period 2π which is not identically zero. Also, determine all such solutions.

4.17 Try to prove the following partial improvements on Theorem 4.4:
(a) If f' is continuous and differentiable on $\mathbf{T}$ except possibly for a finite number of jump discontinuities, then $|c_n| \le M/|n|$ for some constant M.
(b) If f is continuous on $\mathbf{T}$ and has a second derivative everywhere except possibly for a finite number of points, where there are "corners" (i.e., the left-hand and right-hand first derivatives exist but are different from each other), then $|c_n| \le M/n^2$ for some constant M.

4.4 Pointwise convergence

Time is now ripe for the formulation and proof of our most general theorem on the pointwise convergence of Fourier series. We have already mentioned that continuity of the function involved is not sufficient. Now let us assume that f is defined on $\mathbf{T}$ and continuous except possibly for a finite number of finite jumps. This means that f is permitted to be discontinuous at a finite number of points in each period, but at these points we assume that both the one-sided limits exist and are finite. For convenience, we introduce this notation for these limits:

$$f(t_0-) = \lim_{t\nearrow t_0} f(t), \qquad f(t_0+) = \lim_{t\searrow t_0} f(t).$$

In addition, we assume that the "generalized left-hand derivative" $f_L'(t_0)$ exists:

$$f_L'(t_0) = \lim_{h\nearrow 0} \frac{f(t_0+h) - f(t_0-)}{h} = \lim_{u\searrow 0} \frac{f(t_0-u) - f(t_0-)}{-u}.$$

If f happens to be continuous at t_0, this coincides with the usual left-hand derivative; if f has a discontinuity at t_0, we take care to use the left-hand limit instead of just writing $f(t_0)$.

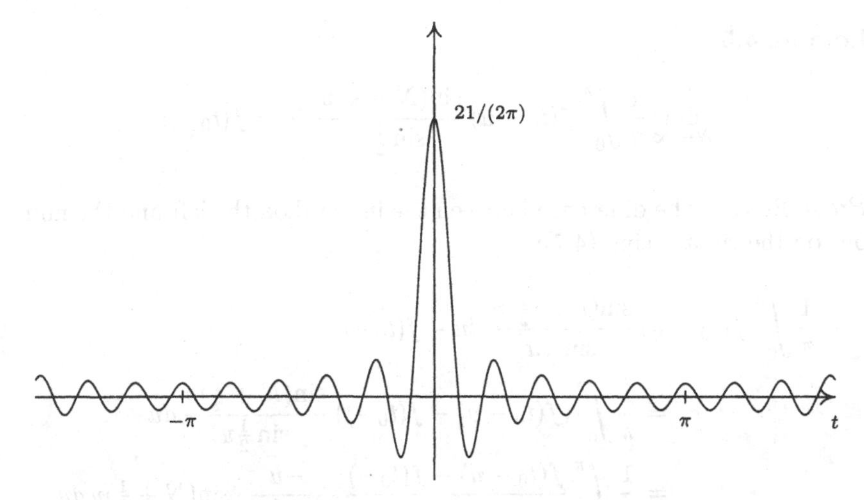

FIGURE 4.3. The graph of D_{10}

Symmetrically, we shall also assume that the "generalized right-hand derivative" exists:

$$f'_R(t_0) = \lim_{h \searrow 0+} \frac{f(t_0 + h) - f(t_0+)}{h}.$$

Intuitively, the existence of these generalized derivatives amounts to the fact that at a jump discontinuity, the graphs of the two parts of the function on either side of the jump have each an end-point tangent direction.

In Sec. 4.2 we proved the following formula for the partial sums of the Fourier series of f:

$$s_N(t) = \frac{1}{2\pi} \int_{-\pi}^{\pi} f(t-u) \frac{\sin(N + \frac{1}{2})u}{\sin \frac{1}{2}u} \, du. \tag{4.6}$$

What complicates matters is that the Dirichlet kernel occurring in the integral is not a positive summation kernel. On the contrary, it takes a lot of negative values, which causes a proof along the lines of Theorem 2.1 to fail completely (see Figure 4.3).

We shall make use of the following formula:

$$\frac{1}{\pi} \int_0^{\pi} \frac{\sin(N + \frac{1}{2})u}{\sin \frac{1}{2}u} \, du = 1. \tag{4.7}$$

This follows directly from the fact that the integrated function is $2\pi D_N(u) = 1 + 2\sum_1^N \cos nu$, where all the cosine terms have integral zero over $[0, \pi]$.

We split the integral (4.6) in two parts, each covering half of the interval of integration, and begin by taking care of the right-hand half:

Lemma 4.5

$$\lim_{N\to\infty} \frac{1}{\pi} \int_0^{\pi} f(t_0 - u) \frac{\sin(N+\frac{1}{2})u}{\sin\frac{1}{2}u} \, du = f(t_0-).$$

Proof. Rewrite the difference between the integral on the left and the number on the right, using (4.7):

$$\begin{aligned} &\frac{1}{\pi} \int_0^{\pi} f(t_0 - u) \frac{\sin(N+\frac{1}{2})u}{\sin\frac{1}{2}u} \, du - f(t_0-) \\ &\qquad = \frac{1}{\pi} \int_0^{\pi} (f(t_0 - u) - f(t_0-)) \frac{\sin(N+\frac{1}{2})u}{\sin\frac{1}{2}u} \, du \\ &\qquad = \frac{1}{\pi} \int_0^{\pi} \frac{f(t_0 - u) - f(t_0-)}{-u} \cdot \frac{-u}{\sin\frac{1}{2}u} \cdot \sin(N+\tfrac{1}{2})u \, du. \end{aligned}$$

The last integrand consists of three factors: The first one is continuous (except for jumps), and it has a finite limit as $u \to 0+$, namely, $f_L'(t_0)$. The second factor is continuous and bounded. The product of the two first factors is thus a function $g(u)$ which is clearly Riemann-integrable on the interval $[0, \pi]$. By the Riemann–Lebesgue lemma we can then conclude that the whole integral tends to zero as N goes to infinity, which proves the lemma. □

In just the same way one can prove that if f has a generalized right-hand derivative at t_0, then

$$\lim_{N\to\infty} \frac{1}{\pi} \int_{-\pi}^{0} f(t_0 - u) \frac{\sin(N+\frac{1}{2})u}{\sin\frac{1}{2}u} \, du = f(t_0+).$$

Taking the arithmetic mean of the two formulae, we have proved the convergence theorem:

Theorem 4.5 *Suppose that f has period 2π, and suppose that t_0 is a point where f has one-sided limiting values and (generalized) one-sided derivatives. Then the Fourier series of f converges for $t = t_0$ to the mean value $\frac{1}{2}(f(t_0+) + f(t_0-))$. In particular, if f is continuous at t_0, the sum of the series equals $f(t_0)$.*

We emphasize that if f is continuous at t_0, the sum of the series is simply $f(t_0)$. At a point where the function has a jump discontinuity, the sum is instead the mean value of the right-hand and left-hand limits.

It is important to realize that the convergence of a Fourier series at a particular point is really dependent only on the local behavior of the function in the neighborhood of that point. This is sometimes called the Riemann localization principle.

Example 4.5. Let us return to Example 4.2 on page 78. Now we finally know that the series

$$\sum_{n=1}^{\infty} \frac{\sin nt}{n}$$

is indeed convergent for all t. If, for example, $t = \pi/2$, we have $\sin nt$ equal to zero for all even values of n, while $\sin(2k+1)t = (-1)^k$. Since f is continuous and has a derivative at $t = \pi/2$, and $f(\pi/2) = \pi/4$, we obtain

$$\frac{\pi}{4} = \sum_{k=0}^{\infty} \frac{(-1)^k}{2k+1} = 1 - \tfrac{1}{3} + \tfrac{1}{5} - \tfrac{1}{7} + \cdots.$$

(In theory, this formula could be used to compute numerical approximations to π, but the series converges so extremely slowly that it is of no practical use whatever.) □

The most comprehensive theorem concerning pointwise convergence of Fourier series of continuous functions was proved in 1966 by Lennart CARLESON. In order to formulate it we first introduce the notion of a *zero set:* a set $E \subset \mathbf{T}$ is called a zero set if, for every $\varepsilon > 0$, it is possible to construct a sequence of intervals $\{\omega_n\}_{n=1}^{\infty}$ on the circle, that together cover the set E and whose total length is less that ε.

Theorem 4.6 (Carleson's theorem) *If f is continuous on $\mathbf{T}$, then its Fourier series converges at all points of $\mathbf{T}$ except possibly for a zero set.*

In fact, it is not even necessary that f be continuous; it is sufficient that $f \in L^2(\mathbf{T})$, which will be explained in Chapter 5. The proof is very complicated.

Carleson's theorem is "best possible" in the following sense:

Theorem 4.7 (Kahane and Katznelson) *If E is a zero set on $\mathbf{T}$, then there exists a continuous function such that its Fourier series diverges precisely for all $t \in E$.*

Exercises

4.18 Define f by letting $f(t) = t \sin t$ for $|t| < \pi$ and $f(t + 2\pi) = f(t)$ for all t. Determine the Fourier series of f and investigate for which values of t it converges to $f(t)$.

4.19 If $f(t) = (t+1)\cos t$ for $-\pi < t < \pi$, what is the sum of the Fourier series of f for $t = 3\pi$? (Note that you do not have to compute the series itself!)

4.20 The function f has period 2π and satisfies

$$f(t) = \begin{cases} t + \pi, & -\pi < t < 0, \\ 0, & 0 \le t \le \pi. \end{cases}$$

(a) Find the Fourier series of f and sketch the sum of the series on the interval $[-3\pi, 3\pi]$.

(b) Sum the series $\displaystyle\sum_{n=1}^{\infty} \frac{1}{(2n-1)^2}$.

4.21 Let $f(x)$ be defined for $-\pi < x < \pi$ by $f(x) = \cos \frac{3}{2}x$ and for other values of x by $f(x) = f(x + 2\pi)$. Determine the Fourier series of f. For all real x, investigate whether the series is convergent. Find its sum for $x = n \cdot \pi/2$, $n = 1, 2, 3$.

4.22 Let α be a complex number but not an integer. Determine the Fourier series of $\cos \alpha t$ ($|t| \le \pi$). Use the result to prove the formula

$$\pi \cot \pi z = \lim_{N\to\infty} \sum_{n=-N}^{N} \frac{1}{z-n} \qquad (z \notin \mathbf{Z})$$

("expansion into partial fractions of the cotangent").

4.5 Formulae for other periods

Here we have collected the formulae for Fourier series of functions with a period different from 2π. It is convenient to have a notation for the *half-period*, so we assume that the period is $2P$, where $P > 0$:

$$f(t + 2P) = f(t) \text{ for all } t \in \mathbf{R}.$$

Put $\Omega = \pi/P$. The number Ω could be called the *fundamental angular frequency*. A linear change of variable in the usual formulae results in the following set of formulae:

$$f(t) \sim \sum_{n\in\mathbf{Z}} c_n\, e^{in\Omega t}, \quad \text{where} \quad c_n = \frac{1}{2P} \int_{-P}^{P} f(t)\, e^{-in\Omega t}\, dt,$$

and, alternatively,

$$f(t) \sim \tfrac{1}{2}\, a_0 + \sum_{n=1}^{\infty} (a_n \cos n\Omega t + b_n \sin n\Omega t), \text{ where } \begin{matrix} a_n \\ b_n \end{matrix} = \frac{1}{P} \int_{-P}^{P} f(t) \begin{matrix} \cos \\ \sin \end{matrix} n\Omega t\, dt.$$

In all cases, the intervals of integration can be changed from $(-P, P)$ to an arbitrary interval of length $2P$. If f is even or odd, we have the special cases

$$f \text{ even} \Rightarrow b_n = 0,\ a_n = \frac{2}{P} \int_0^P f(t) \cos n\Omega t\, dt,$$

$$f \text{ odd} \Rightarrow a_n = 0,\ b_n = \frac{2}{P} \int_0^P f(t) \sin n\Omega t\, dt.$$

All results concerning summability, convergence, differentiability, etc., that we have proved in the preceding sections, will of course hold equally well for any period length.

Exercises

4.23 (a) Determine the Fourier series for the even function f with period 2 that satisfies $f(t) = t$ for $0 < t < 1$.
(b) Determine the Fourier series for the odd function f with period 2 that satisfies $f(t) = t$ for $0 < t < 1$.
(c) Compare the convergence properties of the series obtained in (a) and (b). Illuminate by drawing pictures!

4.24 Find, in the guise of a "complex" Fourier series, a periodic solution with a continuous first derivative on $\mathbf{R}$ of the differential equation $y'' + y' + y = g$, where g has period 4π and $g(t) = 1$ for $|t| < \pi$, $g(t) = 0$ for $\pi < |t| < 2\pi$.

4.25 Determine a solution with period 2 of the differential-difference equation $y'(t) + y(t-1) = \cos^2 \pi t$.

4.26 Compute the Fourier series of the odd function f with period 2 that satifies $f(x) = x - x^2$ for $0 < x < 1$. Use the result to find the sum of the series

$$\sum_{n=0}^{\infty} \frac{(-1)^n}{(2n+1)^3}.$$

4.6 Some worked examples

In this section we give a few more examples of the computational work that may occur in calculating the Fourier coefficients of a function.

Example 4.6. Take $f(t) = t \cos 2t$ for $-\pi < t < \pi$, and assume f to have period 2π. First of all, we try to see if f is even or odd — indeed, it is odd. This means that it should be a good idea to compute the Fourier series in the "real" version; because all a_n will be zero, and b_n is given by the half-range integral

$$b_n = \frac{2}{\pi} \int_0^{\pi} t \cos 2t \sin nt \, dt.$$

The computation is now greatly simplified by using the product formula

$$\sin x \cos y = \tfrac{1}{2}\big(\sin(x+y) + \sin(x-y)\big).$$

Integrating by parts, we get

$$\begin{aligned} b_n &= \frac{1}{\pi} \int t\big(\sin(n+2)t + \sin(n-2)t\big)\, dt \quad (n \neq 2) \\ &= \frac{1}{\pi} \left[t\Big(-\frac{\cos(n+2)t}{n+2} - \frac{\cos(n-2)t}{n-2} \Big) \right]_0^{\pi} \end{aligned}$$

$$+\frac{1}{\pi}\int_0^\pi\left(\frac{\cos(n+2)t}{n+2}+\frac{\cos(n-2)t}{n-2}\right)dt$$
$$=-\frac{1}{\pi}\cdot\pi\left(\frac{\cos(n+2)\pi}{n+2}+\frac{\cos(n-2)\pi}{n-2}\right)+0=-\left(\frac{(-1)^n}{n+2}+\frac{(-1)^n}{n-2}\right)$$
$$=-\frac{2n(-1)^n}{n^2-4}.$$

This computation fails for $n=2$. For this n we get instead

$$b_2=\frac{1}{\pi}\int_0^\pi t(\sin 4t+0)\,dt=\frac{1}{\pi}\left[t\frac{-\cos 4t}{4}\right]+\frac{1}{4\pi}\int_0^\pi \cos 4t\,dt$$
$$=-\tfrac{1}{4}+0=-\tfrac{1}{4}.$$

Noting that $b_1=-\frac{2}{3}$, we can conveniently describe the Fourier series as

$$f(t)\sim-\tfrac{2}{3}\sin t-\tfrac{1}{4}\sin 2t-2\sum_{n=3}^{\infty}\frac{n(-1)^n}{n^2-4}\sin nt.$$

□

Example 4.7. Find the Fourier series of the odd function of period 2 that is described by $f(t)=t(1-t)$ for $0\le t\le 1$. Using the result, find the value of the sum

$$s_1=\sum_{k=0}^{\infty}\frac{(-1)^k}{(2k+1)^3}.$$

Solution. Since the function is odd, we compute a sine series. The coefficients are

$$b_n=2\int_0^1 t(1-t)\sin n\pi t\,dt=\text{(integrations by parts)}=\frac{4(1-(-1)^n)}{n^3\pi^3},$$

which is zero for all even values of n. Writing $n=2k+1$ when n is odd, we get the series

$$f(t)\sim\frac{8}{\pi^3}\sum_{k=0}^{\infty}\frac{\sin(2k+1)\pi t}{(2k+1)^3}.$$

A sketch of the function shows that f is everywhere continuous and has both right- and left-hand derivatives everywhere, which permits us to replace the sign $\sim$ by $=$. In particular we note that if $t=\frac{1}{2}$, then $\sin(2k+1)\pi t=\sin(k+\frac{1}{2})\pi=(-1)^k$, so that

$$\tfrac{1}{4}=f(\tfrac{1}{2})=\frac{8}{\pi^3}\cdot s_1 \qquad\Rightarrow\qquad s_1=\frac{\pi^3}{32}.$$

□

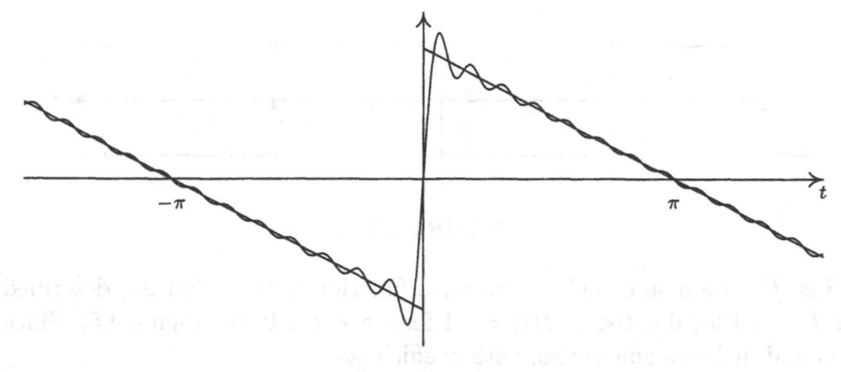

FIGURE 4.4.

Exercises

4.27 Find the Fourier series of f with period 1, when $f(x) = x$ for $1 < x < 2$. Indicate the sum of the series for $x = 0$ and $x = \frac{1}{2}$. Explain your answer!

4.28 Develop into Fourier series the function f given by

$$f(x) = \sin\frac{x}{2}, \quad -\pi < x \le \pi; \quad f(x + 2\pi) = f(x), \; x \in \mathbf{R}.$$

4.29 Compute the Fourier series of period 2π for the function $f(x) = (|x| - \pi)^2$, $|x| \le \pi$, and use it to find the sums

$$\sum_{n=1}^{\infty} \frac{(-1)^{n-1}}{n^2} \quad \text{and} \quad \sum_{n=1}^{\infty} \frac{1}{n^2}.$$

4.7 The Gibbs phenomenon

Let f be a function that satisfies the conditions for pointwise convergence of the Fourier series (Theorem 4.5) and that has a jump discontinuity at a certain point t_0. If we draw a graph of a partial sum of the series, we discover a peculiar behavior: When t approaches t_0, for example, from the left, the graph of $s_n(t)$ somehow grows restless; you might say that it prepares to take off for the jump; and when the jump is accomplished, it overshoots the mark somewhat and then calms down again. Figure 4.4 shows a typical case.

This sort of behavior had already been observed during the nineteenth century by experimental physicists, and it was then believed to be due to imperfection in the measuring apparatuses. The fact that this is not so, but that we are dealing with an actual mathematical phenomenon, was proved by J. W. GIBBS, after whom the behavior has also been named.

The behavior is fundamentally due to the fact that the Dirichlet kernel $D_n(t)$ is restless near $t = 0$. We are going to analyze the matter in detail in one special case and then, using a simple maneuver, show that the same sort of thing occurs in the general case.

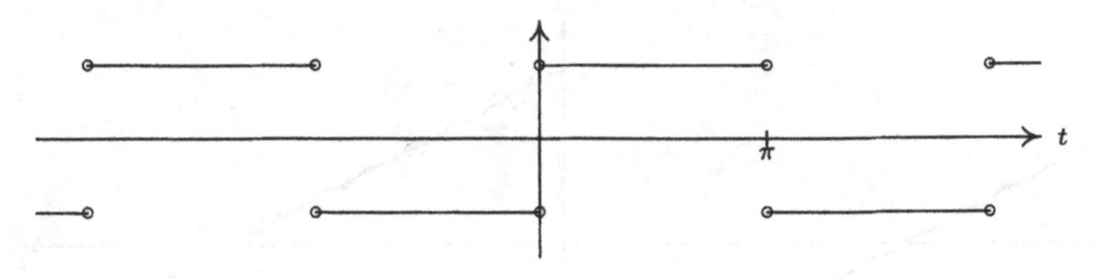

FIGURE 4.5.

Let $f(t)$ be a so-called square-wave function with period 2π, described by $f(t) = 1$ for $0 < t < \pi$, $f(t) = -1$ for $-\pi < t < 0$ (see Figure 4.5). Since f is odd, it has a sine series, with coefficients

$$b_n = \frac{2}{\pi} \int_0^\pi \sin nt\,dt = \frac{2}{\pi}\left[-\frac{\cos nt}{n}\right]_0^\pi = \frac{2}{n\pi}(1-(-1)^n),$$

which is zero if n is even. Thus,

$$f(t) \sim \frac{4}{\pi}\sum_{k=0}^{\infty} \frac{\sin(2k+1)t}{2k+1} = \frac{4}{\pi}\left(\sin t + \frac{\sin 3t}{3} + \frac{\sin 5t}{5} + \cdots\right). \tag{4.8}$$

Because of symmetry we can restrict our study to the interval $(0, \pi/2)$. For a while we dump the factor $4/\pi$ and consider the partial sums of the series in the brackets:

$$S_n(t) = \sin t + \tfrac{1}{3}\sin 3t + \tfrac{1}{5}\sin 5t + \cdots + \frac{1}{2n+1}\sin(2n+1)t.$$

By differentiation we find

$$S_n'(t) = \cos t + \cos 3t + \cdots + \cos(2n+1)t = \tfrac{1}{2}\sum_{k=0}^{n}\left(e^{i(2k+1)t} + e^{-i(2k+1)t}\right)$$

$$= \tfrac{1}{2}e^{-i(2n+1)t}\sum_{k=0}^{2n+1} e^{i2kt} = \tfrac{1}{2}e^{-i(2n+1)t}\,\frac{1-e^{i2(2n+2)t}}{1-e^{i2t}} = \frac{\sin 2(n+1)t}{2\sin t}$$

(compare the method that we used to sum $D_n(t)$). The last formula does not hold for $t = 0$, but it does hold in the half-open interval $0 < t \le \pi/2$. The derivative has zeroes in this interval; they are easily found to be where $2(n+1)t = k\pi$ or $t = \tau_k = (k\pi)/((2(n+1)), k = 1, 2, \ldots, n$. Considering the sign of the derivative between the zeroes one realizes that these points are alternatingly maxima and minima of S_n. More precisely, since $S_n(0) = 0$, integration gives

$$S_n(t) = \int_0^t \frac{\sin 2(n+1)u}{2\sin u}\,du,$$

where the numerator of the integrand oscillates in a smooth fashion between the successive τ_k, while the denominator increases throughout the interval.

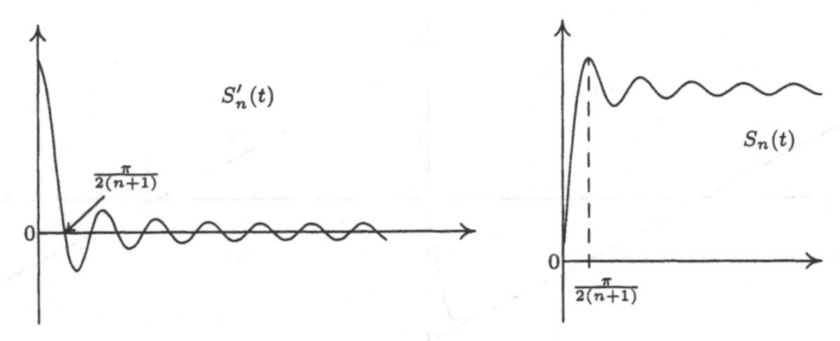

FIGURE 4.6.

This means that the first maximum value, for $t = \tau_1$, is also the largest, and the oscillations in S_n then quiet down as t increases (see Figure 4.6).

It follows that the maximal value of $S_n(t)$ on $]0, \pi/2]$ is given by

$$A_n = S_n(\tau_1) = S_n\left(\frac{\pi}{2(n+1)}\right) = \sum_{k=0}^{n} \frac{1}{2k+1} \sin\frac{(2k+1)\pi}{2(n+1)}.$$

We can interpret the last sum as a Riemann sum for a certain integral: Let $t_k = k\pi/(n+1)$ and $\xi_k = \frac{1}{2}(t_k + t_{k+1})$. Then the points $0 = t_0, t_1, \ldots, t_{n+1} = \pi$ describe a subdivision of the interval $(0, \pi)$, the point ξ_k lies in the subinterval (x_k, x_{k+1}) and, in addition, $\xi_k = (2k+1)\pi/(2(n+1))$. Thus we have

$$A_n = \tfrac{1}{2} \sum_{k=0}^{n} \frac{\sin \xi_k}{\xi_k} \Delta x_k \to \tfrac{1}{2} \int_0^{\pi} \frac{\sin x}{x}\, dx \quad \text{as } n \to \infty.$$

A more detailed scrutiny of the limit process would show that the numbers A_n *decrease* toward the limit.

Now we reintroduce the factor $4/\pi$. We have then established that the partial sums of the Fourier series (4.8) have maximum values that tend to the limit

$$\frac{2}{\pi} \int_0^{\pi} \frac{\sin t}{t}\, dt \approx 1.1789797,$$

and the maximal value of $S_n(t)$ is taken at $t = \pi/(2(n+1))$. On the right-hand side of the maximum, the partial sums oscillate around the value 1 with a decreasing amplitude, up to the point $t = \pi/2$. Because of symmetry, the behavior to the left will be analogous. What we want to stress is the fact that the maximal oscillation does not tend to zero when more terms of the series are added; on the contrary, it stabilizes toward a value that is approximately 9 percent of the total size of the jump. The point where the maximum oscillation takes place moves indefinitely closer to the point of the jump. It is even possible to prove that the Fourier series is actually uniformly convergent to 1 on intervals of the form $[a, \pi - a]$, where $a > 0$.

Now let g be any function with a jump discontinuity at t_0 with the size of the jump equal to $\delta = g(t_0+) - g(t_0-)$, and assume that g satisfies the

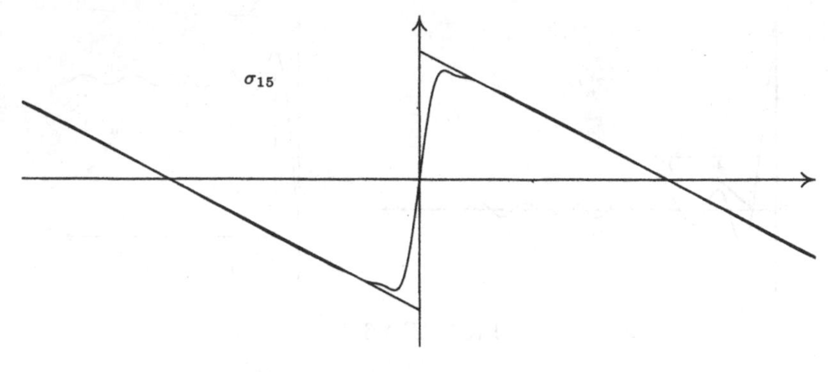

FIGURE 4.7.

conditions of Theorem 4.5 for convergence of the Fourier series in some neighborhood of t_0. Form the function $h(t) = g(t) - \frac{1}{2}\delta f(t - t_0)$, where f is the square-wave function just investigated. Then, $h(t_0+) = h(t_0-)$, so that h is actually continuous at t_0 if one defines $h(t_0)$ in the proper way. Furthermore, h has left- and right-hand derivatives at t_0, and so the Fourier series of h will converge nicely to h in a neighborhood of $t = t_0$. The Fourier series of g can be written as the series of h plus some multiple of a translate of the series of f; the former series is calm near t_0, but the latter oscillates in the manner demonstrated above. It follows that the series of g exhibits on the whole the same restlessness when we approach t_0, as does the series of f when we approach 0. The size of the maximum oscillation is also approximately 9 percent of the size of the whole jump.

If a Fourier series is summed according to Cesàro (Theorem 4.1) or Poisson–Abel (see Sec. 6.3), the Gibbs phenomenon disappears completely. Compare the graphs of $s_{15}(t)$ in Figure 4.4 and $\sigma_{15}(t)$ (for the same f) in Figure 4.7.

4.8 *Fourier series for distributions

We shall here consider the generalized functions of Sec. 2.6 and 2.7 and their Fourier series. Since the present chapter deals with objects defined on $\mathbf{T}$, or, equivalently, periodic phenomena, we begin by considering periodic distributions as such.

In this context, the Heaviside function H is not really interesting. But we can still think of the object $\delta_a(t)$ as a "unit pulse" located at a point $a \in \mathbf{T}$, having the property

$$\int_{\mathbf{T}} \varphi(t)\delta_a(t)\,dt = \varphi(a) \quad \text{if } \varphi \text{ is continuous at } a.$$

The periodic description of the same object consists of a so-called *pulse train* consisting of unit pulses at all the points $a + n \cdot 2\pi$, $n \in \mathbf{Z}$. As an object defined on $\mathbf{R}$, this pulse train could be described by

$$\sum_{n=-\infty}^{\infty} \delta_{a+2\pi n}(t) = \sum_{n=-\infty}^{\infty} \delta(t-a-2\pi n).$$

The convergence of this series is uncontroversial, because at any individual point t at most one of the terms is different from zero.

The derivatives of δ_a can be described using integration by parts, just as in Sec. 2.6, but now the integrals are taken over $\mathbf{T}$ (i.e., over one period). Because everything is periodic, the contributions at the ends of the interval will cancel:

$$\int_{\mathbf{T}} \varphi(t)\delta_a'(t)\,dt = \Big[\varphi(t)\delta_a(t)\Big]_b^{b+2\pi} - \int_{\mathbf{T}} \varphi'(t)\delta_a(t)\,dt = -\varphi'(a).$$

What would be the Fourier series of these distributions? Let us first consider δ_a. The natural approach is to define Fourier coefficients by the formula

$$c_n = \frac{1}{2\pi}\int_{\mathbf{T}} \delta_a(t)e^{-int}\,dt = \frac{1}{2\pi}\cdot e^{-ina}.$$

The series then looks like this:

$$\delta_a(t) \sim \frac{1}{2\pi}\sum_{n\in\mathbf{Z}} e^{-ina}\cdot e^{int}.$$

In particular, when $a=0$, the Fourier coefficients are all equal to $1/(2\pi)$, and the series is

$$\delta(t) \sim \frac{1}{2\pi}\sum_{n\in\mathbf{Z}} e^{int}.$$

By pairing terms with the same values of $|n|$, we can formally rewrite this as

$$\delta(t) \sim \frac{1}{2\pi} + \frac{1}{\pi}\sum_{n=1}^{\infty} \cos nt.$$

Compare the Dirichlet kernel! We might say that δ is the limit of D_N as $N\to\infty$.

These series cannot be convergent in the usual sense, since their terms do not tend to zero. But for certain values of t they can be summed according to Cesàro. Indeed, we can use the result of Exercise 2.16 on page 22. The series for $2\pi\delta_a$ can be written (with $z = e^{i(t-a)}$)

$$\begin{aligned}\sum_{n\in\mathbf{Z}} e^{in(t-a)} &= \sum_{n=0}^{\infty}\left(e^{i(t-a)}\right)^n + \sum_{n=-\infty}^{-1}\left(e^{i(t-a)}\right)^n \\ &= \sum_{n=0}^{\infty} z^n + \sum_{n=1}^{\infty}\left(e^{-i(t-a)}\right)^n = \sum_{n=0}^{\infty} z^n + \overline{z}\sum_{n=0}^{\infty}\overline{z}^n\end{aligned}$$

According to Exercise 2.16, both the series in the last expression can be summed $(C,1)$ if $|z| = 1$ but $z \neq 1$, which is the case if $t \neq a$, and the result will be

$$\frac{1}{1-z} + \frac{\overline{z}}{1-\overline{z}} = \frac{1-\overline{z}+\overline{z}(1-z)}{|1-z|^2} = \frac{1-z\overline{z}}{|1-z|^2} = \frac{1-|z|^2}{|1-z|^2} = 0.$$

If $t = a$, all the terms are ones, and the series diverges to infinity.

Thus the series behaves in a way that is most satisfactory, as it enhances our intuitive image of what δ_a looks like.

Next we find the Fourier series of δ_a'. The coefficients are

$$\begin{aligned} c_n &= \frac{1}{2\pi}\int_{-\pi}^{\pi} \delta_a'(t)e^{-int}\,dt = -\frac{1}{2\pi}\cdot\frac{d}{dt}e^{-int}\bigg|_{t=a} \\ &= -\frac{1}{2\pi}\left(-ine^{-int}\right)\big|_{t=a} = \frac{in}{2\pi}e^{-ina}. \end{aligned}$$

We recognize that the rule in Sec. 4.3 for the Fourier coefficients of a derivative holds true. The summation of the series

$$\delta_a'(t) = \frac{i}{2\pi}\sum_{n\in\mathbf{Z}} ne^{-ian}e^{int} = \frac{i}{2\pi}\sum_{n\in\mathbf{Z}} ne^{in(t-a)}$$

is tougher than that of δ_a itself, because the terms now have moduli that even tend to infinity as $|n| \to \infty$. It can be shown, however, that for $t \neq a$ the series is summable $(C,2)$ to 0.

We give a couple of examples to illustrate the use of these series.

Example 4.8. Consider the function of Example 4.2 on page 78. Its Fourier series can be written

$$f(t) \sim \sum_{n=1}^{\infty}\frac{\sin nt}{n} = \sum_{n\neq 0}\frac{1}{2in}e^{int}.$$

(Notice that the last version is correct — the minus sign in the Euler formula for sin is incorporated in the sign of the n in the coefficient.)

The derivative of f consists of an "ordinary" term $-\frac{1}{2}$, which takes care of the slope between the jumps, and a pulse train that on $\mathbf{T}$ is identified with $\pi\cdot\delta(t)$. This would mean that the Fourier series of the derivative is given by

$$\begin{aligned} f'(t) &= -\tfrac{1}{2} + \pi\delta(t) \sim -\tfrac{1}{2} + \pi\cdot\frac{1}{2\pi}\sum_{n\in\mathbf{Z}} e^{int} \\ &= -\tfrac{1}{2} + \tfrac{1}{2}\sum_{n\in\mathbf{Z}} e^{int} = \tfrac{1}{2}\sum_{n=1}^{\infty}(e^{int}+e^{-int}) = \sum_{n=1}^{\infty}\cos nt. \end{aligned}$$

Notice that this is precisely what a formal differentiation of the original series would yield. □

Example 4.9. Find a 2π-periodic solution of the differential equation $y' + y = 1 + \delta(t)\,(-\pi < t < \pi)$.

Solution. We try a solution of the form $y = \sum c_n e^{int}$. Differentiating this and expanding the right-hand member in Fourier series, we get

$$\sum_{n\in\mathbf{Z}} inc_n e^{int} + \sum_{n\in\mathbf{Z}} c_n e^{int} = 1 + \sum_{n\in\mathbf{Z}} \frac{1}{2\pi}\, e^{int},$$

or

$$c_0 + \sum_{n\neq 0} (in+1)c_n\, e^{int} = \left(1 + \frac{1}{2\pi}\right) + \sum_{n\neq 0} \frac{1}{2\pi}\, e^{int}.$$

Identification of coefficients yields $c_0 = 1 + 1/(2\pi)$ and, for $n \neq 0$, $c_n = 1/(2\pi(1+in))$. A solution should thus be given by

$$y(t) \sim \left(1 + \frac{1}{2\pi}\right) + \frac{1}{2\pi} \sum_{n\neq 0} \frac{e^{int}}{1+in}.$$

By a stroke of luck, it happens that this series has been almost encountered before in the text: in Example 4.1 on page 76 f. we found that

$$f(u) \sim \frac{\sinh \pi}{\pi}\left(1 + \sum_{n\neq 0} \frac{(-1)^n}{1-in}\, e^{inu}\right),$$

where $f(u) = e^u$ for $-\pi < u < \pi$ and f has period 2π. From this we can find that

$$\sum_{n\neq 0} \frac{(-1)^n}{1-in}\, e^{inu} = \frac{\pi}{\sinh \pi}\, f(u) - 1.$$

On the other hand, the series on the left of this equation can be rewritten, using $(-1)^n = e^{in\pi}$ and letting $t = \pi - u$:

$$\sum_{n\neq 0} \frac{(-1)^n}{1-in}\, e^{inu} = \sum_{n\neq 0} \frac{(-1)^n}{1+in}\, e^{-inu} = \sum_{n\neq 0} \frac{e^{in(\pi-u)}}{1+in} = \sum_{n\neq 0} \frac{e^{int}}{1+in}.$$

This means that our solution can be expressed in the following way:

$$y(t) \sim \left(1+\frac{1}{2\pi}\right) + \frac{1}{2\pi}\left(\frac{\pi}{\sinh \pi}\, f(u) - 1\right) = 1 + \frac{1}{2\sinh \pi}\, f(u) = 1 + \frac{f(\pi - t)}{2\sinh \pi}.$$

In particular,

$$y(t) = 1 + \frac{e^{\pi - t}}{2\sinh \pi} = 1 + \frac{e^{\pi}}{2\sinh \pi}\, e^{-t}, \quad 0 < t < 2\pi,$$

since this condition on t is equivalent to $-\pi < \pi - t < \pi$. At the points $t = n \cdot 2\pi$, $y(t)$ has an upward jump of size 1 (check this!).

Let us check the solution by substitution into the equation. Differentiating, we find that $y'(t)$ contains the pulse $\delta(t)$ at the origin, and between jumps one has $y'(t) = -(y(t) - 1)$. This proves that we have indeed found a solution. □

Exercises

4.30 Let f be the even function with period 2π that satisfies $f(t) = \pi - t$ for $0 \le t \le \pi$. Determine f' and f'', and use the result to find the Fourier series of f.

4.31 Let f have period 2π and satisfy

$$f(t) = \begin{cases} e^t, & |t| < \pi/2, \\ 0, & \pi/2 < |t| < \pi. \end{cases}$$

Compute $f' - f$, and then determine the Fourier series of f.

Summary of Chapter 4

Definition
If f is a sufficiently nice function defined on $\mathbf{T}$, we define its *Fourier coefficients* by

$$c_n = \frac{1}{2\pi} \int_{\mathbf{T}} f(t) e^{-int}\,dt \qquad \text{or} \qquad \begin{matrix} a_n \\ b_n \end{matrix} = \frac{1}{\pi} \int_{\mathbf{T}} f(t) \begin{matrix} \cos \\ \sin \end{matrix} nt\,dt.$$

The *Fourier series* of f is the series

$$\sum_{n \in \mathbf{Z}} c_n e^{int}, \qquad \text{resp.} \qquad \tfrac{1}{2} a_0 + \sum_{n=1}^{\infty} (a_n \cos nt + b_n \sin nt).$$

If f has a period other than 2π, the formulae have to be adjusted accordingly. If f is even or odd, the formulae for a_n and b_n can be simplified.

Theorem
If two continuous functions f and g have the same Fourier coefficients, then $f = g$.

Theorem
If f is piecewise continuous on $\mathbf{T}$ and continuous at the point t, then, for this value of t, its Fourier series is summable $(C, 1)$ to the value $f(t)$.

Theorem
If f is continuous on $\mathbf{T}$ and its Fourier coefficients satisfy $\sum |c_n| < \infty$, then its Fourier series converges absolutely and uniformly to $f(t)$ on all of $\mathbf{T}$.

Theorem
If f is differentiable on $\mathbf{T}$, then the Fourier series of the derivative f' can be found by termvise differentiation.

Theorem
If $f \in C^k(\mathbf{T})$, then its Fourier coefficients satisfy $|c_n| \leq M/|n|^k$.

Theorem
If f is continuous except for jump discontinuities, and if it has (generalized) one-sided derivatives at a point t, then its Fourier series for this value of t converges with the sum $\frac{1}{2}(f(t+) + f(t-))$.

Formulae for Fourier series are found on page 251.

Historical notes

JOSEPH FOURIER was not the first person to consider trigonometric series of the kind that came to bear his name. Around 1750, both Daniel Bernoulli and Leonhard Euler were busy investigating these series, but the standard of rigor in mathematics then was not sufficient for a real understanding of them. Part of the problem was the fact that the notion of a function had not been made precise, and different people had different opinions on this matter. For example, a graph pieced together as in Figure 4.2 on page 78 was not considered to represent one function but several. It was not until the times of BERNHARD RIEMANN and KARL WEIERSTRASS that something similar to the modern concept of a function was born. In 1822, when Fourier's great treatise appeared, it was generally regarded as absurd that a series with terms that were smooth and nice trigonometric functions should be able to represent functions that were not everywhere differentiable, or even worse—discontinuous!

The convergence theorem (Theorem 4.5) as stated in the text is a weaker version of a result by the German mathematician J. PETER LEJEUNE-DIRICHLET (1805–59). At the age of 19, the Hungarian LIPÓT FEJÉR (1880–1959) had the bright idea of applying Cesàro summation to Fourier series.

In the twentieth century the really hard questions concerning the convergence of Fourier series were finally resolved, when LENNART CARLESON (1928–) proved his famous Theorem 4.6. The author of this book, then a graduate student, attended the series of seminars in the fall of 1965 when Carleson step by step conquered the obstacles in his way. The final proof consists of 23 packed pages in one of the world's most famous mathematical journals, the *Acta Mathematica.*

Problems for Chapter 4

4.32 Determine the Fourier series of the following functions. Also state what is the sum of the series for all t.
(a) $f(t) = 2 + 7\cos 3t - 4\sin 2t, \quad -\pi < t < \pi.$
(b) $f(t) = |\sin t|, \quad -\pi < t < \pi.$

(c) $f(t) = (\pi - t)(\pi + t), \quad -\pi < t < \pi$.
(d) $f(t) = e^{|t|}, \quad -\pi < t < \pi$.

4.33 Find the cosine series of $f(t) = \sin t$, $0 < t < \pi$.

4.34 Find the sine series of $f(t) = \cos t$, $0 < t < \pi$. Use this series to show that

$$\frac{\pi\sqrt{2}}{16} = \frac{1}{2^2-1} - \frac{3}{6^2-1} + \frac{5}{10^2-1} - \frac{7}{14^2-1} + \cdots.$$

4.35 Let f be the 2π-periodic continuation of the function $H(t-a) - H(t-b)$, where $-\pi < a < b < \pi$. Find the Fourier series of f. For what values of t does it converge? Indicate its sum for for all such $t \in [-\pi, \pi]$.

4.36 Let f be given by $f(x) = -1$ for $-1 < x < 0$, $f(x) = x$ for $0 \le x \le 1$ and $f(x+2) = f(x)$ for all x. Compute the Fourier series of f. State the sum of this series for $x = 10$, $x = 10.5$, and $x = 11$.

4.37 Develop $f(t) = t(t-1)$, $0 < t < 1$, period 1, in a Fourier series. Quote some criterion that implies that the series converges to $f(t)$ for all values of t.

4.38 The function f is defined by $f(t) = t^2$ for $0 \le t \le 1$, $f(t) = 0$ for $1 < t < 2$ and by the statement that it has period 2.
(a) Develop f in a Fourier series with period 2 and indicate the sum of the series in the interval $[0, 5]$.
(b) Compute the value of the sum $s = \sum_{n=1}^{\infty} \dfrac{(-1)^n}{n^2}$.

4.39 Suppose that f is integrable, has period T, and Fourier series

$$f(t) \sim \sum_{n=-\infty}^{\infty} c_n \, e^{2\pi i n t/T}.$$

Determine the Fourier series of the so-called *autocorrelation function* r of f, which is defined by

$$r(t) = \frac{1}{T} \int_0^T f(t+u)\, \overline{f(u)}\, du.$$

4.40 An application to sound waves: Suppose the variation in pressure, p, that causes a sound has period $\frac{1}{262}$ s (seconds), and satisfies

$$p(t) = 1, \quad 0 < t < \tfrac{1}{1048}, \qquad p(t) = -\tfrac{7}{8}, \quad \tfrac{1}{1048} < t < \tfrac{1}{524},$$

$$p(t) = \tfrac{7}{8}, \quad \tfrac{1}{524} < t < \tfrac{3}{1048}, \qquad p(t) = -1, \quad \tfrac{3}{1048} < t < \tfrac{1}{262}.$$

What frequencies can be heard in this sound? Which is the dominant frequency?

4.41 Compute the Fourier series of f, given by

$$f(x) = \left|\sin\frac{x}{2}\right|, \quad -\pi < x \le \pi; \quad f(x+2\pi) = f(x), \; x \in \mathbf{R}.$$

Then find the values of the sums

$$s_1 = \sum_{n=1}^{\infty} \frac{1}{4n^2-1} \quad \text{and} \quad s_2 = \sum_{n=1}^{\infty} \frac{(-1)^n}{4n^2-1}.$$

4.42 Let f be an even function of period 2π described by $f(x) = \cos 2x$ for $0 \le x \le \frac{1}{2}\pi$ and $f(x) = -1$ for $\frac{1}{2}\pi < x \le \pi$. Find its Fourier series and compute the value of the sum

$$s = \sum_{k=1}^{\infty} \frac{(-1)^k}{(2k+1)(2k-1)(2k+3)}.$$

4.43 Find all solutions $y(t)$ with period 2π of the differential-difference equation

$$y'(t) + y(t - \tfrac{1}{2}\pi) - y(t - \pi) = \cos t, \qquad -\infty < t < \infty.$$

4.44 Let f be an even function with period 4 such that $f(x) = 1-x$ for $0 \le x \le 1$ and $f(x) = 0$ for $1 < x \le 2$. Find its Fourier series and compute

$$s = \sum_{k=0}^{\infty} \frac{1}{(2k+1)^2}.$$

4.45 Let α be a real number but not an integer. Define $f(x)$ by putting $f(x) = e^{i\alpha x}$ for $-\pi < x < \pi$ and $f(x + 2\pi) = f(x)$. By studying its Fourier series, prove the following formulae:

$$\frac{\pi}{\sin \pi\alpha} = \frac{1}{\alpha} + \sum_{n=1}^{\infty} \frac{2(-1)^n \alpha}{\alpha^2 - n^2}. \qquad \left(\frac{\pi}{\sin \pi\alpha}\right)^2 = \sum_{n=-\infty}^{\infty} \frac{1}{(\alpha - n)^2}.$$

4.46 Compute the Fourier series of the 2π-periodic function f given by $f(x) = x^3 - \pi^2 x$ for $-\pi < x < \pi$. Find the sum

$$s = \sum_{n=1}^{\infty} \frac{(-1)^{n+1}}{(2n-1)^3}.$$

4.47 Let f be a 2π-periodic function with ("complex") Fourier coefficients c_n ($n \in \mathbf{Z}$). Assume that for an integer $k > 0$ it holds that

$$\sum_{n \in \mathbf{Z}} |n|^k \, |c_n| < \infty.$$

Prove that f is of class C^k, i.e., that the kth derivative of f is continuous.

4.48 Find the Fourier series of f with period 2 which is given for $|x| < 1$ by $f(x) = 2x^2 - x^4$. The result can be used to find the value of the sum

$$\zeta(4) = \sum_{n=1}^{\infty} \frac{1}{n^4}.$$

5

L^2 Theory

5.1 Linear spaces over the complex numbers

We assume that the reader is familiar with vector spaces, where the scalars are real numbers. It is also possible, and indeed very fruitful, to allow scalars to be *complex* numbers. Thus, we consider now a set V, whose elements can be added to each other, and also can be multiplied by complex numbers, to produce new elements in V; and these operations are to obey the same rules of calculation as in ordinary real vector spaces. Simple examples of such *complex vector spaces* are given by $\mathbf{C}^n$, which consists of n-tuples $\mathbf{z} = (z_1, z_2, \ldots, z_n)$ of complex numbers. Another example is the set of all complex-valued functions f defined on an interval $[a, b]$.

It turns out that elementary complex linear algebra can be developed in almost exact parallel to its real counterpart. Linear dependence, subspaces, and dimension can be defined word for word in the same way, with the understanding that all scalars occurring in the process are complex numbers. This gives a new significance to the notion of dimension. For example, $\mathbf{C}$ is a *one-dimensional* complex vector space. It could also be considered as a real vector space, but then the dimension is 2. In the same manner, $\mathbf{C}^n$ has *complex* dimension n but *real* dimension $2n$.

When we reach the notions of scalar product and distance, we must, however, modify the details. The old formula for the (standard) scalar product of two vectors $\mathbf{z}$ and $\mathbf{w}$ in n-space has the form $\langle \mathbf{z}, \mathbf{w} \rangle = z_1 w_1 + z_2 w_2 + \cdots + z_n w_n$. This cannot be allowed to hold any more, because it does not work in a reasonable way.

Example 5.1. Consider the vector $\mathbf{z} = (1, i) \in \mathbf{C}^2$. With the usual formula for the scalar product we would get $|\mathbf{z}|^2 = 1 \cdot 1 + i \cdot i = 1 - 1 = 0$. The length of the vector would be 0, which is no good. Still more strange would be the case for the vector $\mathbf{w} = (1, 2i)$, which would not even have real length. □

By considering the one-dimensional space $\mathbf{C}$, we can get an idea for a more suitable definition. The vector, or number, $z = x+iy$ is normally identified with the vector (x, y) in $\mathbf{R}^2$. The length of this vector is the number $|z| = \sqrt{x^2 + y^2} = \sqrt{z\overline{z}}$. If the scalar product of two complex numbers z and w is defined to be $\langle z, w \rangle = z\overline{w}$, we get the formula $|z|^2 = \langle z, z \rangle$. This means that we have to modify a few rules for the scalar product, but that price is well worth paying. Just to indicate that the rules of computation are no longer exactly the same, we choose to use the name *inner product* instead of scalar product.

Definition 5.1 *Let V be a complex vector space. An* inner product on V *is a complex-valued function $\langle u, v \rangle$ of u and $v \in V$ having the following properties:*

(1) $\langle u, v \rangle = \overline{\langle v, u \rangle}$ *(Hermitian symmetry)*
(2) $\langle \alpha u + \beta v, w \rangle = \alpha \langle u, w \rangle + \beta \langle v, w \rangle$ *(Linearity in the first argument)*
(3) $\langle u, u \rangle \geq 0$ *(Positive-definiteness)*
(4) $\langle u, u \rangle = 0 \;\Rightarrow\; u = \mathbf{0}$

Combining rules 1 and 2 we find the rule
(5) $\langle u, \alpha v + \beta w \rangle = \overline{\alpha} \langle u, v \rangle + \overline{\beta} \langle u, w \rangle$.

Example 5.2. In $\mathbf{C}^n$, we can define an inner product (using natural notation) by the formula

$$\langle \mathbf{z}, \mathbf{w} \rangle = z_1 \overline{w}_1 + z_2 \overline{w}_2 + \cdots + z_n \overline{w}_n.$$

□

Example 5.3. Let $C(a, b)$ be the set of continuous, complex-valued functions defined on the compact interval $[a, b]$, and put

$$\langle f, g \rangle = \int_a^b f(x)\, \overline{g(x)}\, dx.$$

The fact that this is an inner product is almost trivial, except possibly for condition 4. That implication follows from the fact that if a *continuous* function is non-negative on an interval and its integral is 0, then the function must indeed be identically 0 (see textbooks on calculus). □

Example 5.4. More generally, let w be a fixed continuous function on $[a, b]$ such that $w(x) > 0$, and put

$$\langle f, g\rangle = \int_a^b f(x)\,\overline{g(x)}\,w(x)\,dx.$$

□

A complex vector space with a chosen inner product is called an *inner product space.* Because of rules 3 and 4 it is natural to define a measure of the size of a vector in such a space, corresponding to the length in the real case. One prefers to use the word *norm* instead of length and write

$$\text{the norm of } u = \|u\| = \sqrt{\langle u, u\rangle}.$$

In the case described in Example 5.3 we thus have

$$\|f\|^2 = \int_a^b |f(x)|^2\,dx.$$

The following inequalities are wellknown in the real case, and they hold just as well in the new setting:

Theorem 5.1

$$|\langle u, v\rangle| \le \|u\|\,\|v\| \qquad \textit{(Cauchy–Schwarz inequality)}$$
$$\|u + v\| \le \|u\| + \|v\| \qquad \textit{(Triangle inequality)}$$

Proof. (a) If $u = 0$, then both members are 0 and the statement is true. Thus, let us assume that $u \ne 0$. Put $\alpha = -\langle v, u\rangle/\langle u, u\rangle$. Then,

$$\begin{aligned} 0 &\le \|\alpha u + v\|^2 = \langle \alpha u + v, \alpha u + v\rangle = \alpha\overline{\alpha}\langle u, u\rangle + \alpha\langle u, v\rangle + \overline{\alpha}\langle v, u\rangle + \langle v, v\rangle \\ &= \frac{\langle v, u\rangle \cdot \langle u, v\rangle}{\langle u, u\rangle^2}\langle u, u\rangle - \frac{\langle v, u\rangle}{\langle u, u\rangle}\langle u, v\rangle - \frac{\langle u, v\rangle}{\langle u, u\rangle}\langle v, u\rangle + \langle v, v\rangle \\ &= -\frac{|\langle u, v\rangle|^2}{\langle u, u\rangle} + \langle v, v\rangle. \end{aligned}$$

Rearranging this we arrive at the Cauchy–Schwarz inequality.

(b) The triangle inequality is proved using Cauchy–Schwarz:

$$\begin{aligned} \|u + v\|^2 &= \langle u + v, u + v\rangle = \|u\|^2 + \langle u, v\rangle + \langle v, u\rangle + \|v\|^2 \\ &= \|u\|^2 + 2\,\mathrm{Re}\,\langle u, v\rangle + \|v\|^2 \le \|u\|^2 + 2|\langle u, v\rangle| + \|v\|^2 \\ &\le \|u\|^2 + 2\|u\|\,\|v\| + \|v\|^2 = \big(\|u\| + \|v\|\big)^2. \end{aligned}$$

Since both $\|u + v\|$ and $\|u\| + \|v\|$ are non-negative numbers, the triangle inequality follows. □

Two vectors u and v are called *orthogonal* with respect to the chosen inner product, if $\langle u, v\rangle = 0$. A vector u is *normed* if $\|u\| = 1$; and an *orthonormal*, or simply ON, set of vectors consists of normed and pairwise orthogonal vectors.

Example 5.5. $C(\mathbf{T})$ consists of the continuous complex-valued functions on the unit circle $\mathbf{T}$, and let the inner product be defined by

$$\langle f, g\rangle = \frac{1}{2\pi}\int_{\mathbf{T}} f(x)\,\overline{g(x)}\,dx.$$

Let $\varphi_k(x) = e^{ikx}$, $k \in \mathbf{Z}$. Then we have

$$\langle \varphi_m, \varphi_n\rangle = \frac{1}{2\pi}\int_{\mathbf{T}} e^{imx}\,e^{-inx}\,dx = \frac{1}{2\pi}\int_{-\pi}^{\pi} e^{i(m-n)x}\,dx = \begin{cases} 0, & m \neq n \\ 1, & m = n \end{cases}$$

The sequence $\{\varphi_k\}_{k=-\infty}^{\infty}$ is thus an orthonormal set in $C(\mathbf{T})$. □

Just as in the case of real spaces one can show that a set of non-zero, pairwise orthogonal vectors is linearly independent. In a space with finite (complex) dimension N, an orthonormal set can thus contain at most N vectors. On the other hand, there always exists such a set. Starting from an arbitrary basis in the space, one can use the Gram–Schmidt process to construct an ON basis, working precisely as in the real case.

Example 5.6. As a reminder of the Gram–Schmidt process, we construct an orthonormal basis in the space of polynomials of degree at most 2, with the inner product

$$\langle f, g\rangle = \int_0^1 f(x)\overline{g(x)}(1+x)\,dx.$$

The "raw material" consists of the vectors u_0, u_1, u_2 defined by

$$u_0(x) = 1, \quad u_1(x) = x, \quad u_2(x) = x^2.$$

As a tentative first vector in the new basis we choose $v_0 = u_0$, which is to be adjusted to have unit norm. Since

$$\|v_0\|^2 = \langle v_0, v_0\rangle = \int_0^1 1\cdot 1\cdot(1+x)\,dx = \tfrac{3}{2},$$

the normed vector is described by

$$\varphi_0(x) = \frac{v_0(x)}{\|v_0\|} = \sqrt{\tfrac{2}{3}}\,.$$

Next we adjust u_1 so that it is orthogonal to φ_0. This is done by subtracting a certain multiple of φ_0 according to the following recipe:

$$v_1 = u_1 - \langle u_1, \varphi_0\rangle\,\varphi_0.$$

The inner product is

$$\langle u_1, \varphi_0\rangle = \int_0^1 x\cdot\sqrt{\tfrac{2}{3}}(1+x)\,dx = \tfrac{5}{6}\sqrt{\tfrac{2}{3}}\,.$$

Thus,

$$v_1(x) = x - \tfrac{5}{6}\sqrt{\tfrac{2}{3}} \cdot \sqrt{\tfrac{2}{3}} = x - \tfrac{5}{9}.$$

The squared norm of v_1 is

$$\|v_1\|^2 = \langle v_1, v_1 \rangle = \int_0^1 \left(x - \tfrac{5}{9}\right)^2 (x+1)\,dx = \tfrac{13}{108}\,.$$

Thus the second normed basis vector is

$$\varphi_1(x) = \frac{x - \frac{5}{9}}{\sqrt{\frac{13}{108}}} = 6\sqrt{\tfrac{3}{13}}\left(x - \tfrac{5}{9}\right).$$

The third vector is obtained by the following steps:

$$v_2 = u_2 - \langle u_2, \varphi_0\rangle\varphi_0 - \langle u_2, \varphi_1\rangle\varphi_1, \qquad \langle u_2, \varphi_0\rangle = \tfrac{7}{36}\sqrt{6}, \quad \langle u_2, \varphi_1\rangle = \tfrac{34}{585}\sqrt{39},$$

$$v_2 = x^2 - \tfrac{68}{65}\,x + \tfrac{5}{26}\,, \qquad \langle v_2, v_2\rangle = \tfrac{21}{2600}\,,$$

$$\varphi_2 = \tfrac{10}{21}\sqrt{546}\left(x^2 - \tfrac{68}{65}\,x + \tfrac{5}{26}\right).$$

It is obvious that the computations quickly grow very involved, not least because the norms turn out to be rather hideous numbers. Mostly, therefore, one is satisfied with just orthogonal sets, instead of orthonormal sets. □

There is also the following theorem, which should be proved by the reader:

Theorem 5.2 *If $\varphi_1, \varphi_2, \ldots, \varphi_N$ is an ON basis in an N-dimensional inner product space V, then every $u \in V$ can be written as $u = \sum_{j=1}^{N} \langle u, \varphi_j\rangle\varphi_j$, and furthermore one has*

$$\|u\|^2 = \sum_{j=1}^{N} |\langle u, \varphi_j\rangle|^2 \quad \text{(theorem of Pythagoras)}.$$

For the inner product of two vectors one also has the following formula:

$$\langle u, v\rangle = \sum_{j=1}^{N} \langle u, \varphi_j\rangle\overline{\langle v, \varphi_j\rangle}.$$

Exercises

5.1 Let $\mathbf{u} = (1-2i,\ 3,\ 2+i)$ and $\mathbf{v} = (i,\ 1-3i,\ 0)$ be vectors in $\mathbf{C}^3$. Compute $\|\mathbf{u}\|$, $\|\mathbf{v}\|$, and $\langle \mathbf{u}, \mathbf{v}\rangle$.

5.2 Let $C(-1,1)$ be as in Example 5.3 above, and let f and $g \in C(-1,1)$ be described by $f(x) = x+2$, $g(x) = ix + x^2$. Compute $\langle f, g\rangle$.

5.3 Are the functions 1, x, x^2, $\ldots$, x^n, $\sin x$, $\cos x$, e^x linearly independent, considered as vectors in $C(0,1)$? (n is some positive integer.)

5.4 Apply the Gram–Schmidt orthogonalization procedure to construct an orthogonal basis in the subspace of $C(0,1)$ spanned by the polynomials 1, x and x^2.

5.5 Orthogonalize the following sets of vectors:
(a) the vectors $(1,2,3)$, $(3,1,4)$ and $(2,1,1)$ in $\mathbf{C}^3$;
(b) the functions 1, x, x^2 i $C(-1,1)$;
(c) the functions e^{-x}, xe^{-x}, x^2e^{-x} in $C(0,\infty)$.

5.6 Prove the following formula for the inner product:

$$4\langle u,v\rangle = \|u+v\|^2 + i\|u+iv\|^2 - \|u-v\|^2 - i\|u-iv\|^2.$$

5.2 Orthogonal projections

The reader should recall the following fact from real linear algebra: Let $\{\varphi_k\}_{k=1}^N$ be an orthonormal set in the space V, and let u be an arbitrary vector in V. The *orthogonal projection* of u on to the subspace of V spanned by $\{\varphi_k\}_{k=1}^N$ is the vector

$$P_N(u) = \langle u,\varphi_1\rangle\varphi_1 + \langle u,\varphi_2\rangle\varphi_2 + \cdots + \langle u,\varphi_N\rangle\varphi_N = \sum_{k=1}^N \langle u,\varphi_k\rangle\varphi_k.$$

This definition works just as well in the complex case, and the projection thus defined has also the least-squares property described in the following theorem.

Theorem 5.3 (Least squares approximation) *Let $\{\varphi_k\}_{k=1}^N$ be an orthonormal set in an inner product space V and let u be a vector in V. Among all the linear combinations $\Phi = \sum_{k=1}^N \gamma_k\varphi_k$, the one that minimizes the value of $\|u-\Phi\|$ is the one with the coefficients γ_k equal to $\langle u,\varphi_k\rangle$, i.e., $\Phi = P_N(u)$.*

Proof. For brevity, we write $\langle u,\varphi_k\rangle = c_k$. We get

$$\|u-\Phi\|^2 = \langle u-\Phi, u-\Phi\rangle = \left\langle u - \sum_{k=1}^N \gamma_k\varphi_k,\, u - \sum_{k=1}^N \gamma_k\varphi_k\right\rangle$$
$$= \langle u,u\rangle - \left\langle u, \sum_{k=1}^N \gamma_k\varphi_k\right\rangle - \left\langle \sum_{k=1}^N \gamma_k\varphi_k, u\right\rangle + \left\langle \sum_{k=1}^N \gamma_k\varphi_k, \sum_{k=1}^N \gamma_k\varphi_k\right\rangle$$
$$= \langle u,u\rangle - \sum_{k=1}^N \overline{\gamma}_k\langle u,\varphi_k\rangle - \sum_{k=1}^N \gamma_k\langle\varphi_k,u\rangle + \sum_{k=1}^N\sum_{j=1}^N \gamma_k\overline{\gamma}_j\langle\varphi_k,\varphi_j\rangle$$
$$= \langle u,u\rangle - \sum_{k=1}^N (\overline{\gamma}_k c_k + \gamma_k\overline{c}_k) + \sum_{k=1}^N \gamma_k\overline{\gamma}_k$$

$$= \|u\|^2 + \sum_{k=1}^{N}(\gamma_k\overline{\gamma}_k - \overline{\gamma}_k c_k - \gamma_k \overline{c}_k + c_k\overline{c}_k) - \sum_{k=1}^{N}|c_k|^2$$
$$= \|u\|^2 + \sum_{k=1}^{N}|\gamma_k - c_k|^2 - \sum_{k=1}^{N}|c_k|^2.$$

In the last expression, we note that u is given in the formulation of the theorem; also, since the φ_k are given, the $c_k = \langle u, \varphi_k\rangle$ are given. But we have the γ_k at our disposal, to minimize the value of the expression. In order to make it as small as possible, we choose the numbers γ_k to be equal to c_k, which proves the assertion. □

A couple of consequences of the proof are worth emphasizing. When a vector u is approximated by the orthogonal projection $P_N(u)$, then the "error" $u - P_N(u)$ is called the *residual.* An important fact is that *the residual is orthogonal to the projection.* As for the "size" of the residual, we have the identity ("Pythagoras' theorem")

$$\|u - P_N(u)\|^2 = \|u\|^2 - \|P_N(u)\|^2.$$

If this is written out in full, it becomes

$$\left\|u - \sum_{k=1}^{N}\langle u, \varphi_k\rangle\varphi_k\right\|^2 = \|u\|^2 - \sum_{k=1}^{N}|\langle u, \varphi_k\rangle|^2. \tag{5.1}$$

Since the left-hand member of these equations is non-negative, we also have

$$\|P_N(u)\|^2 \le \|u\|^2, \quad \text{or} \quad \|P_N(u)\| \le \|u\|,$$

or, indeed,

$$\sum_{k=1}^{N}|\langle u, \varphi_k\rangle|^2 \le \|u\|^2.$$

If the ON set contains infinitely many vectors, we can let $N \to \infty$ and get what is known as the BESSEL inequality:

$$\sum_{k=1}^{\infty}|\langle u, \varphi_k\rangle|^2 \le \|u\|^2.$$

Let us now concentrate on the case when the space V has infinite dimension. A system $\{\varphi_j\}_{j=1}^{\infty}$ in V is said to be *complete* in V if, for every $u \in V$ and every $\varepsilon > 0$, there exists a linear combination $\sum_{j=1}^{N} a_j\varphi_j$ such that

$$\left\|u - \sum_{j=1}^{N} a_j\varphi_j\right\| < \varepsilon.$$

This means that every u in V can be approximated arbitrarily closely, in the sense of the norm, by linear combinations of the elements in the set $\{\varphi_j\}$. The theorem above shows that the *best* approximation must be given by the "infinite projection" $\sum_{j=1}^{\infty}\langle u, \varphi_j\rangle\varphi_j$ of u. A condition that is equivalent to completeness is given by the following theorem:

Theorem 5.4 *The ON system $\{\varphi_j\}_{j=1}^{\infty}$ is complete in V if and only if for every $u \in V$ it holds that*

$$\|u\|^2 = \sum_{j=1}^{\infty} |\langle u, \varphi_j\rangle|^2$$

(*the* PARSEVAL formula *or the* completeness relation).

The proof follows from (5.1).

As a corollary we also have the following:

Theorem 5.5 (Expansion theorem) *If the ON system $\{\varphi_j\}_{j=1}^{\infty}$ is complete in V, then every $u \in V$ can be written as $u = \sum_{j=1}^{\infty}\langle u, \varphi_j\rangle\varphi_j$, where the series converges in the sense of the norm (i.e., $\left\|u - \sum_{j=1}^{N}\langle u, \varphi_j\rangle\varphi_j\right\| \to 0$ as $N \to \infty$).*

Furthermore, one has the following theorem, which generalizes the usual formula (in finite dimension) for computing the inner product in an ON basis:

Theorem 5.6 *If the ON system $\{\varphi_j\}_{j=1}^{\infty}$ is complete in V, then*

$$\langle u, v\rangle = \sum_{j=1}^{\infty}\langle u, \varphi_j\rangle\overline{\langle v, \varphi_j\rangle}$$

for all $u, v \in V$.

Proof. Let $P_n(u)$ be the projection of u on to the subspace spanned by the n first φ's:

$$P_n(u) = \sum_{j=1}^{n}\langle u, \varphi_j\rangle\varphi_j\,.$$

By Theorem 5.2 we have

$$\langle P_n(u), P_n(v)\rangle = \sum_{j=1}^{n}\langle u, \varphi_j\rangle\overline{\langle v, \varphi_j\rangle}.$$

Using the triangle and Cauchy–Schwarz inequalities, we get

$$|\langle u, v\rangle - \langle P_n(u), P_n(v)\rangle|$$

$$\begin{aligned}
&= |\langle u,v\rangle - \langle u,P_n(v)\rangle + \langle u,P_n(v)\rangle - \langle P_n(u),P_n(v)\rangle| \\
&\le |\langle u, v-P_n(v)\rangle| + |\langle u-P_n(u), P_n(v)\rangle| \\
&\le \|u\|\,\|v-P_n(v)\| + \|u-P_n(u)\|\,\|P_n(v)\| \\
&\le \|u\|\,\|v-P_n(v)\| + \|u-P_n(u)\|\,\|v\|.
\end{aligned}$$

In the last part we also used Bessel's inequality. Now we know, because of completeness, that $\|v-P_n(v)\| \to 0$ as $n \to \infty$, and similarly for u, which implies that the final member of the estimate tends to zero. Then also the first member must tend to zero, and so

$$\langle u,v\rangle = \lim_{n\to\infty} \langle P_n(u),P_n(v)\rangle = \sum_{j=1}^{\infty} \langle u,\varphi_j\rangle\overline{\langle v,\varphi_j\rangle},$$

and the proof is complete. □

Remark. Using an estimate of the same kind as in the proof, one can see that $\langle u,v\rangle$ is a *continuous* function of u and v in the sense that if $u_n \to u$ and $v_n \to v$ (in the sense of the norm), then $\langle u_n,v_n\rangle \to \langle u,v\rangle$. □

In the interest of simplicity, we have throughout this section been working with ortho*normal* systems. In practice one is often satisfied with using *orthogonal* systems, since the normalizing factors can be quite unwieldy numbers. In such a case, our formulae have to be somewhat modified.

The projection of u on to an orthogonal set of vectors $\{\varphi_j\}_{j=1}^N$ is given by

$$P_N(u) = \sum_{j=1}^{N} \frac{\langle u,\varphi_j\rangle}{\langle \varphi_j,\varphi_j\rangle}\,\varphi_j = \sum_{j=1}^{N} \frac{\langle u,\varphi_j\rangle}{\|\varphi_j\|^2}\,\varphi_j.$$

The other formulae must be adjusted in the same vein: every φ_j that occurs has to be divided by its norm, and this holds whether φ_j is "free" or is part of an inner product. For example, the Parseval formula takes the form

$$\|u\|^2 = \sum_{j=1}^{\infty} \frac{|\langle u,\varphi_j\rangle|^2}{\|\varphi_j\|^2},$$

and the formula for inner products is

$$\langle u,v\rangle = \sum_{j=1}^{\infty} \frac{\langle u,\varphi_j\rangle\overline{\langle v,\varphi_j\rangle}}{\|\varphi_j\|^2}.$$

Exercises

5.7 Determine the polynomial p of degree at most 1 that minimizes $\int_0^2 |e^x - p(x)|^2\,dx$. (Hint: first find an orthogonal basis for a suitably chosen space of polynomials of degree ≤ 1.)

5.8 Determine the constants a and b in order to minimize the integral $\int_{-1}^{1} |ax + bx^2 - \sin \pi x|^2 \, dx$.

5.9 Find the polynomial $p(x)$ of degree at most 2 that minimizes the integral

$$\int_{-\pi/2}^{\pi/2} |\sin x - p(x)|^2 \cos x \, dx.$$

5.3 Some examples

We have already seen the finite-dimensional inner-product spaces $\mathbf{C}^n$. A generalization of these spaces can be constructed in the following manner. Let M be an arbitrary set (with finitely or infinitely many elements). Let $l^2(M)$ be the set of all functions $a : M \to \mathbf{C}$ such that

$$\sum_{x \in M} |a(x)|^2 < \infty. \tag{5.2}$$

The fact that this defines a linear space can be proved in the following way. Because of the inequality

$$|p\overline{q}| \le \tfrac{1}{2}(|p|^2 + |q|^2), \tag{5.3}$$

that holds for all complex numbers p and q, the following estimate is true:

$$\begin{aligned} |\lambda a + \mu b|^2 &= (\lambda a + \mu b)(\overline{\lambda}\overline{a} + \overline{\mu}\overline{b}) = |\lambda|^2|a|^2 + |\mu|^2|b|^2 + 2\mathrm{Re}\,\{\lambda\overline{\mu}a\overline{b}\} \\ &\le |\lambda|^2|a|^2 + |\mu|^2|b|^2 + 2|\lambda a||\mu b| \le 2\big(|\lambda|^2|a|^2 + |\mu|^2|b|^2\big). \end{aligned}$$

Using this, and the comparison test for positive series, one finds that if $\sum |a(x)|^2$ and $\sum |b(x)|^2$ are both convergent, then $\sum |\lambda a(x) + \mu b(x)|^2$ is also convergent. This means that linear combinations of elements in $l^2(M)$ also belong to $l^2(M)$. Using (5.3) we also see that if both a and b are members of $l^2(M)$, then the series

$$\sum_{x \in M} a(x)\,\overline{b(x)}$$

will converge absolutely. This expression can be taken as the definition of an inner product $\langle a, b \rangle$. The square of the norm of a is then given by the left member of (5.2).

If, as an example, we choose $M = \mathbf{N} = \{0, 1, 2, \ldots\}$, we can write a_n instead of $a(n)$ and get the inner product $\sum_{n=0}^{\infty} a_n \overline{b_n}$. This certainly looks like a natural generalization of $\mathbf{C}^n$.

When we gave examples of inner products in function spaces in Sec. 5.1, we assumed, for convenience, that all the functions were continuous. It is, however, often desirable to be able to work with more general functions.

One such class is the class of Riemann-integrable functions, which consists of functions that can be approximated in a certain way by so-called step functions; the class includes, for example, functions with a finite number of jumps. In order to get a really efficient theory, one should actually go still further and allow *measurable* functions in the sense of LEBESGUE. However, this is a rather complicated step, and in this text we shall confine ourselves to more "ordinary" functions. At one point (p. 121), we shall, however, mention how the Lebesgue functions obtrude upon us.

In what follows, it will often be of no interest whether an interval contains its endpoints or not. For simplicity, we shall write (a,b), which can be interpreted at will to mean either $[a,b]$ or $[a,b[$, etc.

Let I be an interval, bounded or unbounded, and let $w : I \to]0,\infty[$ be a positive, continuous, real-valued function on I. Let finally p be a number ≥ 1. The set $L^p(I,w)$ is defined to consist of all (Lebesgue-measurable) functions f such that

$$\int_I |f(x)|^p \, w(x)\, dx < \infty.$$

(The integral may be improper in one way or another, without this being indicated when we write it.) It can be proved that this defines a linear space: if f and g belong to it, then the same goes for all linear combinations $\alpha f + \beta g$. The proof is very simple in the case $p = 1$:

$$\int_I |\alpha f + \beta g|\, w\, dx \leq |\alpha| \int_I |f|\, w\, dx + |\beta| \int_I |g|\, w\, dx.$$

For $p = 2$, it can be done in a way that is analogous to what was done for $l^2(M)$ above (integrals replacing sums). For other values of p it is more difficult, and we skip it here.

The space $L^p(I,w)$ is called the *Lebesgue space* with *weight function* w and *exponent* p. If the weight is identically 1 on all of I, one simply writes $L^p(I)$. For $f \in L^p(I,w)$ one can define a *norm*

$$\|f\|_{p,w} = \left(\int_I |f(x)|^p \, w(x)\, dx \right)^{1/p},$$

that can be used to introduce a notion of *distance* $\|f - g\|_{p,w}$ between two functions f and g. This, in turn, gives rise to a notion of *convergence*: we say that a sequence of functions f_n converges to g in $L^p(I,w)$ if $\|f_n - g\|_{p,w} \to 0$ as $n \to \infty$.

In order for this to make sense, it is necessary to modify slightly what is meant by *equality* between functions. Two functions f and g are considered to be equal if their mutual distance is zero; explicitly this means that

$$\int_I |f(x) - g(x)|^p \, w(x)\, dx = 0.$$

This does not necessarily imply that $f(x) = g(x)$ for all x, it means only that they are equal "almost everywhere": the set where they differ is a so-called zero set (see the text preceding Theorem 4.6 on page 89). When working with "nice" functions, which are continuous except for a finite number of jumps, this means that the function values are actually equal except possibly for at a finite number of points.

We shall almost exclusively consider the case $p = 2$. It can be proved that only in this case can there exist an inner product $\langle f, g\rangle$ in $L^p(I, w)$ such that the norm is recovered through the formula $\|f\|^2 = \langle f, f\rangle$. This inner product must be defined by

$$\langle f, g\rangle = \int_I f(x)\,\overline{g(x)}\,w(x)\,dx.$$

Let us now look at a few important cases.

Example 5.7. Let $I = (-\pi, \pi)$ or, equivalently, interpret I as the unit circle $\mathbf{T}$, and take $w(x) = 1$. The inner product here is almost the same as in Example 5.5, page 108. There we showed that the functions $\varphi_n(x) = e^{inx}$ are orthogonal in $C(\mathbf{T})$, and this holds equally well in $L^2(\mathbf{T})$. Because the factor $1/(2\pi)$ is now missing in the inner product, they are no longer normed, however, but the norm of each φ_n is now $\sqrt{2\pi}$.

The formula for projections on to an orthogonal set of vectors has the form

$$P_N(f) = \sum_{n=1}^{N} \frac{\langle f, \varphi_n\rangle}{\langle \varphi_n, \varphi_n\rangle}\,\varphi_n = \sum_{n=1}^{N} \frac{\langle f, \varphi_n\rangle}{\|\varphi_n\|^2}\,\varphi_n = \sum_{n=1}^{N} c_n \varphi_n,$$

where

$$c_n = \frac{\langle f, \varphi_n\rangle}{\|\varphi_n\|^2} = \frac{1}{2\pi}\int_{\mathbf{T}} f(x)\,\overline{e^{inx}}\,dx = \frac{1}{2\pi}\int_{\mathbf{T}} f(x)\,e^{-inx}\,dx.$$

We recognize our good old Fourier coefficients. □

Example 5.8. In the same space as in the preceding example we study the system consisting of *firstly* all the functions $\varphi_n(x) = \cos nx$, $n = 0, 1, 2, \ldots$, *secondly* all the functions $\psi_n(x) = \sin nx$, $n = 1, 2, 3, \ldots$. Using some suitable trigonometric formula (or Euler's formulae), it is easily proved that all these functions are mutually orthogonal, and furthermore $\langle \varphi_0, \varphi_0\rangle = 2\pi$, while all the other members of the system have the square of the norm equal to π. Here, too, the coefficients in the orthogonal projections turn out to be the well-known classical Fourier coefficients. □

Example 5.9. In the space $L^2(0, \pi)$, the functions $\varphi_n(x) = \cos nx$, $n = 0, 1, 2, \ldots$ form an orthogonal set, and $\psi_n(x) = \sin nx$, $n = 1, 2, 3, \ldots$ another orthogonal set (check it!). However, as a rule, a φ_m is not orthogonal to a ψ_n. □

Examples 5.7 and 5.8 provide a new viewpoint for considering Fourier series. Such series can be regarded as representations with respect to an orthogonal system. The general results in Sections 5.1–2 concerning such representations are thus valid for our Fourier expansions. As an example, we formulate what Theorem 5.3 purports in the case of "real" Fourier series.

Theorem 5.7 *Let $f \in L^2(\mathbf{T})$ and let N be a fixed non-negative integer. Among all trigonometric polynomials of the form*

$$P_N(x) = \tfrac{1}{2}\alpha_0 + \sum_{n=1}^{N}(\alpha_n \cos nx + \beta_n \sin nx),$$

the polynomial that minimizes the value of the integral

$$\int_{\mathbf{T}} |f(x) - P_N(x)|^2\, dx$$

is the one where the coefficients are the usual Fourier coefficients of f, viz., (using the ordinary notation) where $\alpha_n = a_n$ and $\beta_n = b_n$.

We have also, for example, Bessel's inequality, which can be written, in complex and real guise, respectively, as

$$\sum_{n\in\mathbf{Z}} |c_n|^2 \le \frac{1}{2\pi}\int_{\mathbf{T}} |f(x)|^2\, dx, \quad \tfrac{1}{2}|a_0|^2 + \sum_{n=1}^{\infty}(|a_n|^2 + |b_n|^2) \le \frac{1}{\pi}\int_{\mathbf{T}} |f(x)|^2\, dx.$$

The question of completeness will be addressed in the next section.

Remark. At this point, the reader will have been confronted by a number of different notions of convergence for sequences of functions. There is *pointwise* convergence and *uniform* convergence, and now we have also various versions of L^p convergence. It is natural to ask whether there are any sensible connections between all these notions.

It is trivial that uniform convergence implies pointwise convergence. If one works on a finite interval (a, b) with weight 1, it is also easy to see that uniform convergence implies convergence in $L^p(a, b)$. Indeed, suppose that $f_n \to f$ uniformly on (a, b). Then $(p \ge 1)$

$$\|f_n - f\|_p = \left(\int_a^b |f_n(t) - f(t)|^p\, dt\right)^{1/p} \le \left(\int_a^b \sup_{(a,b)} |f_n(t) - f(t)|^p\, dt\right)^{1/p}$$

$$= \sup_{(a,b)} |f_n(t) - f(t)| \left(\int_a^b dt\right)^{1/p} = C \sup_{(a,b)} |f_n(t) - f(t)| \to 0$$

as $n \to \infty$.

In the converse direction, L^p convergence does not even imply pointwise convergence. This is shown by the following example; for simplicity it is formulated in $L^1(0, 1)$, but the same functions can be used to prove the same thing in all $L^p(0, 1)$, and they are easily rescaled to suit other intervals.

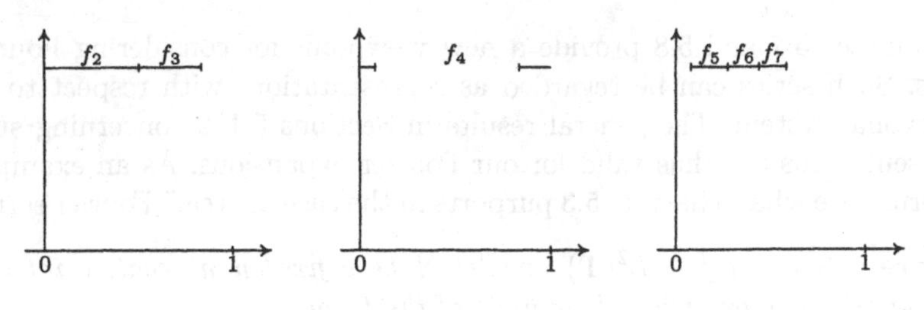

FIGURE 5.1.

Let $f_1(t) = 1$ on the interval $I = (0, 1)$. Then $\|f_1\|_1 = 1$. Let then f_2 be 1 on $(0, \frac{1}{2})$ and 0 on the rest of I; f_3 is set to 1 on $(\frac{1}{2}, \frac{5}{6})$ and 0 otherwise; f_4 is 1 on the intervals $(\frac{5}{6}, 1)$ and $(0, \frac{1}{12})$, and 0 otherwise. In general: f_n is equal to 1 on one or two intervals with total length equal to $1/n$, and, in a manner of speaking, f_n starts where f_{n-1} ends; and if you exceed the right-hand end of I you put the remainder on the far left of I and start again moving to the right (see Figure 5.1). Then it holds that $\|f_n\|_1 = 1/n$, and so $f_n \to 0$ in the sense of norm convergence in $L^1(I)$.

Since the series

$$\sum_{n=1}^{\infty} \frac{1}{n}$$

is divergent, in the sense that its partial sums tend to plus infinity, the "piling of intervals" described above must run to an infinite number of "turns." If x is an arbitrary point in I, there will thus exist arbitrarily large values of n such that $f_n(x) = 1$, but also arbitrarily large values of n for which $f_n(x) = 0$. In other words, the sequence of numbers $\{f_n(x)\}_{n=1}^{\infty}$ has no limit at all, which means that the sequence of functions $\{f_n\}$ does not converge pointwise anywhere.

The functions in this example are close to zero "in the mean": when n is large, $f_n(x)$ is equal to zero on most of the interval. For this reason, L^p convergence is often called *convergence in the mean*, more precisely in the L^p mean. □

Exercises

5.10 Find the three first orthonormal polynomials with respect to the inner product

$$\langle f, g \rangle = \int_0^1 f(x)\, \overline{g(x)}\, x\, dx,$$

by orthogonalizing the polynomials 1, x, x^2.

5.11 Solve the same problem as the preceding, when

$$\langle f, g \rangle = \int_{\mathbf{R}} f(x)\, \overline{g(x)}\, e^{-x^2}\, dx.$$

5.12 Determine the polynomial of degree ≤ 2, that is the best approximation to $f(x) = \sqrt{|x|}$ in the space $L^2(I, w)$, where $I = [-1, 1]$, $w(x) = 1$.

5.13 Determine the complex numbers c_j ($j = 0, 1, 2, 3$) in order to make the integral $\int_0^\pi |c_0 + c_1 \cos x + c_2 \cos 2x + c_3 \cos 3x - \cos^4 x|^2\, dx$ as small as possible.

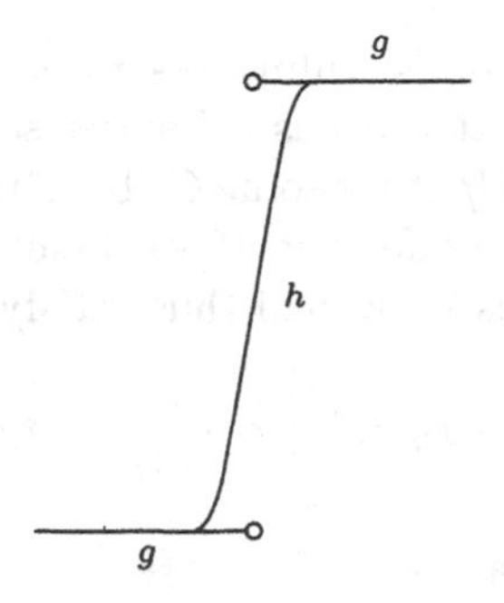

FIGURE 5.2.

5.4 The Fourier system is complete

We are now prepared to prove the following result, which, in a way, crowns the classical Fourier theory.

Theorem 5.8 *The two orthogonal systems* $\{e^{int}\}_{n\in\mathbf{Z}}$ *and* $\{\cos nt,\ n \geq 0;\ \sin nt,\ n \geq 1\}$ *are each complete in* $L^2(\mathbf{T})$.

Proof. We want to prove that if $f \in L^2(\mathbf{T})$, then f can be approximated arbitrarily well by linear combinations of elements from one of the systems. Since the elements of one system are simple linear combinations of elements from the other system, it does not matter which of the systems we consider. Our proof will be incomplete in one way: we only show that functions f that happen to be piecewise continuous can be approximated; but this is really just due to the fact that we have not said anything about what a more general f might look like.

Thus we assume that f is piecewise continuous. It is then bounded on $\mathbf{T}$, so that $|f(t)| \leq M$ for some constant M and all t. Now let $\varepsilon > 0$ be given. By the definition of the Riemann integral, there exists a *step function* (i.e., a piecewise constant function) g such that

$$\int_{\mathbf{T}} |f(t) - g(t)|\, dt < \frac{\varepsilon^2}{2M}.$$

Clearly we can choose g in such a way that $|g(t)| \leq M$, and then it follows that

$$\|f - g\|^2 = \int_{\mathbf{T}} |f(t) - g(t)|^2\, dt \leq \int_{\mathbf{T}} 2M|f(t) - g(t)|\, dt \leq 2M \cdot \frac{\varepsilon^2}{2M} = \varepsilon^2,$$

and so $\|f - g\| < \varepsilon$.

In the next step of the proof we round off the corners of the step function g to obtain a C^2 function h such that $\|g - h\| < \varepsilon$. At this point, the author appeals to the reader's willingness to accept that this is possible (see Figure 5.2). The function h has a Fourier series $\sum \gamma_n e^{int}$, where the coefficients satisfy $|\gamma_n| \le C/n^2$ for some C (by Theorem 4.4), which implies that the series converges uniformly. If we take N sufficiently large, the partial sum $s_N(t;h)$ of this series will thus satisfy

$$|h(t) - s_N(t;h)| < \frac{\varepsilon}{\sqrt{2\pi}}, \quad t \in \mathbf{T}.$$

From this we conclude that

$$\|h - s_N(\cdot\,;h)\|^2 = \int_{\mathbf{T}} |h(t) - s_N(t;h)|^2\,dt \le \frac{\varepsilon^2}{2\pi} \cdot 2\pi = \varepsilon^2,$$

so that $\|h - s_N(\cdot\,;h)\| \le \varepsilon$. Finally, let s_N be the corresponding partial sum of the Fourier series of the function f that we started with. Because of the Approximation theorem (Theorem 5.3), it is certainly true that $\|f - s_N\| \le \|f - s_N(\cdot\,;h)\|$. Time is now ripe for combining all our approximations in this manner:

$$\begin{aligned}\|f - s_N\| &\le \|f - s_N(\cdot\,;h)\| = \|f - g + g - h + h - s_N(\cdot\,;h)\| \\ &\le \|f - g\| + \|g - h\| + \|h - s_N(\cdot\,;h)\| < \varepsilon + \varepsilon + \varepsilon = 3\varepsilon.\end{aligned}$$

This means that f can be approximated to within 3ε by a certain linear combination of the functions e^{int}, and since ε can be chosen arbitrarily small we have proved the theorem. □

As a consequence we now have the Parseval formula and the formula for the inner product (which is also often called Parseval's formula, sometimes qualified as the *polarized* Parseval formula). For the "complex" system these formulae take the form

$$\frac{1}{2\pi}\int_{\mathbf{T}} |f(t)|^2\,dt = \sum_{n\in\mathbf{Z}} |c_n|^2, \qquad \frac{1}{2\pi}\int_{\mathbf{T}} f(t)\,\overline{g(t)}\,dt = \sum_{n\in\mathbf{Z}} c_n\,\overline{d_n},$$

and for the "real" system,

$$\begin{aligned}\frac{1}{\pi}\int_{\mathbf{T}} |f(t)|^2\,dt &= \tfrac{1}{2}|a_0|^2 + \sum_{n=1}^{\infty}(|a_n|^2 + |b_n|^2), \\ \frac{1}{\pi}\int_{\mathbf{T}} f(t)\,\overline{g(t)}\,dt &= \tfrac{1}{2}a_0\overline{\alpha_0} + \sum_{n=1}^{\infty}(a_n\overline{\alpha_n} + b_n\overline{\beta_n}).\end{aligned}$$

(The reader will have to figure out independently how the letters on the right correspond to the functions involved.)

Example 5.10. In Sec. 4.1, we saw that the odd function f with period 2π that is described by $f(t) = (\pi - t)/2$ for $0 < t < \pi$ has Fourier series

$\sum_1^\infty (\sin nt)/n$. Parseval's formula looks like this

$$\sum_{n=1}^{\infty} \frac{1}{n^2} = \sum_{n=1}^{\infty} b_n^2 = \frac{1}{\pi} \int_{\mathbf{T}} (f(t))^2\,dt = \frac{2}{\pi} \int_0^{\pi} \frac{(\pi - t)^2}{4}\,dt = \frac{1}{2\pi} \left[\frac{(\pi - t)^3}{-3} \right]_0^{\pi}$$
$$= \frac{1}{-6\pi}\,(0 - \pi^3) = \frac{\pi^2}{6}.$$

This provides yet another way of finding the value of $\zeta(2)$. □

Example 5.11. For $f(t) = t^2$ on $|t| \le \pi$ we had $a_0 = 2\pi^2/3$, $a_n = 4(-1)^n/n^2$ for $n \ge 1$ and all $b_n = 0$. Parseval's formula becomes

$$\frac{1}{\pi} \int_{-\pi}^{\pi} t^4\,dt = \tfrac{1}{2} \left(\frac{2\pi^2}{3} \right)^2 + 16 \sum_{n=1}^{\infty} \frac{1}{n^4},$$

which can be solved for $\zeta(4) = \pi^4/90$. □

Suppose that f is defined only on the interval $(0, \pi)$. Then f can be extended to an odd function on $(-\pi, \pi)$ by defining $f(t) = -f(-t)$ for $-\pi < t < 0$ (if $f(0)$ happens to be defined already, this value may have to be changed to 0, but changing the value of a function at one point does not matter when dealing with Fourier series). The extended function can be expanded in a series containing only sine terms. Since ordinary Fourier series present a complete system in $L^2(-\pi, \pi)$, f can be approximated as closely as we want in this space by partial sums of this series. But then f is also approximated in $L^2(0, \pi)$ by the same partial sums (for the square of the norm in this space is exactly one half of the norm in $L^2(-\pi, \pi)$ for odd functions, and, for that matter, even functions). We interpret this to say that the system $\psi_n(t) = \sin nt$, $n \ge 1$, is a complete orthogonal system on the interval $(0, \pi)$ (the orthogonality was pointed out in Example 5.9).

In an analogous way, a function on $(0, \pi)$ can be extended to an even function and be approximated by the partial sums of a cosine series. This shows that the orthogonal system $\varphi_n(t) = \cos nt$, $n \ge 0$, is also complete on $(0, \pi)$.

We see that a function on $(0, \pi)$ can be represented by either a sine series or a cosine series, whichever is suitable, and both these series converge to the function in the norm of $L^2(0, \pi)$. This turns out to be useful in applications to problems for differential equations.

Remark. The reader may now be asking the following question: suppose that $\{c_n\}_{n \in \mathbf{Z}}$ is a sequence of numbers such that $\sum |c_n|^2$ converges. Does there then exist some $f \in L^2(\mathbf{T})$ having these numbers as its Fourier coefficients? The answer is yes — provided we admit functions that are Lebesgue-measurable but not necessarily Riemann-integrable. If we do this, we actually have a bijective mapping between $L^2(\mathbf{T})$ and the space $l^2(\mathbf{Z})$, so that $f \in L^2(\mathbf{T})$ corresponds to the sequence of its Fourier coefficients, considered as an element of $l^2(\mathbf{Z})$. □

As an application of Parseval we can prove the following nice theorem.

Theorem 5.9 *Let $f \in L^2(\mathbf{T})$. If the Fourier series of f is integrated term by term over a finite interval (a, b), the series obtained is convergent with the sum $\int_a^b f(t)\,dt$.*

Note that we do not assume that the Fourier series of f is convergent in itself!

Proof. Let us first assume that the interval (a, b) is contained within a period, for instance, $-\pi \le a < b \le \pi$. We define a function g on $\mathbf{T}$ by letting $g(t) = 1$ for $a < t < b$ and 0 otherwise. We compute the Fourier coefficients of g:

$$\widehat{g}(n) = \frac{1}{2\pi}\int_a^b e^{-int}\,dt = \left[\frac{e^{-int}}{-2\pi i n}\right]_a^b = \frac{i}{2\pi n}\left(e^{-inb} - e^{-ina}\right), \ n \ne 0;$$

$$\widehat{g}(0) = \frac{1}{2\pi}\int_a^b 1\,dt = \frac{b-a}{2\pi}.$$

If now $f \sim \sum c_n e^{int}$, the polarized Parseval relation takes the form

$$\frac{1}{2\pi}\int_{\mathbf{T}} f(t)\,\overline{g(t)}\,dt = \sum_{n\ne 0} c_n \overline{\frac{i}{2\pi n}\left(e^{-inb} - e^{-ina}\right)} + c_0\,\frac{b-a}{2\pi}$$

$$= \frac{1}{2\pi}\left(\sum_{n\ne 0} c_n\,\frac{e^{inb} - e^{ina}}{in} + c_0(b-a)\right).$$

But the integral on the left is nothing but $\int_a^b f(t)\,dt$ divided by 2π, and the terms in the sum on the right are just what you get if you integrate $c_n e^{int}$ from a to b. After multiplication by 2π we have the assertion for this case.

If the interval (a, b) is longer than 2π, it can be subdivided into pieces, each of them shorter than 2π, and then we can use the case just proved on each piece. When the results are added, the contributions from the subdivision points will cancel (convergent series can be added termwise!), and the result follows also in the general case. □

If we choose $a = 0$ and let the upper limit of integration be variable and equal to t, the theorem gives a formula for the primitive functions (antiderivatives) of f:

$$F(t) = \int_0^t f(u)\,du + K = \sum_{n\ne 0} c_n\,\frac{e^{int} - 1}{in} + c_0 t + K,$$

where K is some constant. We shall rewrite this constant; first notice that, by the Cauchy–Schwarz inequality for sums,

$$\sum_{n\ne 0}\frac{|c_n|}{|n|} \le \left(\sum_{n\ne 0}|c_n|^2\right)^{1/2}\left(\sum_{n\ne 0}\frac{1}{n^2}\right)^{1/2} \le \left(\frac{\|f\|}{2\pi}\right)^{1/2}\left(\frac{\pi^2}{3}\right)^{1/2} < \infty,$$

which means that the series $\sum_{n\neq 0} c_n/(in)$ converges absolutely with sum equal to some number K_1. Write $C = K - K_1$ and use the fact that convergent series can be added term by term, and we find that

$$F(t) = \sum_{n\neq 0} \frac{c_n}{in} e^{int} + c_0 t + C.$$

This is the simplest form of the "formally" integrated Fourier series of f, and we have thus shown that this integration is permitted and indeed results in a series convergent for all t.

In general, the integrated series is no longer a Fourier series; this happens only if $c_0 = 0$, i.e., if f has mean value 0.

Exercises

5.14 Using the result of Exercise 4.13 on page 84, find the value of the sum

$$\sum_{n=1}^{\infty} \frac{1}{(4n^2-1)^2}.$$

5.15 We reconnect to Exercise 4.22 on page 90. By studying the series established there on the interval $(0, \pi/2)$, prove the formula

$$\sum_{k=0}^{\infty} \frac{(-1)^k}{(2k+1)((2k+1)^2-\alpha^2)} = \frac{\pi}{4\alpha^2}\left(\frac{1}{\cos\frac{1}{2}\alpha\pi} - 1\right), \quad \alpha \notin \mathbf{Z}.$$

5.16 Let f be a continuous real-valued function on $0 < x < \pi$ such that $f(0) = f(\pi) = 0$ and $f' \in L^2(0,\pi)$.
(a) Prove that $\int_0^\pi (f(x))^2\,dx \leq \int_0^\pi (f'(x))^2\,dx$.
(b) For what functions does equality hold?
(Hint: extend f to an odd function on $(-\pi,\pi)$.)

5.5 Legendre polynomials

A number of "classical" ON systems in various L^2 spaces consist of *polynomials*. Polynomials are very practical functions because their values always can be computed exactly using elementary operations (addition and multiplication), which makes them immediately accessible to computers. If you want a value of a function such as e^x or $\cos x$, the effective calculation always has to be performed using some sort of approximation, and this approximation often consists of some polynomial (or a combination of polynomials).

The success of such approximations depends fundamentally on the validity of the following theorem.

Theorem 5.10 (The Weierstrass approximation theorem) *An arbitrary continuous function f on a compact interval K can be approximated uniformly arbitrarily well by polynomials.*

In greater detail, the assertion is the following: If K is compact, $f : K \to \mathbf{C}$ is continuous and ε is any positive number, then there exists a polynomial $P(x)$ such that $|f(x) - P(x)| < \varepsilon$ for all $x \in K$.

A proof can be conducted along the following lines (which are best understood by a reader who is familiar with slightly more than the barest elements about power series expansions of analytic functions): By a linear change of variable, the interval can be assumed to be, say, $K = [0, \pi]$. On this interval, f can be represented by a cosine series (which involves, if you like, considering f to be extended to an even function). The Fejér sums σ_n of this series converge uniformly to f, according to the remark following Theorem 4.1. We can then choose n to make $\sup_K |f(x) - \sigma_n(x)| < \varepsilon/2$. The function σ_n is a finite linear combination of functions of the form $\cos kx$. These can be developed in Maclaurin series, each converging to its cosine function uniformly on every compact set, in particular on K. Take partial sums of these series and construct a polynomial P, such that $\sup_K |\sigma_n(x) - P(x)| < \varepsilon/2$. Using the triangle inequality one sees that P is a polynomial with the required property.

Now we make things concrete. Let the interval be $K = [-1, 1]$, so that we live in the space $L^2(-1, 1)$, where the inner product is given by

$$\langle f, g \rangle = \int_{-1}^{1} f(x)\, \overline{g(x)}\, dx.$$

If we orthogonalize the polynomials $1, x, x^2, \ldots,$ according to the Gram–Schmidt procedure with respect to this inner product, the result is a sequence $\{P_n\}$ of polynomials of degree $0, 1, 2, \ldots$. They are traditionally scaled by the condition $P_n(1) = 1$. The polynomials obtained are called LEGENDRE polynomials. The first few Legendre polynomials are

$$P_0(x) = 1,\ P_1(x) = x,\ P_2(x) = \tfrac{1}{2}(3x^2 - 1),\ P_3(x) = \tfrac{1}{2}(5x^3 - 3x).$$

We notice a few simple facts that are easily seen to be universally valid. The polynomial P_n has degree exactly n; for odd n it contains only odd powers of x and for even n only even powers of x. An *arbitrary* polynomial $p(x)$ of degree n can, in a unique way, be written as a linear combination of $P_0, \ldots, P_n$, with the coefficient in front of P_n different from zero. We illustrate this with the example $p(x) = x^2 + 3x$:

$$x^2 + 3x = \tfrac{2}{3}(P_2(x) + \tfrac{1}{2}) + 3P_1(x) = \tfrac{2}{3}P_2(x) + 3P_1(x) + \tfrac{1}{3}P_0(x).$$

We saw in Sec. 5.3 that uniform convergence implies L^2-convergence on bounded intervals. By Theorem 5.10, continuous functions can be uniformly approximated by polynomials; these can be rewritten as linear combinations of Legendre polynomials, and thus a continuous function on $[-1,1]$ can be approximated arbitrarily well by such expressions in the sense of L^2. Just as in the proof of Theorem 5.8 it follows that the Legendre polynomials make up a complete orthogonal system in $L^2(-1,1)$. For historical reasons, they are not normed; instead one has $\|P_n\|^2 = 2/(2n+1)$ (see Exercise 5.17).

The following so-called RODRIGUES formula holds for the Legendre polynomials:

$$P_n(x) = \frac{1}{2^n n!} D^n\big((x^2-1)^n\big). \tag{5.4}$$

Example 5.12. Find the polynomial $p(x)$ of degree at most 4 that minimizes the value of

$$\int_{-1}^{1} |\sin \pi x - p(x)|^2\, dx.$$

Solution. By the general theory, the required polynomial can be obtained as the orthogonal projection of $f(x) = \sin \pi x$ onto the first five Legendre polynomials:

$$p(x) = \sum_{k=0}^{4} \frac{\langle f, P_k\rangle}{\langle P_k, P_k\rangle} P_k(x).$$

Since f is an odd function, its inner products with even functions are zero. Thus the sum reduces to just two terms:

$$p(x) = \frac{\langle f, P_1\rangle}{\langle P_1, P_1\rangle} P_1(x) + \frac{\langle f, P_3\rangle}{\langle P_3, P_3\rangle} P_3(x).$$

The denominators are taken from a table, $\langle P_k, P_k\rangle = 2/(2k+1)$, and the numerators are computed (taking advantage of symmetries):

$$\langle f, P_1\rangle = \int_{-1}^{1} (\sin \pi x)\cdot x\, dx = 2\int_0^1 x \sin \pi x\, dx = \frac{2}{\pi},$$

$$\langle f, P_3\rangle = 2\int_0^1 \tfrac{1}{2}(5x^3 - 3x)\sin \pi x\, dx = \frac{2\pi^2 - 30}{\pi^3}.$$

Putting everything together, we arrive at

$$p(x) = \frac{315 - 15\pi^2}{2\pi^3}\, x - \frac{525 - 35\pi^2}{2\pi^3}\, x^3 \approx 2.6923x - 2.8956x^3.$$

In Figure 5.3 we can see the graphs of f and p. For comparison, we have also included the Taylor polynomial $T(x)$ of degree 3, that approximates $f(x)$ near $x = 0$. It is clear that T and p serve quite different purposes: T is a very good approximation when we are close to the origin, but quite

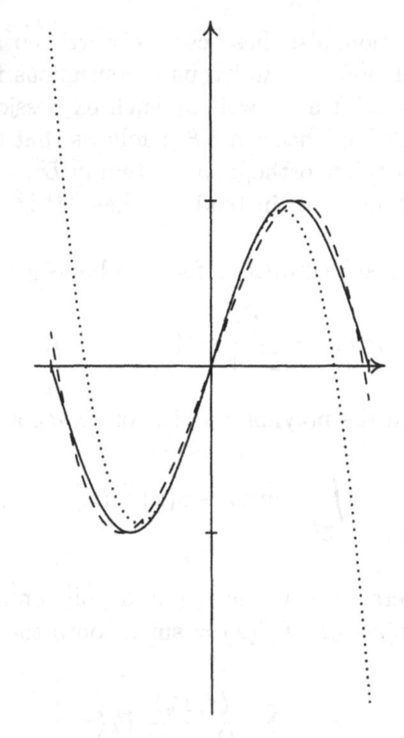

FIGURE 5.3. Solid line: f, dashed: p, dotted: T

worthless away from that point, whereas p is a reasonable approximation over the whole interval. □

We have standardized the situation in this section by choosing the interval to be $(-1, 1)$. By a simple linear change of variable everything can be transferred to an arbitrary finite interval (a, b). Here, the polynomials Q_n make up a complete orthogonal system, if we let

$$Q_n(x) = P_n\left(\frac{2x - (a+b)}{b-a}\right),$$

and the norm is given by

$$\|Q_n\|^2 = \int_a^b |Q_n(x)|^2 \, dx = \frac{b-a}{2n+1}.$$

When solving problems, use the formula collection in Appendix C, page 254.

Exercises

5.17 Show that the polynomials defined by (5.4) are orthogonal and that $\|P_n\|^2 = \dfrac{2}{2n+1}$. (Hint: write down $\langle P_m, P_n \rangle$, where $m \leq n$, and, integrating by parts, move differentiations from P_n to P_m. It is also possible to keep track of the leading coefficients.)

5.18 Find the best approximations with polynomials of degree at most 3, in the sense of $L^2(-1,1)$, to the functions (a) $H(x)$ (the Heaviside function), (b) $1/(1+x^2)$. Draw pictures!

5.19 Compare the result of Exercise 5.4, page 110, with what is said in the text about Legendre polynomials on an interval (a, b).

5.20 Let $p_0(x) = 1$, $p_1(x) = x$, and define p_n for $n \geq 2$ by the recursion formula $(n+1)p_{n+1}(x) = (2n+1)x\,p_n(x) - n\,p_{n-1}(x)$ for $n = 1, 2, \ldots$. Prove that p_n is the same as P_n.

5.21 *Prove that $u(x) = P_n(x)$, as defined by Rodrigues' formula, satisfies the differential equation

$$(1-x^2)u''(x) - 2x\,u'(x) + n(n+1)\,u(x) = 0.$$

5.6 Other classical orthogonal polynomials

In this section we collect data concerning some orthogonal systems that have been studied ever since the nineteenth century, because they occur, for instance, in the study of problems for differential equations. Proofs may sometimes be supplied by an interested reader; see the exercises at the end of the section. When solving problems in the field, a handbook containing the formulae should of course be consulted. A small collection of such formulae is found on page 254 f.

First we consider the L^2 space on the semi-infinite interval $(0, \infty)$ with the weight function $w(x) = e^{-x}$, which means that the inner product is given by

$$\langle f, g \rangle = \int_0^\infty f(x)\,\overline{g(x)}\,e^{-x}\,dx.$$

The LAGUERRE polynomials $L_n(x)$ can be defined by a so-called Rodrigues formula (where D denotes differentiation with respect to x):

$$L_n(x) = \frac{e^x}{n!}\,D^n\bigl(x^n\,e^{-x}\bigr). \tag{5.5}$$

It is not hard to see that L_n is actually a polynomial of degree n; it is somewhat more laborious to check that $\langle L_m, L_n \rangle = \delta_{mn}$; indeed these polynomials are not only orthogonal but even normed. See Exercises 5.22–23.

Next we take the interval to be the whole axis $\mathbf{R}$ and the weight to be $w(x) = e^{-x^2}$. Thus the inner product is

$$\langle f, g \rangle = \int_{\mathbf{R}} f(x)\,\overline{g(x)}\,e^{-x^2}\,dx.$$

The HERMITE polynomials $H_n(x)$ can be defined by

$$H_n(x) = (-1)^n\,e^{x^2}\,D^n\bigl(e^{-x^2}\bigr). \tag{5.6}$$

The facts that these functions are actually polynomials of degree equal to the index n and that they are orthogonal with respect to the considered inner product are left to the reader in Exercises 5.24–25. The polynomials are not normed; instead one has $\|H_n\|^2 = n!\, 2^n \sqrt{\pi}$.

Finally, we return to a finite interval, taken to be $(-1, 1)$, and let the weight function be $1/\sqrt{1-x^2}$:

$$\langle f, g \rangle = \int_{-1}^{1} f(x)\, \overline{g(x)}\, \frac{dx}{\sqrt{1-x^2}}\,.$$

Orthogonal polynomials are defined by the formula $T_n(x) = \cos(n \arccos x)$. They are called CHEBYSHEV polynomials (which can be spelled in various ways: Čebyšev, Chebyshev, Tschebyschew, Tchebycheff, and Чебышев are a few variants seen in the literature, the last one being (more or less) the original). That the formula actually defines polynomials is most easily recognized after the change of variable $x = \cos\theta$, $0 \le \theta \le \pi$, which gives the formula $T_n(\cos\theta) = \cos n\theta$, and it is well known that $\cos n\theta$ can be expressed as a polynomial in $\cos\theta$. In the case $n = 2$, for example, one has $\cos 2\theta = 2\cos^2\theta - 1$, which means that $T_2(x) = 2x^2 - 1$. The orthogonality is proved by making the same change of variable in the integral $\langle T_m, T_n \rangle$. One also finds that $\|T_0\|^2 = \pi$ and $\|T_n\|^2 = \pi/2$ for $n > 0$.

It can be proved that the polynomials named after Laguerre, Hermite, and Chebyshev actually constitute *complete* orthogonal systems in their respective spaces.

We round off with a couple of examples.

Example 5.13. Find the polynomial $p(x)$ of degree at most 2, that minimizes the value of the integral

$$\int_0^\infty |x^3 - p(x)|^2\, e^{-x}\, dx.$$

Solution. The norm occurring in the problem belongs together with the Laguerre polynomials. These even happen to be orthonormal (not merely orthogonal), which means that the wanted polynomial must be

$$p(x) = \langle f, L_0 \rangle\, L_0(x) + \langle f, L_1 \rangle\, L_1(x) + \langle f, L_2 \rangle\, L_2(x),$$

where $f(x) = x^3$. From a handbook we fetch $L_0(x) = 1$, $L_1(x) = 1 - x$, $L_2(x) = 1 - 2x + \frac{1}{2}x^2$. When computing the inner products it is convenient to notice that $\int_0^\infty x^n\, e^{-x}\, dx = n!$. We get

$$\langle f, L_0 \rangle = \int_0^\infty x^3\, e^{-x}\, dx = 3! = 6,$$

$$\langle f, L_1 \rangle = \int_0^\infty x^3(1-x)\,e^{-x}\,dx = 3! - 4! = 6 - 24 = -18,$$
$$\langle f, L_2 \rangle = \int_0^\infty x^3(1-2x+\tfrac{1}{2}\,x^2)\,e^{-x}\,dx = 3! - 2\cdot 4! + \tfrac{1}{2}\cdot 5! = 18.$$

Thus, $p(x) = 6L_0(x) - 18L_1(x) + 18L_2(x) = 6 - 18(1-x) + 18(1-2x+\frac{1}{2}\,x^2) = 6 - 18x + 9x^2$. □

Example 5.14. Let $f(x) = \sqrt{1-x^2}$ for $|x| \le 1$. Find the polynomial $p(x)$, of degree at most 3, that minimizes the value of the integral

$$\int_{-1}^{1} |f(x) - p(x)|^2 \,\frac{dx}{\sqrt{1-x^2}}.$$

Solution. This inner product belongs with the Chebyshev polynomials. Because of "odd-even" symmetry, these have the property that a polynomial of even index contains only terms of even degree, and similarly for odd indices. Since f is an even function and the inner product itself has "even" symmetry, the wanted polynomial will only contain terms of even degree:

$$p(x) = \frac{\langle f, T_0 \rangle}{\langle T_0, T_0 \rangle}\,T_0(x) + \frac{\langle f, T_2 \rangle}{\langle T_2, T_2 \rangle}\,T_2(x).$$

The data required are taken from a handbook: $T_0(x) = 1$, $T_2(x) = 2x^2 - 1$, and the denominators are found above (and in the handbook), so all that remains to be computed are the numerators:

$$\langle f, T_0 \rangle = 2\int_0^1 \sqrt{1-x^2}\cdot 1 \cdot \frac{dx}{\sqrt{1-x^2}} = 2,$$
$$\langle f, T_2 \rangle = 2\int_0^1 \sqrt{1-x^2}\cdot(2x^2-1)\cdot\frac{dx}{\sqrt{1-x^2}} = 2\int_0^1 (2x^2-1)\,dx = -\tfrac{2}{3}.$$

Substituting we get

$$p(x) = \frac{2}{\pi}\cdot 1 - \frac{2/3}{\pi/2}\,(2x^2-1) = \frac{1}{3\pi}(6 - 8x^2 + 4) = \frac{2}{3\pi}\,(5 - 4x^2).$$

□

Exercises

5.22 Show that the formula (5.5) defines a polynomial of degree n.

5.23 Show that the Laguerre polynomials are orthonormal. (Hint: the same as for Exercise 5.17.)

5.24 Show that the formula (5.6) gives a polynomial of degree n.

5.25 Show that the Hermite polynomials are orthogonal and $\|H_n\|^2 = n!\,2^n\,\sqrt{\pi}$.

5.26 Expand $e^{x/3}$ in a Laguerre series; i.e., determine the coefficients c_n in the formula

$$e^{x/3} \sim \sum_{n=0}^{\infty} c_n L_n(x), \qquad x \geq 0.$$

(The formula $\int_0^\infty e^{-at} t^n \, dt = n!/a^{n+1}$ may come in handy.)

5.27 Let $f(x) = e^{x^2}(H(x+1) - H(x-1))$ Approximate f with a polynomial $p(x)$ of degree at most 3 so as to minimize the expression

$$\int_{-\infty}^{\infty} |f(x) - p(x)|^2 e^{-x^2} \, dx.$$

5.28 Let $f(t) = \sqrt{1-t^2}$ for $|t| \leq 1$. Find a polynomial $p(t)$ of degree at most 3 that minimizes

$$\int_{-1}^{1} |f(t) - p(t)|^2 \frac{dt}{\sqrt{1-t^2}}.$$

5.29 Approximate $f(x) = |x|$ on the interval $(-1, 1)$ with a polynomial of degree ≤ 3, *first* with the weight function 1 and *secondly* with weight $1/\sqrt{1-x^2}$. *Thirdly*, do the same with weight function $(1-x^2)$ (but here you'll have to construct your own orthogonal polynomials). Compare the approximating polynomials obtained in the three cases. Draw pictures! Comment!

Summary of Chapter 5

In this chapter we studied vector spaces where the scalars are the complex numbers. Practically all results from real linear algebra remain valid in this case. The only important exception to this is the appearance of the *inner product.*

A typical example of an inner product space is given by the set $L^2(I, w)$, where I is an interval on the real axis and w is a *weight function*, i.e., a function such that $w(x) > 0$ on I; the inner product in $L^2(I, w)$ is defined by

$$\langle f, g \rangle = \int_I f(x)\overline{g(x)}\, w(x)\, dx.$$

With the *norm* defined by $\|u\| = \sqrt{\langle u, u \rangle}$, a notion of *distance* is given by $\|u - v\|$.

Theorem
In an inner product space the inequalities

$$|\langle u, v \rangle| \leq \|u\| \|v\|, \qquad \|u + v\| \leq \|u\| + \|v\|$$

are valid.

With respect to an inner product, one defines *orthogonality* and *orthonormal* sets (or ON sets), as in the real case. Gram–Schmidt's method can be used to construct such sets.

If $\{\varphi_k\}_{k=1}^N$ is an ON set, the *orthogonal projection* of a $u \in V$ on to the subspace spanned by this set is the vector

$$P(u) = \sum_{k=1}^{N} \langle u, \varphi_k \rangle \varphi_k.$$

Theorem
If $\{\varphi_1, \varphi_2, \ldots, \varphi_N\}$ is an ON basis in an N-dimensional inner product space V, then every $u \in V$ can be written as $u = \sum\limits_{j=1}^{N} \langle u, \varphi_j \rangle \varphi_j$, and furthermore one has

$$\|u\|^2 = \sum_{j=1}^{N} |\langle u, \varphi_j \rangle|^2 \quad \text{(theorem of Pythagoras).}$$

For the inner product of two vectors one also has the following formula:

$$\langle u, v \rangle = \sum_{j=1}^{N} \langle u, \varphi_j \rangle \overline{\langle v, \varphi_j \rangle}.$$

Theorem
Let $\{\varphi_k\}_{k=1}^N$ be an orthonormal set in an inner product space V and let u be a vector in V. Among all the linear combinations $\Phi = \sum_{k=1}^N \gamma_k \varphi_k$, the one that minimizes the value of $\|u - \Phi\|$ is the orthogonal projection of u on to the subspace spanned by the ON set, i.e., $\Phi = P(u)$. Also, it holds that

$$\left\| u - \sum_{k=1}^{N} \langle u, \varphi_k \rangle \varphi_k \right\|^2 = \|u\|^2 - \sum_{k=1}^{N} |\langle u, \varphi_k \rangle|^2.$$

Theorem
If $\{\varphi_k\}_{k=1}^\infty$ is an ON set in V and $u \in V$, then

$$\sum_{k=1}^{\infty} |\langle u, \varphi_k \rangle|^2 \le \|u\|^2 \qquad \text{(Bessel's inequality).}$$

If every element in V can be approximated arbitrarily closely by linear combinations of the elements of $\{\varphi_k\}$, then this set is said to be *complete* in V.

Theorem
The system $\{\varphi_k\}_{k=1}^\infty$ is complete in V if and only if for every $u \in V$ it holds that

$$\|u\|^2 = \sum_{k=1}^{\infty} |\langle u, \varphi_k \rangle|^2$$

(the PARSEVAL formula or the completeness relation).

Theorem
If the system $\{\varphi_k\}_{k=1}^{\infty}$ is complete in V, then every $u \in V$ can be written as $u = \sum_{k=1}^{\infty} \langle u, \varphi_k \rangle \varphi_k$ where the series converges in the sense of the norm, i.e., $\left\| u - \sum_{k=1}^{N} \langle u, \varphi_k \rangle \varphi_k \right\| \to 0$ as $N \to \infty$).

Theorem
If the system $\{\varphi_k\}_{k=1}^{\infty}$ is complete in V, then

$$\langle u, v \rangle = \sum_{k=1}^{\infty} \langle u, \varphi_k \rangle \overline{\langle v, \varphi_k \rangle}$$

for all $u, v \in V$.

If the set $\{\varphi_k\}$ is not ON but merely orthogonal, all these formulae must be adjusted by dividing each occurrence of a φ_k by its norm.

Theorem
The two orthogonal systems $\{e^{int}\}_{n \in \mathbf{Z}}$ and $\{\cos nt,\ n \geq 0;\ \sin nt,\ n \geq 1\}$ are each complete in $L^2(\mathbf{T})$.

As a consequence of this, Parseval's identities hold for ordinary Fourier series (with conventional notation):

$$\frac{1}{2\pi} \int_{\mathbf{T}} |f(t)|^2 \, dt = \sum_{n \in \mathbf{Z}} |c_n|^2, \qquad \frac{1}{2\pi} \int_{\mathbf{T}} f(t) \, \overline{g(t)} \, dt = \sum_{n \in \mathbf{Z}} c_n \, \overline{d_n};$$

$$\frac{1}{\pi} \int_{\mathbf{T}} |f(t)|^2 \, dt = \tfrac{1}{2} |a_0|^2 + \sum_{n=1}^{\infty} (|a_n|^2 + |b_n|^2),$$

$$\frac{1}{\pi} \int_{\mathbf{T}} f(t) \, \overline{g(t)} \, dt = \tfrac{1}{2} a_0 \overline{\alpha_0} + \sum_{n=1}^{\infty} (a_n \overline{\alpha_n} + b_n \overline{\beta_n}).$$

Theorem
The Fourier series of a function $f \in L^2(\mathbf{T})$ can always be integrated term by term over any bounded interval (a, b). The series obtained by this operation is always convergent, regardless of the convergence of the original Fourier series.

Theorem
(The Weierstrass approximation theorem) An arbitrary continuous function f on a compact interval K can be approximated uniformly arbitrarily well using polynomials.

Historical notes

The insight that certain notions of geometry, such as orthogonality and projections, can be fruitfully applied to sets of functions dawned upon mathematicians in various situations during the nineteenth century. Round the turn of the century, the Swedish mathematician IVAR FREDHOLM (1866–1927) treated certain problems for linear integral equations in a way that made obvious analogies with problems for systems of linear equations. Building on these ideas, the German DAVID HILBERT (1862–1943) and the Pole STEFAN BANACH (1892–1945) introduced notions such as Hilbert and Banach spaces. The L^p spaces mentioned in the present text are all Banach spaces (if one uses the Lebesgue integral in the definitions); and in particular L^2 spaces are Hilbert spaces. The latter spaces are infinite-dimensional, complex-scalar counterparts of ordinary Euclidean spaces, with a concept of distance that is coupled to an inner product.

Parseval's formula, which can be seen as a counterpart of the theorem of Pythagoras, is named after an obscure French amateur mathematician, MARC-ANTOINE PARSEVAL DES CHÊNES (1755–1836).

ADRIEN-MARIE LEGENDRE (1752–1833) was an influential French mathematician who worked in many areas. EDMOND LAGUERRE (1834-86) and CHARLES HERMITE (1822–1901) were also French. Hermite is most famous for his proof that the number π is transcendental. PAFNUTY LVOVICH CHEBYSHEV (1821–94) founded the great Russian mathematical tradition that lives on to this day.

Problems for Chapter 5

5.30 Determine the Fourier series of the function

$$f(x) = \begin{cases} \cos x, & 0 < x < \pi, \\ -\cos x, & -\pi < x < 0. \end{cases}$$

Also compute the sum $S = \displaystyle\sum_{n=1}^{\infty} \frac{n^2}{(4n^2-1)^2}$.

5.31 Let f be the even function with period 2π described by $f(x) = \sin \frac{3}{2} x$ for $0 < x < \pi$. Using the Fourier series of f, find the values of the sums

$$s_1 = \sum_{n=1}^{\infty} \frac{1}{4n^2-9}, \quad s_2 = \sum_{n=1}^{\infty} \frac{(-1)^n}{4n^2-9}, \quad s_3 = \sum_{n=1}^{\infty} \frac{1}{(4n^2-9)^2}.$$

5.32 Use the result of Problem 4.46 on page 103 to compute the value of

$$\sum_{n=1}^{\infty} \frac{1}{(2n-1)^6}.$$

5.33 Use the result of Problem 4.48 on page 103 to compute the value of $\zeta(8)$.

5.34 Find a polynomial $p(x)$ of degree at most 1 that minimizes the integral

$$\int_0^1 (p(x) - x^2)^2 (1+x)\, dx.$$

5.35 Let Q be the square $\{(x,y) : |x| \le \pi,\ |y| \le \pi\}$, and let $L^2(Q)$ denote the set of functions $f : Q \to \mathbf{C}$ that satisfy $\iint_Q |f(x,y)|^2\,dx\,dy < \infty$. In this space we define an inner product by the formula

$$\langle f, g\rangle = \iint_Q f(x,y)\,\overline{g(x,y)}\,dx\,dy.$$

Define the functions $\varphi_{mn} \in L^2(Q)$ by $\varphi_{mn}(x,y) = e^{i(mx+ny)}$, $m,n \in \mathbf{Z}$. Show that these functions are orthogonal with respect to $\langle\cdot,\cdot\rangle$, and determine their norms in $L^2(Q)$.

5.36 Expand the function $f(x) = e^{-ax}$ in a Fourier–Hermite series:

$$f(x) \sim \sum_{n=0}^{\infty} c_n H_n(x).$$

5.37 Expand $f(x) = x^3$, $x \ge 0$, in a Fourier–Laguerre series:

$$f(x) \sim \sum_{n=0}^{\infty} c_n L_n(x).$$

5.38 Prove this formula for Legendre polynomials:

$$(2n+1)P_n(x) = P'_{n+1}(x) - P'_{n-1}(x), \quad n \ge 1.$$

5.39 A function $f(x)$, defined on $(-1,1)$, can be expanded in a Fourier–Legendre series:

$$f(x) \sim \sum_{n=0}^{\infty} c_n P_n(x).$$

What does the Parseval formula look like for this expansion?

5.40 Determine the distance in $L^2(-1,1)$ from $\sin \pi x$ to the subspace spanned by 1, x, x^2. (The distance is the norm of the residual.)

5.41 The functions 1 and $\sqrt{3}\,(2x-1)$ constitute an orthogonal system in the space $L^2(0,1)$. Find the linear combination of these that is the best approximation of $\cos x$ in $L^2(0,1)$.

5.42 f is continuous on the interval $[0,1]$. Moreover,

$$\int_0^1 f(x)\,x^n\,dx = 0, \quad n = 0,1,2,\ldots,.$$

Prove that $f(x) = 0$ for all x in $[0,1]$.

5.43 Determine the coefficients a_k, $k = 0,1,2,3$, so that the integral

$$\int_{-1}^{1} |a_0 + a_1 x + a_2 x^2 + a_3 x^3 - x^4|^2\,dx$$

is made as small as possible.

5.44 Let $f(x) = \cos \pi x$. Let V be the space of continuous functions on the interval $[-1, 1]$ with the inner product

$$\langle u, v \rangle = \int_{-1}^{1} u(x)\,\overline{v(x)}\,dx.$$

M is the subspace in V consisting of polynomials of degree at most 3. Find the orthogonal projection of f onto M.

5.45 Determine the numbers a, b och c so as to make the expression

$$\int_{-1}^{1} \left| \cos \frac{\pi x}{2} - (a + bx + cx^2) \right|^2 dx$$

as small as possible.

5.46 Let $f(x) = \operatorname{sgn} x = 2H(x) - 1$ for $x \in \mathbf{R}$. Approximate f with a third-degree polynomial in the sense of Hermite.

5.47 Let $f(x) = (1 - x^2)^{3/2}$. Find a polynomial $P(x)$ of degree at most 3 that minimizes

$$\int_{-1}^{1} \frac{|f(x) - P(x)|}{\sqrt{1 - x^2}}\,dx.$$

6

Separation of variables

6.1 The solution of Fourier's problem

We now return, at last, to the problem stated in Sec. 1.4: heat conduction in a rod of finite length, with its end points kept at temperature 0. The mathematical formulation of the problem was this:

$$\begin{aligned} &\text{(E)} \quad u_{xx} = u_t, \quad 0 < x < \pi, \quad t > 0; \\ &\text{(B)} \quad u(0,t) = u(\pi,t) = 0, \quad t > 0; \\ &\text{(I)} \quad u(x,0) = f(x), \quad 0 < x < \pi. \end{aligned} \tag{6.1}$$

We had found the following solutions of the homogeneous sub-problem consisting of the conditions (E) and (B):

$$u(x,t) = \sum_{n=1}^{N} b_n \, e^{-n^2 t} \sin nx. \tag{6.2}$$

Then we asked two questions: can we allow $N \to \infty$ in this sum? And can the coefficients be chosen so that (I) is also satisfied? Now we can answer these questions.

Let $f(x)$ be the initial values for $0 < x < \pi$. By defining $f(x) = -f(-x)$ for $-\pi < x < 0$ we get an *odd* function. It can be expanded in a Fourier series, which is a sine series with coefficients

$$b_n = \frac{2}{\pi} \int_0^{\pi} f(x) \sin nx \, dx.$$

The coefficients are bounded, $|b_n| \le M$, they even tend to zero as $n \to \infty$. This implies that the series

$$u(x,t) = \sum_{n=1}^{\infty} b_n\, e^{-n^2 t} \sin nx \tag{6.3}$$

converges very nicely as soon as $t > 0$. If $a > 0$, we can estimate the terms like this for $t \ge a$:

$$|b_n\, e^{-n^2 t} \sin nx| \le M\, e^{-n^2 a} \le M\, e^{-na} = M_n,$$

and $\sum M_n$ is a convergent geometric series. The considered series then converges uniformly in the region $t \ge a$. If it is differentiated termwise with respect to t once, or with respect to x twice, the new series are also uniformly convergent in the same region (check this!). According to the theorem on differentiation of series, the function u, defined by (6.3), is differentiable to the extent needed, and since all the partial sums satisfy (E)+(B), so will the sum.

To check that the initial values are right is somewhat more tricky. If f happens to be so nice that $\sum |b_n| < \infty$, then we are home; for in this case the series will actually converge uniformly in the closed set $0 \le x \le \pi$, $t \ge 0$, and so the sum is continuous in this set, making

$$\lim_{t \searrow 0} u(x,t) = u(x,0) = \sum_{n=1}^{\infty} b_n \sin nx = f(x)$$

(cf. Theorem 4.2, page 83). This holds, say, if the odd, 2π-periodic extension of f belongs to C^2; it even holds under weaker assumptions, but this is harder to prove.

If $f \in L^2(0,\pi)$, we can alternatively study convergence in the L^2 sense. Let v_t be the restriction of u to time t, i.e., $v_t(x) = u(x,t)$, $0 < x < \pi$ (here the subscript $_t$ does not stand for a derivative). The function v_t has Fourier coefficients $b_n\, e^{-n^2 t}$, and by Parseval we have

$$\|v_t\|^2 = \int_0^{\pi} |v_t(x)|^2\, dx = \frac{\pi}{2} \sum_{n=1}^{\infty} |b_n|^2\, e^{-2n^2 t} \le \frac{\pi}{2} \sum_{n=1}^{\infty} |b_n|^2 = \|f\|^2 < \infty.$$

Thus, v_t also belongs to $L^2(0,\pi)$ for each $t > 0$. Now we investigate what happens if $t \searrow 0$:

$$\|f - v_t\|^2 = \frac{\pi}{2} \sum_{n=1}^{\infty} |b_n|^2 (1 - e^{-n^2 t})^2 = \Phi(t).$$

The series defining $\Phi(t)$ converges uniformly on $t \ge 0$ and its terms are continuous functions of t. Thus $\Phi(t)$ is continuous on the right for $t = 0$, and

$$\lim_{t \searrow 0} \Phi(t) = \Phi(0) = 0,$$

which means that $\|f - v_t\| \to 0$ as t decreases to 0. The solution u thus has the L^2-limit f, which is our way of saying that

$$\lim_{t \searrow 0} \int_0^\pi |u(x,t) - f(x)|^2 \, dx = 0.$$

The terms of the series representing the solution consist of sine functions, multiplied by exponentially decreasing factors. The higher the frequency of the sine factor, the faster does the term containing it tend to zero – small fluctuations in the temperature along the rod are faster to even out than fluctuations of longer period. As time goes by, the temperature of the entire rod will approach zero – which should be expected, considering the physical experiment that we have attempted to describe with our model.

Remark. For $t > 0$, the series in (6.3) can actually be differentiated an indefinite number of times with respect to both variables. What happens to the term $b_n e^{-n^2 t} \sin nx$ when it is differentiated is that one or more factors n come out, that sin and cos may interchange and also the sign may change. But, for $t \geq a > 0$, the resulting term can always be estimated by an expression of the form $M n^P e^{-n^2 t} \leq M n^P e^{-na} = Q_n$, and it is easy to see that $\sum Q_n < \infty$ (apply the ratio test). We conclude that all functions such as (6.3) are indeed of class C^∞; they are "infinitely smooth." □

Exercises

6.1 Find the solution of Fourier's problem when (a) $f(x) = \sin^3 x$ for $0 < x < \pi$; (b) $f(x) = \cos 3x$ for $0 < x < \pi$.

6.2 Find a solution to the following modified Fourier problem (heat conduction in a rod of length 1; a is a positive constant):

$$u_t = \frac{1}{a^2} u_{xx}, \quad 0 < x < 1, \ t > 0;$$
$$u(0,t) = u(1,t) = 0, \ t > 0; \quad u(x,0) = f(x), \ 0 < x < 1.$$

6.2 Variations on Fourier's theme

In this section we perform some slight variations on the theme that has just been concluded. Later on in the chapter we shall indicate the possibility of more far-reaching variations.

Example 6.1. Let us study the problem of heat conduction in a *completely isolated* rod, where there is no exchange of heat with the surroundings, not even at the end points. As before, the rod is represented by the interval $[0, \pi]$, and the temperature at the point x at time t is denoted by $u(x,t)$. Within the mathematical model that gives rise to the heat equation, the flow of heat is assumed to run from warmer to colder areas in such a way that the velocity of the flow is proportional to the gradient of the temperature (and having the opposite direction). The mathematical formulation of the condition that no heat shall flow past the end points is then that the gradient of the temperature be zero at these points; in the one-dimensional case this condition is simply $u_x(0,t) = u_x(\pi,t) = 0$. If the temperature of the rod at time 0 is called $f(x)$, we have the following problem:

$$\begin{aligned} &\text{(E)} \quad u_{xx} = u_t, \quad 0 < x < \pi, \quad t > 0; \\ &\text{(B)} \quad u_x(0,t) = u_x(\pi,t) = 0, \quad t > 0; \\ &\text{(I)} \quad u(x,0) = f(x), \quad 0 < x < \pi. \end{aligned} \tag{6.4}$$

This problem is largely similar to the previous one, and we attack it by the same means (cf. Sec. 1.4). Thus we start by looking for nontrivial solutions of the homogeneous sub-problem (E)+(B), and we try to find solutions having the form $u(x,t) = X(x)T(t)$. Substituting into (E) leads, just as before, to the separated conditions

$$X''(x) + \lambda X(x) = 0, \qquad T'(t) + \lambda T(t) = 0.$$

To satisfy (B) without having u identically zero we must also have $X'(0) = X'(\pi) = 0$. This leaves us with the following boundary value problem for X:

$$X''(x) + \lambda X(x) = 0, \quad 0 < x < \pi; \qquad X'(0) = X'(\pi) = 0. \tag{6.5}$$

Just as in Sec. 1.4, we look through the different cases according to the value of λ. It will be sufficient for us to give account of "basis vectors," so we omit scalar factors that can always be adjoined.

For all $\lambda < 0$ one finds that the only possible solution is $X(x) \equiv 0$ (the reader should check this). If $\lambda = 0$, the equation is $X''(x) = 0$ with solutions $X(x) = A + Bx$. The boundary conditions are satisfied if $B = 0$. This means that we have the solutions $X(x) = X_0(x) = A = \text{constant}$. For the same value of λ, the T-equation also has the solutions $T = \text{constants}$; as a "basis vector" we can choose

$$u_0(x,t) = X_0(x)\,T_0(t) = \tfrac{1}{2}. \tag{6.6}$$

When $\lambda > 0$ we can put $\lambda = \omega^2$ with $\omega > 0$. Thus we have $X'' + \omega^2 X = 0$ with solutions $X(x) = A\cos\omega x + B\sin\omega x$ and $X'(x) = -\omega A\sin\omega x + \omega B\cos\omega x$. The condition $X'(0) = 0$ directly gives $B = 0$, and then $X'(\pi) = 0$ means that $0 = -\omega A\sin\omega\pi$. This can be satisfied with $A \neq 0$ precisely if ω is a (positive) integer. Thus, for $\lambda = n^2$ we have the solution $X(x) = X_n(x) = \cos nx$ and multiples of this function. The corresponding equation for T is solved by $T_n(t) = e^{-n^2 t}$. In addition to (6.6), the problem (E)+(B) thus has the solutions

$$u_n(x,t) = X_n(x)\,T_n(t) = e^{-n^2 t}\cos nx, \quad n = 1, 2, 3, \ldots.$$

By homogeneity, series of the form

$$u(x,t) = \tfrac{1}{2}\, a_0 + \sum_{n=1}^{\infty} a_n\, e^{-n^2 t} \cos nx \tag{6.7}$$

are solutions of (E)+(B), provided they converge nicely enough. It remains to be seen if it is possible to choose the constants a_n so that (I) can be satisfied. Direct substitution of $t = 0$ in the solution would give

$$f(x) = u(x,0) = \tfrac{1}{2}\, a_0 + \sum_{n=1}^{\infty} a_n \cos nx, \quad 0 < x < \pi.$$

We can see that if f is extended to an even function on the interval $(-\pi, \pi)$ and we let the a_n be the Fourier coefficients of this function, then we ought to have a solution to the whole problem.

And, just as in the preceding section, everything works excellently if we know, for example, that $\sum |a_n| < \infty$.

It can be noted that the solution (6.7) has the property that all terms except for the first one tend rapidly to zero when t tends to infinity. One is left with the term $\frac{1}{2}\, a_0$. As we have seen, this is equal to the mean value of f, and this is in accordance with the intuitive feeling for what ought to happen in the physical situation: a completely isolated rod will eventually assume a constant temperature, which is precisely the mean of the initial temperature distribution. □

Example 6.2. Let us now modify the original problem in a few different ways. We let the rod be the interval $(0, 2)$, and the end points are kept each at a constant temperature, but these are different at the two ends. To be specific, say that $u(0,t) = 2$ and $u(2,t) = 5$. Let us take the initial temperature to be given by $f(x) = 1 - x^2$. The whole problem is

$$\begin{array}{lll} \text{(E)} & u_{xx} = u_t, & 0 < x < 2, \quad t > 0; \\ \text{(B)} & u(0,t) = 2, \quad u(2,t) = 5, & t > 0; \\ \text{(I)} & u(x,0) = 1 - x^2, & 0 < x < 2. \end{array} \tag{6.8}$$

Here, separation of variables cannot be applied directly; an important feature of that method is making use of the homogeneity of the conditions, enabling us to add solutions to each other to obtain other solutions. For this reason, we now start by *homogenizing* the problem in the following way. Since the boundary values are constants, independent of time, it should be possible to write $u(x,t) = v(x,t) + \varphi(x)$, where $\varphi(x)$ should be chosen to make v the solution of a modified problem with homogeneous boundary conditions. Substitution into (E) gives

$$v_{xx}(x,t) + \varphi''(x) = v_t(x,t),$$

so it is desirable to have $\varphi''(x) = 0$. If we can also achieve $\varphi(0) = 2$ and $\varphi(2) = 5$, we would get $v(0,t) = v(2,t) = 0$.

Thus we are faced with this simple problem for an ordinary differential equation:

$$\varphi''(x) = 0; \quad \varphi(0) = 2, \ \varphi(2) = 5.$$

The unique solution is easily found to be $\varphi(x) = \frac{3}{2}\,x + 2$. Substituting this into the initial condition of the original problem, we have

$$1 - x^2 = u(x,0) = v(x,0) + \varphi(x) = v(x,0) + \tfrac{3}{2}\,x + 2.$$

We collect all the conditions to be satisfied by v:

$$\begin{aligned} &(\mathrm{E}') \quad v_{xx} = v_t, \qquad 0 < x < 2, \quad t > 0; \\ &(\mathrm{B}') \quad v(0,t) = 0, \quad v(2,t) = 0, \qquad t > 0; \\ &(\mathrm{I}') \quad v(x,0) = -x^2 - \tfrac{3}{2}\,x - 1, \quad 0 < x < 2. \end{aligned} \tag{6.9}$$

This problem is essentially of the sort considered and solved in Sec. 1.4 and 6.1. A slight difference is the fact that the length of the rod is 2 instead of π, but the only consequence of this is that the sine functions in the solution will be adapted to this interval (as in Sec. 4.5). The reader is urged to perform all the steps that lead to the following formula for "general" solutions of (E′)+(B′):

$$v(x,t) = \sum_{n=1}^{\infty} a_n \exp\left(-\frac{n^2\pi^2}{4}\,t\right) \sin\frac{n\pi}{2}\,x.$$

Next, the coefficients are adapted to (I′):

$$a_n = \frac{2}{2}\int_0^2 \left(-x^2 - \tfrac{3}{2}\,x - 1\right) \sin\frac{n\pi}{2}\,x\,dx = \frac{16(-1)^n - 2}{n\pi} + \frac{16(1 - (-1)^n)}{n^3\pi^3}.$$

Finally, we put together the answer to the original problem:

$$u(x,t) = \tfrac{3}{2}\,x + 2 + \sum_{n=1}^{\infty} \left(\frac{16(-1)^n - 2}{n\pi} + \frac{16(1 - (-1)^n)}{n^3\pi^3}\right) e^{-n^2\pi^2 t/4} \sin\frac{n\pi}{2}\,x.$$

As time goes by, the temperature along the rod will stabilize at the distribution given by the function $\varphi(x)$. This is called the *stationary* distribution of the problem. □

Example 6.3. In our next variation we consider a rod with a built-in source of heat. The length of the rod is again π, and we assume that at the point with coordinate x there is generated an amount of heat per unit of time and unit of length along the rod, described by the function $\sin(x/2)$. It can be shown that this leads to the following modification of the heat equation:

$$(\mathrm{E}) \qquad u_t = u_{xx} + \sin\frac{x}{2}, \qquad 0 < x < \pi, \quad t > 0.$$

We also assume that both ends are kept at temperature 0 for $t > 0$ and that the initial temperature along the rod is 1:

$$\text{(B)}\quad u(0,t) = u(\pi,t) = 0,\ t > 0; \qquad \text{(I)}\quad u(x,0) = 1,\ 0 < x < \pi.$$

Here there is an inhomogeneity in the equation itself. We try to amend this by using the same trick as in Example 2: put $u(x,t) = v(x,t) + \varphi(x)$ and substitute into (E) and (B). (Do it!) We conclude that it would be very nice to have

$$\varphi''(x) = -\sin\frac{x}{2}, \qquad \varphi(0) = \varphi(\pi) = 0.$$

The first condition implies that φ must be of the form $\varphi(x) = 4\sin(x/2) + Ax+B$, and the boundary conditions force us to take $B = 0$ and $A = -4/\pi$. As a consequence, v shall be a solution of the problem

$$\begin{aligned}
&(\mathrm{E}') \quad v_{xx} = v_t, \qquad 0 < x < \pi, \quad t > 0;\\
&(\mathrm{B}') \quad v(0,t) = 0, \quad v(\pi,t) = 0, \qquad t > 0;\\
&(\mathrm{I}') \quad v(x,0) = 1 - 4\sin(x/2) + (4x)/\pi, \quad 0 < x < \pi.
\end{aligned} \tag{6.10}$$

The reader is asked to complete the calculations; the answer is

$$u(x,t) = 4\sin\frac{x}{2} - \frac{4}{\pi}x + \frac{2}{\pi}\sum_{n=1}^{\infty}\frac{1-(-1)^n(4n^2-5)}{n(4n^2-1)}e^{-n^2t}\sin nx.$$

□

Example 6.4. We leave the heat equation and turn to the wave equation. We shall solve the problem of the *vibrating string.*

Imagine a string (a violin string or guitar string), stretched between the points 0 and π of an x-axis. The point with coordinate x at time t has a position deviating from the equilibrium by the amount $u(x,t)$. If the string is homogeneous, its vibrations are small and considered to be at right angles to the x-axis, gravitation can be disregarded; and the units of mass, length, and time are suitably chosen, then the function u will satisfy the wave equation in the simple form $u_{xx} = u_{tt}$. The fact that the string is anchored at its ends means that $u(0,t) = u(\pi,t) = 0$. At time $t = 0$, every point of the string is located at a certain position and has a certain speed of movement. We want to find $u(x,t)$ for $t > 0$ and all the interesting values of x. This is collected into a problem of the following appearance:

$$\begin{aligned}
&(\mathrm{E}) \quad u_{xx} = u_{tt}, \qquad 0 < x < \pi, \quad t > 0;\\
&(\mathrm{B}) \quad u(0,t) = u(\pi,t) = 0, \quad t > 0;\\
&(\mathrm{I}_1) \quad u(x,0) = f(x), \quad 0 < x < \pi,\\
&(\mathrm{I}_2) \quad u_t(x,0) = g(x), \quad 0 < x < \pi;
\end{aligned} \tag{6.11}$$

Again, (E) and (B) are homogeneous conditions. The usual attempt $u(x,t) = X(x)\,T(t)$ this time leads up to this set of coupled problems:

$$\begin{cases} X''(x) + \lambda X(x) = 0, \\ X(0) = X(\pi) = 0; \end{cases} \qquad T''(t) + \lambda T(t) = 0.$$

The X problem is familiar by now: it has nontrivial solutions exactly for $\lambda = n^2$ $(n = 1, 2, 3, \ldots)$, viz., multiples of $X_n(x) = \sin nx$. For these values of λ, the T problem is solved by $T_n(t) = a_n \cos nt + b_n \sin nt$. Because of homogeneity we obtain the following solutions of the sub-problem (E)+(B):

$$u(x,t) = \sum_{n=1}^{\infty} X_n(x)\, T_n(t) = \sum_{n=1}^{\infty} (a_n \cos nt + b_n \sin nt) \sin nx. \qquad (6.12)$$

Letting $t = 0$ in order to investigate (I_1), we get

$$f(x) = u(x,0) = \sum_{n=1}^{\infty} a_n \sin nx.$$

Termwise differentiation with respect to t and then substitution of $t = 0$ gives for the second initial condition (I_2) that

$$g(x) = u_t(x,0) = \sum_{n=1}^{\infty} n b_n \sin nx.$$

Thus, if we choose a_n to be the sine coefficients of (the odd extension of) f, and choose b_n so that nb_n are the corresponding coefficients of g, then the series (6.12) ought to represent the wanted solution.

As we saw already in Sec. 1.3, the wave equation may have rather irregular, non-smooth solutions. This is reflected by the fact that the series in (6.12) can converge quite "badly." See, for example, the solution of Exercise 6.7, which is, after all, an attempt at a quite natural situation. If we allow distributions as derivatives, as indicated in Sec. 2.6–7, the mathematical troubles go away. It should also be borne in mind that the conditions of Exercise 6.7 are not physically realistic: a string does not really have thickness 0 and cannot really take on the shape of an angle. □

Remark. The typical term in the sum (6.12) can be rewritten in the form $A_n \sin(nt + \alpha_n) \sin nx$. Its musical significance is the *nth partial tone* in the sound emitted by the string. (The first partial is often called the fundamental.) Figure 6.1 illustrates in principle the shapes of the string that correspond to different values of n. These are also called the *modes of vibrations* of the string. □

Remark. Of considerable musical importance is the fact that the nth partial also vibrates *in time* with a frequency that is the nth multiple of the fundamental. This is what was noted already by PYTHAGORAS: if the length of the string is halved (making it vibrate in the same manner as the whole string would vibrate in the second mode), one hears a note sounding one octave higher. The successive partials are illustrated in Figure 6.2. The accidental ♭ stands for lowering the pitch by slightly more than a (tempered) semi-tone, while ♯ and ♯ indicate raising the pitch by slightly less or respectively more, than a semi-tone. Partial number 7 is

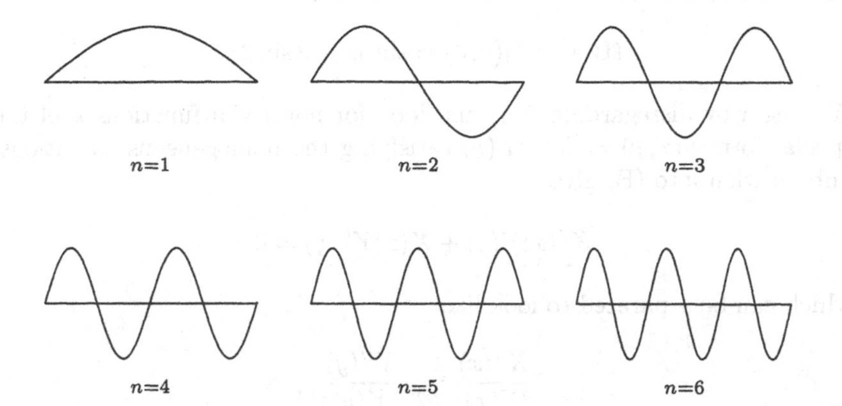

FIGURE 6.1.

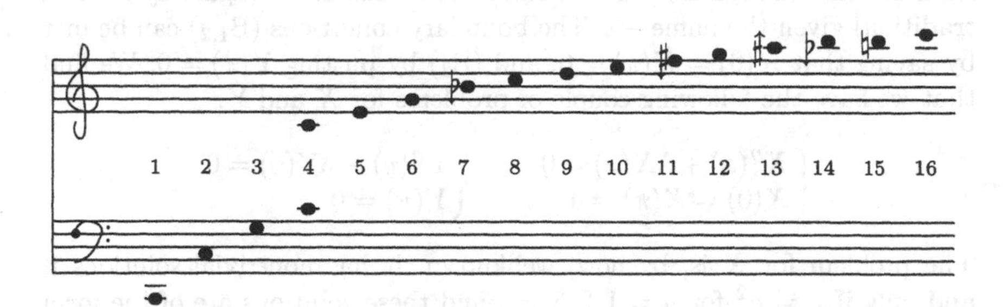

FIGURE 6.2.

wellknown to musicians (especially brass players) as an ugly pitch that is to be avoided in normal music. Partials 11 and 13 are also bad approximations of the pitches indicated in the figure, but they are so high up that they cause relatively little trouble in normal playing. □

We round off this section with a problem for the Laplace equation in a square. This sort of problem is called a DIRICHLET problem: the Laplace equation in a region of the plane, with values prescribed on the boundary of the region.

Example 6.5. Find $u(x,y)$ that solves $u_{xx} + u_{yy} = 0$, $0 < x < \pi$, $0 < y < \pi$, with boundary conditions $u(x,0) = \sin 3x - 3\sin 2x$ for $0 < x < \pi$, $u(x,\pi) = u(0,y) = u(\pi,y) = 0$, $0 < x,y < \pi$.

Solution. Draw a picture! We have a *homogeneous equation*,

$$\text{(E)} \qquad u_{xx} + u_{yy} = 0,$$

together with *three homogeneous boundary conditions*,

$$(\mathrm{B}_{1,2,3}) \qquad u(0,y) = u(\pi,y) = 0, \quad u(x,\pi) = 0,$$

and one *non-homogeneous boundary condition,*

$$(\mathrm{B}_4) \qquad u(x,0) = \sin 3x - 3\sin 2x.$$

We begin by disregarding (B_4) and look for nontrivial functions u of the special form $u(x,y) = X(x)Y(y)$ satisfying the homogeneous conditions. Substitution into (E) gives

$$X''(x)Y(y) + X(x)Y''(y) = 0,$$

which can be separated to look like

$$\frac{X''(x)}{X(x)} = -\frac{Y''(y)}{Y(y)},$$

and by the same argument as in preceding cases we conclude that the two sides of this equation must be constant. This constant is (again by force of tradition) given the name $-\lambda$. The boundary conditions $(\mathrm{B}_{1,2})$ can be met by saying that $X(0) = X(\pi) = 0$, and (B_3) by putting $Y(\pi) = 0$. We find that we have the following couple of problems for X and Y*:*

$$\begin{cases} X''(x) + \lambda X(x) = 0 \\ X(0) = X(\pi) = 0 \end{cases} \qquad \begin{cases} Y''(y) - \lambda Y(y) = 0 \\ Y(\pi) = 0 \end{cases}$$

The problem for X is, by now, wellknown. It has nontrivial solutions if and only if $\lambda = n^2$ for $n = 1, 2, 3, \ldots$, and these solutions are of the form $X_n(x) = \sin nx$. For the same values of λ, the Y problem is solved by $Y_n(y) = Ae^{ny} + Be^{-ny}$, where A and B shall be chosen to meet the condition $Y_n(\pi) = 0$. This is done by letting $B = -Ae^{2n\pi}$. We thus have the solutions

$$u_n(x,y) = A_n\left(e^{ny} - e^{n(2\pi-y)}\right)\sin nx, \quad n = 1, 2, 3, \ldots,$$

of the homogeneous conditions (E) and $(\mathrm{B}_{1,2,3})$. Because of the homogeneity, sums of these solutions are again solutions. A "general" solution is given by

$$u(x,y) = \sum_{n=1}^{\infty} A_n\left(e^{ny} - e^{n(2\pi-y)}\right)\sin nx.$$

We now have to choose the coefficients A_n to meet the remaining condition (B_4). The reader should check the computations that lead to the final result

$$u(x,y) = \frac{-3}{1-e^{4\pi}}\left(e^{2y} - e^{2(2\pi-y)}\right)\sin 2x + \frac{1}{1-e^{6\pi}}\left(e^{3y} - e^{3(2\pi-y)}\right)\sin 3x.$$

□

In the last example, the boundary condition was homogeneous on three of the edges of the square. A general Dirichlet problem for a square might be taken care of by solving four problems of this kind, with non-homogeneous boundary values on one edge at a time, and adding the solutions.

In the exercises the reader will have the opportunity to apply the basic ideas of the method of separation of variables to a variety of problems. In all cases, the success of the method is coupled to the fact that one reaches a problem for an ordinary differential equation together with boundary conditions. This problem turns out to have nontrivial solutions only for certain values of the "separation constant," and these solutions are a sort of building blocks out of which the solutions are constructed. This sort of ODE problem is called a STURM–LIOUVILLE problem and will be considered in Sec. 6.4 for its own sake.

It is even possible to treat partial differential equations with more than two independent variables in much the same way.

Exercises

6.3 Find a solution of the heat problem $u_t = u_{xx}$ for $0 < x < \pi$, $t > 0$, such that $u_x(0,t) = u_x(\pi,t) = 0$ for $t > 0$ and $u(x,0) = \frac{1}{2}(1 + \cos 3x)$ for $0 < x < \pi$.

6.4 Determine a solution of the problem

$$\begin{cases} u_{xx} = t\,u_t, & 0 < x < \pi, \ t > 1; \\ u(0,t) = u(\pi,t) = 0, & t > 1, \\ u(x,1) = \sin x + 2\sin 3x, & 0 < x < \pi. \end{cases}$$

6.5 Find a solution of the non-homogeneous heat conduction problem

$$\begin{cases} u_{xx} = u_t + \sin x, & 0 < x < \pi, \ t > 0; \\ u(0,t) = u(\pi,t) = 0, & t > 0; \\ u(x,0) = \sin x + \sin 2x, & 0 < x < \pi. \end{cases}$$

6.6 Solve the following problem for the vibrating string:

$$\begin{cases} u_{xx} = u_{tt}, & 0 < x < \pi, \ t > 0; \\ u(x,0) = 3\sin 2x, \quad u_t(x,0) = 5\sin 3x, & 0 < x < \pi; \\ u(0,t) = u(\pi,t) = 0, & t > 0. \end{cases}$$

6.7 The plucked string: a point on the string is pulled from its resting position and then released with no initial speed. If the string is plucked at its middle point, what tones are heard? In a mathematical formulation: Solve the problem (6.11), when f is given by

$$f(x) = ax, \ 0 \le x \le \tfrac{1}{2}\pi, \quad f(x) = a(\pi - x), \ \tfrac{1}{2}\pi \le x \le \pi,$$

and $g(x) = 0$.

6.8 Find $u(x,t)$ if

$$\begin{cases} u_{xx}(x,t) = u_{tt}(x,t), & 0 < x < 1, \ t > 0; \\ u(0,t) = u(1,t) = 0, & t > 0; \\ u(x,0) = \sin 3\pi x, \quad u_t(x,0) = \sin \pi x \cos^2 \pi x, & 0 < x < 1. \end{cases}$$

6.9 Find a solution of the following problem (one-dimensional heat conduction with loss of heat to the surrounding medium) for $h > 0$ constant: $u_t = u_{xx} - hu$, $0 < x < \pi$, $t > 0$, together with $u(0,t) = 0$, $u(\pi,t) = 1$ for $t > 0$ and $u(x,0) = 0$ for $0 < x < \pi$.

6.10 A Dirichlet problem: $u_{xx} + u_{yy} = 0$ for $0 < x, y < 1$, $u(x,0) = u(x,1) = 0$, $u(0,y) = 0$ and $u(1,y) = \sin^3 \pi y$.

6.11 Find a solution $u = u(x,t)$ of this problem:

$$\begin{cases} u_{xx} + \frac{1}{4}u = u_t, & 0 < x < \pi, \ t > 0; \\ u(0,t) = 0, \ u(\pi,t) = 1, & t > 0; \\ u(x,0) = 0, & 0 < x < \pi. \end{cases}$$

6.3 The Dirichlet problem in the unit disk

We shall study a problem for the Laplace equation in two dimensions. Let $u = u(x,y)$ be a function defined in an open, connected set Ω in $\mathbf{R}^2$. The Laplace equation in Ω is

$$\Delta u := \frac{\partial^2 u}{\partial x^2} + \frac{\partial^2 u}{\partial y^2} = 0, \quad (x,y) \in \Omega.$$

The solutions of this equation are called *harmonic functions* in Ω. The DIRICHLET problem is the task of finding all such functions with prescribed values on the boundary $\partial\Omega$. We shall study this problem in the case when Ω is the unit disk $\mathbf{D} : x^2 + y^2 < 1$, so that the boundary $\partial\Omega$ is the unit circle $\mathbf{T} : x^2 + y^2 = 1$. Concisely, the problem is

$$\Delta u(x,y) = 0, \ (x,y) \in \mathbf{D}; \ \ u(x,y) = g(x,y) = \text{known function}, \ (x,y) \in \mathbf{T}. \tag{6.13}$$

We shall describe two lines of attack: first a method that requires knowledge of the elementary theory of analytic (or holomorphic) functions, then a different approach involving separation of variables.

Method 1. Interpret (x,y) as a complex number $z = re^{i\theta}$. The boundary function g can then conveniently be considered as a function of the polar coordinate θ, so that we are looking for harmonic functions $u = u(z) = u(re^{i\theta})$ for $r < 1$ with boundary values

$$u(e^{i\theta}) = \lim_{r \nearrow 1} u(re^{i\theta}) = g(\theta), \quad -\pi < \theta \leq \pi.$$

The unit disk $\mathbf{D}$ is simply connected. By the theory of analytic functions, every harmonic function u in $\mathbf{D}$ has a conjugate-harmonic partner v such that the expression $f(z) = u(z) + iv(z)$ is analytic in $\mathbf{D}$. An analytic function in the unit disk is the sum of a power series:

$$u(z) + iv(z) = f(z) = \sum_{n=0}^{\infty} A_n z^n = \sum_{n=0}^{\infty} A_n r^n e^{in\theta}$$

$$= \sum_{n=0}^{\infty} (B_n + iC_n)\, r^n \,(\cos n\theta + i \sin n\theta)$$

$$= \sum_{n=0}^{\infty} r^n \big((B_n \cos n\theta - C_n \sin n\theta) + i(C_n \cos n\theta + B_n \sin n\theta)\big),$$

where the A_n are Taylor coefficients with real parts B_n and imaginary parts C_n. Taking the real part of the whole equation one sees that $u(z)$ must be representable by a series of the form

$$u(z) = u(re^{i\theta}) = \sum_{n=0}^{\infty} r^n \big(B_n \cos n\theta - C_n \sin n\theta\big),$$

and, conversely, one realizes (by reading the equation backward) that all such series represent harmonic functions (provided the corresponding series $\sum A_n z^n$ converges in $\mathbf{D}$).

In order to make the formula neater, we switch letters: put $a_n = B_n$ and $b_n = -C_n$ for $n \geq 1$, $a_0 = 2B_0$, and note that the value of C_0 is immaterial (since $\sin 0\theta = 0$ for all θ), and we get

$$u(re^{i\theta}) = \tfrac{1}{2}\, a_0 + \sum_{n=1}^{\infty} r^n \big(a_n \cos n\theta + b_n \sin n\theta\big).$$

Method 2. We want to find solutions of the Laplace equation in the region described in polar coordinates by $r < 1$, $-\pi \leq \theta \leq \pi$. First we transform the equation into polar coordinates. Using the chain rule one finds that $\Delta u = 0$ (for $r > 0$) is equivalent to

$$r \frac{\partial}{\partial r}\left(r \frac{\partial u}{\partial r}\right) + \frac{\partial^2 u}{\partial \theta^2} = 0$$

(the computations required are usually carried through in calculus textbooks as examples of the chain rule). We then proceed to find nontrivial solutions of the special form $u(r, \theta) = R(r)\,\Theta(\theta)$, i.e., solutions that are products of one function of r and one function of θ. Substitution into the equation results in

$$r \frac{\partial}{\partial r}\left(r \frac{dR}{dr}\,\Theta(\theta)\right) + R(r)\,\frac{d^2\Theta}{d\theta^2} = 0 \quad \iff \quad r(rR')'\Theta = -R\Theta''.$$

We divide by $R\Theta$ and get

$$\frac{1}{R}\, r\,(rR')' = -\frac{\Theta''}{\Theta}.$$

The left-hand member in this equation is independent of θ and the right-hand member is independent of r. Just as in Sec. 1.4 we can conclude that both members must then be constant, and this constant value is called λ. The situation splits into the two ordinary differential equations

$$\Theta'' + \lambda\Theta = 0, \qquad r\big(rR'\big)' - \lambda R = 0.$$

In addition, there are a couple of "boundary conditions": in order that the function $u = R\Theta$ be uniquely determined in $\mathbf{D}$, the function $\Theta(\theta)$ must have period 2π. Furthermore, since u shall have a finite value at the origin, we demand that $R(r)$ have a finite limit $R(0+)$ as $r \searrow 0$.

We begin with the problem for Θ:

$$\begin{cases} \Theta''(\theta) + \lambda\Theta(\theta) = 0, \\ \Theta(\theta + 2\pi) = \Theta(\theta) \text{ for all } \theta. \end{cases}$$

As in Sec. 1.4, we work through the cases $\lambda < 0$, $\lambda = 0$ and $\lambda > 0$. In the first case there are no periodic solutions (except for $\Theta(\theta) \equiv 0$). In the case $\lambda = 0$, all *constant* functions will do: $\Theta_0(\theta) = A_0$. When $\lambda > 0$, let $\lambda = \omega^2$ with $\omega > 0$, and we find the solutions $\Theta_\omega(\theta) = A_\omega \cos\omega\theta + B_\omega \sin\omega\theta$. These have period 2π precisely if ω is a positive integer: $\omega = n$, $n = 1, 2, 3, \ldots$. Summarizing, we have found interesting solutions of the Θ problem precisely when $\lambda = n^2$, $n = 0, 1, 2, \ldots$.

For these λ we solve the R-problem. When $\lambda = 0$, the equation becomes

$$r\big(rR'\big)' = 0 \iff rR' = C \iff R' = \frac{C}{r} \iff R = C\ln r + D.$$

The value $R(0+)$ exists only if $C = 0$. In this case we thus have a solution $u = u_0 = \Theta_0(\theta)R_0(r) = A_0 \cdot D = \text{constant}$. For reasons that will presently become evident we denote this constant by $\frac{1}{2}a_0$.

When $\lambda = n^2$ with $n > 0$ we have a so-called Euler equation:

$$r\big(rR'\big)' = n^2 R \iff r^2R'' + rR' - n^2R = 0.$$

To solve it, we change the independent variable by putting $r = e^s$, which results in

$$\frac{d^2R}{ds^2} - n^2\,R = 0,$$

which has the solutions $R = Ce^{ns} + De^{-ns} = Cr^n + Dr^{-n}$. We must take $D = 0$ to ascertain that $R(0+)$ exists. Piecing together with Θ we arrive at the solution

$$\begin{aligned} u(re^{i\theta}) = u_n(re^{i\theta}) = \Theta_n(\theta)\,R_n(r) &= C_n\,r^n\,(A_n\cos n\theta + B_n\sin n\theta) \\ &= r^n\,(a_n\cos n\theta + b_n\sin n\theta). \end{aligned}$$

The Laplace equation is homogeneous. Assuming convergence for the series, the solutions of $\Delta u = 0$ in the unit disk should be representable by series of the form

$$u(re^{i\theta}) = \tfrac{1}{2}\,a_0 + \sum_{n=1}^{\infty} r^n\big(a_n\cos n\theta + b_n\sin n\theta\big).$$

This is the same form for harmonic functions in $\mathbf{D}$ as obtained by "Method 1" above.

The formal solutions that we have obtained can, of course, also be written in "complex" form. Via Euler's formulae we find

$$u(r,\theta) = u(re^{i\theta}) = \sum_{n\in\mathbf{Z}} c_n r^{|n|} e^{in\theta}. \tag{6.14}$$

Notice that the exponents on r have a modulus sign.

Now we turn to the boundary condition. As $r \nearrow 1$, we wish that the values of the solution approach a prescribed function $g(\theta)$. For simplicity, we assume that g is continuous. Let the numbers c_n be the Fourier coefficients of g:

$$c_n = \frac{1}{2\pi}\int_{\mathbf{T}} g(\theta)\, e^{-in\theta}\, d\theta,$$

and, using these coefficients, form the function $u(r,\theta)$ as in (6.14). By Lemma 4.1, there exists a number M such that $|c_n| \le M$. Using the Weierstrass M-test we can easily conclude that the series defining u converges absolutely and uniformly in every inner closed circular disk $r \le r_0$, where $r_0 < 1$, and this still holds after differentiations with respect to r as well as θ. According to the theorem on differentiation of series, termwise differentiation is thus possible, and since each term of the series satisfies the Laplace equation and this equation is homogeneous, the sum function u will also satisfy the same equation. Now we turn to the boundary condition. The uniform convergence in $r \le r_0$ allows us to interchange the order of sum and integral in the following formula:

$$\begin{aligned} u(r,\theta) &= \sum_{n\in\mathbf{Z}} c_n\, r^{|n|}\, e^{in\theta} = \sum_{n\in\mathbf{Z}} r^{|n|}\, e^{in\theta}\,\frac{1}{2\pi}\int_{\mathbf{T}} g(t)\, e^{-int}\, dt \\ &= \int_{\mathbf{T}} \left(\frac{1}{2\pi}\sum_{n\in\mathbf{Z}} r^{|n|}\, e^{in(\theta-t)}\right) g(t)\, dt. \end{aligned}$$

The sum in brackets can be computed explicitly – it is made up of two geometric series:

$$\begin{aligned} \sum_{n\in\mathbf{Z}} r^{|n|}\, e^{ins} &= \sum_{n=-\infty}^{-1} r^{-n}\, e^{ins} + \sum_{n=0}^{\infty} r^n\, e^{ins} \\ &= \sum_{n=1}^{\infty} r^n\, e^{-ins} + \sum_{n=0}^{\infty} r^n\, e^{ins} = \frac{r\,e^{-is}}{1 - r\,e^{-is}} + \frac{1}{1 - r\,e^{is}} \\ &= \frac{(1 - re^{is})re^{-is} + 1 - re^{-is}}{|1 - r\,e^{is}|^2} = \frac{1 - r^2}{1 + r^2 - 2r\cos s}. \end{aligned}$$

We define the POISSON kernel to be the function

$$P_r(s) = P(s,r) = \frac{1}{2\pi} \sum_{n \in \mathbf{Z}} r^{|n|} e^{ins} = \frac{1}{2\pi} \cdot \frac{1 - r^2}{1 + r^2 - 2r \cos s}.$$

This gives the following formula for u:

$$u(r,\theta) = \int_{\mathbf{T}} P_r(\theta - t)\, g(t)\, dt = \int_{\mathbf{T}} P_r(t)\, g(\theta - t)\, dt.$$

The Poisson kernel has some interesting properties:

1. $P_r(s) = P_r(-s) \geq 0$ for $r < 1$, $s \in \mathbf{T}$.
2. $\int_{\mathbf{T}} P_r(s)\, ds = 1$ for $r < 1$.
3. If $\delta > 0$, then

$$\lim_{r \nearrow 1} \int_{\delta}^{\pi} P_r(s)\, ds = 0.$$

The proofs of 1 and 2 are simple (2 follows by integrating the series term by term, which is legitimate). The property 3 can be shown thus: since $P_r(s)$ is decreasing as s goes from 0 to π, we have $P_r(s) \leq P_r(\delta)$ on the interval, and

$$\int_{\delta}^{\pi} P_r(s)\, ds \leq P_r(\delta) \int_{\delta}^{\pi} ds = (\pi - \delta) P_r(\delta) \to 0 \text{ as } r \nearrow 1.$$

This sort of properties of a collection of functions should be familiar to the reader. They actually amount to the fact that P_r is a positive summation kernel of the kind studied in Sec. 2.4. The only difference is the fact that the present kernel is "numbered" by a variable r that tends to 1, instead of using an integer N tending to infinity. Theorem 2.1 can be used, and we get the result that *we have constructed a solution of the Dirichlet problem with boundary values* $g(\theta)$ *in the sense that*

$$\lim_{r \nearrow 1} u(r,\theta) = g(\theta)$$

at all points θ *where* g *is continuous.*

In addition, the solution is actually unique. This can be proved using a technique similar to that employed at the end of Sec. 4.2. First one proves that the problem with boundary values identically zero has only the solution identically zero, and then this is applied to the difference of two solutions corresponding to the same boundary values.

Remark. We have here touched upon another method of summing series that may not be convergent. It is called POISSON or ABEL summation. For a numerical series $\sum_{n=0}^{\infty} a_n$ it consists in forming the function

$$f(r) = \sum_{n=0}^{\infty} a_n r^n, \qquad 0 < r < 1.$$

If this function exists in the interval indicated, and if it has a limit as $r \nearrow 1$, then this limit is called the *Poisson* or *Abel* sum of the series. It can be proved that this method sums a convergent series to its ordinary sum. It is also a quite powerful method: it is stronger than Cesàro summation in the sense that every Cesàro summable series is also Abel summable; and there exist series summable by Abel that are not summable by Cesàro, not even after any number of iterations. □

Example 6.6. Find a solution of the Dirichlet problem in the disk having boundary values $u(1,\theta) = \cos 4\theta - 1$. Express the solution in rectangular coordinates!

Solution. It is immediately seen that in polar coordinates the solution must be

$$u(r,\theta) = -1 + r^4 \cos 4\theta.$$

We rewrite the cosine to introduce cos and sin of the single value θ:

$$\begin{aligned} u &= -1 + r^4(\cos^2 2\theta - \sin^2 2\theta) \\ &= -1 + r^4((\cos^2\theta - \sin^2\theta)^2 - (2\sin\theta\cos\theta)^2) \\ &= -1 + r^4(\cos^4\theta - 2\cos^2\theta\sin^2\theta + \sin^4\theta - 4\sin^2\theta\cos^2\theta) \\ &= -1 + x^4 - 6x^2y^2 + y^4. \end{aligned}$$

□

Exercises

6.12 Find a solution of the Dirichlet problem in the unit disk such that $u(e^{i\theta}) = 2 + \cos 3\theta + \sin 4\theta$.

6.13 Find a solution of the same problem such that $u(x,y) = x^4 + y^4$ for $x^2 + y^2 = 1$.

6.14 Solve the Dirichlet problem with boundary values $u(1,\theta) = \sin^3\theta$.

6.15 Perform the details of the proof of the uniqueness of the solution of Dirichlet's problem.

6.4 Sturm–Liouville problems

In our solutions of the problems in the preceding sections, a central role was played by a boundary value problem for an ordinary differential equation containing a parameter λ. This problem proved to have nontrivial solutions for certain values of λ, but these values had the character of being "exceptional". The situation seems loosely similar to a kind of problem that the reader should have been faced with in a seemingly completely different context, namely linear algebra: eigenvalue problems for an operator or a matrix. We shall see that this similarity is really not loose at all!

We start with a few definitions. Let V be a space with an inner product $\langle\cdot,\cdot\rangle$. A linear mapping A, defined in some subspace $\mathcal{D}_A$ of V and having its values in V, is called an *operator* on V. Notice that this definition is slightly different from the one that is common in the case of finite-dimensional spaces: we do not demand that the domain of definition of the operator be the entire space V. We write $V \supseteq \mathcal{D}_A \xrightarrow{A} V$ or $A : \mathcal{D}_A \to V$. The image of a vector $u \in \mathcal{D}_A$ is written $A(u)$ or, mostly, simply Au.

Definition 6.1 *An operator $A : \mathcal{D}_A \to V$ is said to be* symmetric, *if*

$$\langle Au, v\rangle = \langle u, Av\rangle \quad \textit{for all } u, v \in \mathcal{D}_A.$$

Example 6.7. Let $V = L^2(\mathbf{T})$, $\mathcal{D}_A = V \cap C^2(\mathbf{T})$ and let A be the operator $-D^2$, so that $Au = -u''$. Since $u \in C^2(\mathbf{T})$, the image Au is a continuous function and thus belongs to V. We have

$$\begin{aligned}\langle Au, v\rangle &= -\int_{\mathbf{T}} u''(x)\,\overline{v(x)}\,dx = -\big[u'(x)\,\overline{v(x)}\big]_{-\pi}^{\pi} + \int_{\mathbf{T}} u'(x)\,\overline{v'(x)}\,dx \\ &= \big[u(x)\,\overline{v'(x)}\big]_{-\pi}^{\pi} - \int_{\mathbf{T}} u(x)\,\overline{v''(x)}\,dx = \langle u, Av\rangle.\end{aligned}$$

The integrated parts are zero, because all the functions are periodic and thus have the same values at $-\pi$ and π. □

Definition 6.2 *An operator $A : \mathcal{D}_A \to V$ is said to have an* eigenvalue λ, *if there exists a vector $u \in \mathcal{D}_A$ such that $u \neq 0$ and $Au = \lambda u$. Such a vector u is called an* eigenvector, *more precisely, an eigenvector belonging to the eigenvalue λ. The set of eigenvectors belonging to a particular eigenvalue λ (together with the zero vector) make up the* eigenspace *belonging to λ.*

Example 6.8. We return to the situation in Example 6.7. If $u(x) = a\cos nx + b\sin nx$, where a and b are arbitrary constants and n is an integer ≥ 0, then clearly $Au = n^2u$. In this situation we have thus the eigenvalues $\lambda = 0, 1, 4, 9, \ldots$. For $\lambda = 0$, the eigenspace has dimension 1 (it consists of the constant functions), for the other eigenvalues the dimension is 2. (The fact that this is the complete story of the eigenvalues of this operator was shown in Sec. 6.3, "Method 2.") □

For symmetric operators on a finite-dimensional space there is a spectral theorem, which is a simple adjustment to the case of complex scalars of the theorem from real linear algebra: If A is a symmetric operator defined on all of $\mathbf{C}^n$ (for example), then there is an orthogonal basis for $\mathbf{C}^n$, consisting of eigenvectors for A. The proof of this can be performed as a replica of the corresponding proof for the real case (if anything, the complex case is rather easier to do than a purely "real" proof). In infinite dimensions things are more complicated, but in many cases similar results do hold there as well.

First we give a couple of simple results that do not depend on dimension.

Lemma 6.1 *A symmetric operator has only real eigenvalues, and eigenvectors corresponding to different eigenvalues are orthogonal.*

Proof. Suppose that $Au = \lambda u$ and $Av = \mu v$, where $u \neq 0$ and $v \neq 0$. Then we can write

$$\lambda\langle u, v\rangle = \langle \lambda u, v\rangle = \langle Au, v\rangle = \langle u, Av\rangle = \langle u, \mu v\rangle = \overline{\mu}\langle u, v\rangle. \tag{6.15}$$

First, choose $v = u$, so that also $\mu = \lambda$, and we have that $\lambda\|u\|^2 = \overline{\lambda}\|u\|^2$. Because of $u \neq 0$ we conclude that $\lambda = \overline{\lambda}$, and thus λ is real. It follows that all eigenvalues must be real. But then we can return to (6.15) with the information that μ is also real, and thus $(\lambda - \mu)\langle u, v\rangle = 0$. If now $\lambda - \mu \neq 0$, then we must have that $\langle u, v\rangle = 0$, which proves the second assertion. □

Regrettably, it is not easy to prove in general that there are "sufficiently many" eigenvectors (to make it possible to construct a "basis," as in finite dimensions). We shall here mention something about one situation where this does hold, the study of which was initiated by STURM and LIOUVILLE during the nineteenth century. As special cases of this situation we shall recognize some of the boundary value problems studied in this text, starting in Sec. 1.4.

We settle on a compact interval $I = [a, b]$. Let $p \in C^1(I)$ be a real-valued function such that $p(a) \neq 0 \neq p(b)$; let $q \in C(I)$ be another real-valued function; and let $w \in C(I)$ be a *positive* function on the same interval (i.e., $w(x) > 0$ for $x \in I$). We are going to study the ordinary differential equation

$$\text{(E)} \qquad (pu')' + qu + \lambda wu = 0 \iff$$
$$\frac{d}{dx}\left(p(x)\,\frac{du}{dx}\right) + q(x)\,u(x) + \lambda w(x)u(x) = 0, \qquad x \in I.$$

Here, λ is a parameter and u the "unknown" function. Furthermore, we shall consider *boundary conditions*, initially of the form

$$\text{(B)} \qquad A_0u(a) + A_1u'(a) = 0, \quad B_0u(b) + B_1u'(b) = 0.$$

Here, A_j and B_j are real constants such that $(A_0, A_1) \neq (0, 0) \neq (B_0, B_1)$.

Remark. If we take $p(x) = w(x) = 1$, $q(x) = 0$, $A_0 = B_0 = 1$ and $A_1 = B_1 = 0$, we recover the problem studied in Sec. 1.4. □

The problem (E)+(B) is called a *regular Sturm–Liouville problem.* We introduce the space $L^2(I, w)$, where w is the function occurring in (E). This means that we have an inner product

$$\langle u, v\rangle = \int_I u(x)\,\overline{v(x)}\,w(x)\,dx.$$

In particular, all functions $u \in C(I)$ will belong to $L^2(I, w)$, since the interval is compact.

We define an operator A by the formula

$$Au = -\frac{1}{w}((pu')' + qu),$$
$$\mathcal{D}_A = \{u \in C^2(I) : Au \in L^2(I, w) \text{ and } u \text{ satisfies (B)}\}.$$

Then, (E) can be written simply as $Au = \lambda u$. The problem of finding nontrivial solutions of the problem (E)+(B) has been rephrased as the problem of finding eigenvectors of the operator A. (The fact that $\mathcal{D}_A$ is a linear space is a consequence of the homogeneity of the boundary conditions.)

The symmetry of A can be shown as a slightly more complicated parallel of Example 6.7 above. On the one hand,

$$\langle Au, v\rangle = -\int_a^b \frac{1}{w}((pu')' + qu)\,\overline{v}\,w\,dx = -\int_a^b ((pu')' + qu)\overline{v}\,dx$$
$$= -\int_a^b (pu')'\overline{v}\,dx - \int_a^b q\,u\,\overline{v}\,dx = -\big[p\,u'\,\overline{v}\big]_a^b + \int_a^b (p\,u'\overline{v'} - q\,u\,\overline{v})\,dx.$$

On the other hand (using the fact that p, q and w are real-valued),

$$\langle u, Av\rangle = \int_a^b u\cdot\left(-\frac{1}{\overline{w}}\right)\overline{((pv')' + qv)}w\,dx = -\int_a^b u(p\overline{v}')'\,dx - \int_a^b u\,q\,\overline{v}\,dx$$
$$= -\big[u\,p\,\overline{v}'\big]_a^b + \int_a^b (u'\,p\,\overline{v'} - u\,q\,\overline{v})\,dx.$$

We see that

$$\langle Au, v\rangle - \langle u, Av\rangle = \big[pu\overline{v}' - pu'\overline{v}\big]_a^b = \left[p(x)\begin{vmatrix} u(x) & u'(x) \\ \overline{v(x)} & \overline{v'(x)}\end{vmatrix}\right]_{x=a}^{x=b}.$$

But the determinant in this expression, for $x = a$, must be zero: indeed, we assume that both u and v satisfy the boundary condition (B) at a, which means that

$$\begin{cases} A_0 u(a) + A_1 u'(a) = 0, \\ A_0\overline{v(a)} + A_1\overline{v'(a)} = 0. \end{cases}$$

This can be considered to be a homogeneous linear system of equations with (the real numbers) A_0 and A_1 as unknowns, and it has a nontrivial solution (since we assume that $(A_0, A_1) \neq (0, 0)$). Thus the determinant is zero. In the same way it follows that the determinant is zero at $x = b$. We conclude then that

$$\langle Au, v\rangle = \langle u, Av\rangle,$$

so that A is symmetric.

In this case, the symmetry is achieved by the fact that a certain substitution of values results in zero at each end of the interval. Clearly, this is not necessary. An operator can be symmetric for other reasons, too. We shall not delve deeper into this in this text, but refer the reader to texts on ordinary differential equations.

For the case we have sketched above, the following result holds.

Theorem 6.1 (Sturm–Liouville's theorem) *The operator A, belonging to the problem* (E)+(B), *has infinitely many eigenvalues, which can be arranged in an increasing sequence:*

$$\lambda_1 < \lambda_2 < \lambda_3 < \cdots, \quad \text{where } \lambda_n \to \infty \text{ as } n \to \infty.$$

The eigenspace of each eigenvalue has dimension 1, and if φ_n is an eigenvector corresponding to λ_n, then $\{\varphi_n\}_{n=1}^{\infty}$ is a complete orthogonal system in $L^2(I, w)$.

This can be rewritten to refer directly to the differential equation problem:

Theorem 6.2 *The problem* (E)+(B) *has solutions for an infinite number of values of the parameter λ, which can be arranged in an increasing sequence:*

$$\lambda_1 < \lambda_2 < \lambda_3 < \cdots, \quad \text{where } \lambda_n \to \infty \text{ as } n \to \infty.$$

For each of these values of λ, the solutions make up a one-dimensional space, and if φ_n is a non-zero solution corresponding to λ_n, the set $\{\varphi_n\}_{n=1}^{\infty}$ is a complete orthogonal system in $L^2(I, w)$.

Proofs can be found in texts on ordinary differential equations.

It is of considerable interest that one gets a *complete* orthogonal system. We already know this to be true in a couple of special cases. First we have the problem

$$u''(x) + \lambda u(x) = 0, \quad 0 < x < \pi; \qquad u(0) = u(\pi) = 0, \tag{6.16}$$

that we first met already in Sec. 1.4; here the eigenfunctions are $\varphi_n(x) = \sin nx$, and according to Sec. 5.4 these are complete in $L^2(0, \pi)$ (with weight function 1). Secondly, we have seen this problem, treated in Example 6.1 of Sec. 6.2:

$$u''(x) + \lambda u(x) = 0, \quad 0 < x < \pi; \qquad u'(0) = u'(\pi) = 0. \tag{6.17}$$

There we found the eigenfunctions $\varphi_0(x) = \frac{1}{2}$ and $\varphi_n(x) = \cos nx$, and we have seen that they are also complete in $L^2(0, \pi)$.

In the exercises, the reader is invited to investigate a few more problems that fall within the conditions of Theorem 6.2.

If the assumptions are changed, the results may deviate from those of Theorem 6.2. We have already seen this in Examples 6.7 and 6.8 of the present section. There, we studied the operator $-D^2$ on $\mathbf{T}$, which corresponds to the problem

$$u''(x) + \lambda u(x) = 0, \quad -\pi \le x \le \pi; \qquad u(-\pi) = u(\pi), \quad u'(-\pi) = u'(\pi).$$

The boundary conditions are of a different character from (B): they mean that u and u' have periodic extensions with period 2π (so that they can truly be considered to be functions on the unit circle $\mathbf{T}$). They are also commonly called *periodic* boundary conditions. (In contrast, the conditions considered in (B) are said to be *separated:* the values at a and b have no connection with each other.) In this case the eigenspaces (except for one) have dimension 2. If we choose orthogonal bases in each of the eigenspaces and pool all these together, the result is again a complete system in the relevant space, which is $L^2(\mathbf{T})$.

Yet another few examples are given in the next section. In one of these cases it happens that the function p goes to zero at the ends of the compact interval; in others the interval is no longer compact. It can be finite, but open, and one or more of the functions p, q, and w may have singularities at the ends; the interval may also be a half-axis or even the entire real line. All these situations give rise to what are known as *singular* Sturm–Liouville problems, and they sometimes occur when treating classical situations for partial differential equations.

Exercises

6.16 Determine a complete orthogonal system in $L^2(0, \pi)$ consisting of solutions of the problem

$$u''(x) + \lambda\, u(x) = 0, \quad 0 < x < \pi; \qquad u(0) = u'(\pi) = 0.$$

6.17 The same problem, but with boundary conditions

$$u(0) = u(\pi) + u'(\pi) = 0.$$

6.18 Show that the problem

$$\frac{d}{dx}\left(\sqrt{1-x^2}\,\frac{du}{dx}\right) + \frac{\lambda}{\sqrt{1-x^2}}\,u(x) = 0, \quad -1 < x < 1$$

has the eigenvalues $\lambda = n^2$ ($n = 0, 1, 2, \ldots$) and eigenfunctions $T_n(x) = \cos(n \arccos x)$ for $\lambda = n^2$. (You are not expected to prove that these are *all* the eigenvalues and eigenfunctions of the problem.)

6.5 Some singular Sturm–Liouville problems

Some celebrated problems in classical physics lead up to problems for ordinary differential equations that are similar to the problems considered in the preceding section. We review some of these problems here, partly because of their historical interest, but also because they have solutions that are polynomials that we met in Sec. 5.5–6.

The Legendre polynomials are solutions of the following singular Sturm–Liouville problem. Let $I = [-1, 1]$ and study the problem

$$\frac{d}{dx}\left((1-x^2)\,u'(x)\right) + \lambda u(x) = 0, \quad -1 < x < 1,$$

with no boundary conditions at all (except that $u(x)$ should be defined in the closed interval). Here we can identify $p(x) = 1 - x^2$, $q(x) = 0$ and $w(x) = 1$. Since $p(x) = 0$ at both ends of the interval, the corresponding operator A will be symmetric if one takes $\mathcal{D}_A = \{u \in C^2(I) : Au \in L^2(I)\}$ (the reader should check this, which is not difficult). It can be proved that this problem has eigenvalues $\lambda = n(n+1)$, $n = 0, 1, 2, \ldots$, and that the eigenfunctions are actually (multiples of) the Legendre polynomials.

Remark. The origin of this problem is the three-dimensional Laplace equation in spherical coordinates (r, θ, ϕ), defined implicitly by

$$\begin{cases} x = r\sin\phi\cos\theta & r \geq 0, \\ y = r\sin\phi\sin\theta & 0 \leq \phi \leq \pi, \\ z = r\cos\phi & -\pi < \theta \leq \pi. \end{cases}$$

In these coordinates, the equation takes the form

$$\Delta u \equiv \frac{\partial^2 u}{\partial r^2} + \frac{2}{r}\frac{\partial u}{\partial r} + \frac{1}{r^2}\frac{\partial^2 u}{\partial \phi^2} + \frac{\cot\phi}{r^2}\frac{\partial u}{\partial \phi} + \frac{1}{r^2\sin^2\phi}\frac{\partial^2 u}{\partial \theta^2} = 0.$$

This can also be written as

$$r\left(ru\right)_{rr} + \frac{1}{\sin\phi}\left(\sin\phi\, u_\phi\right)_\phi + \frac{1}{\sin^2\phi}u_{\theta\theta} = 0. \tag{6.18}$$

A function $f(x, y, z)$ is said to be *homogeneous of degree n*, if $f(tx, ty, tz) = t^n f(x, y, z)$ for $t > 0$. This means that f is completely determined by its values on, say, the unit sphere, so that it can be written $f(x, y, z) = r^n g(\phi, \theta)$ for a certain function g. We now look for solutions u_n of (6.18) that are homogeneous of degree n; these solutions are called *spherical harmonics*. Write

$$u_n(x, y, z) = r^n S_n(\phi, \theta),$$

where S_n is called a *spherical surface harmonic*. Substitution into (6.18) and subsequent division by r^n gives

$$(n+1)nS_n + \frac{1}{\sin\phi}\frac{\partial}{\partial\phi}\left(\sin\phi\frac{\partial S_n}{\partial\phi}\right) + \frac{1}{\sin^2\phi}\frac{\partial^2 S_n}{\partial\theta^2} = 0.$$

Now we specialize once more and restrict ourselves to solutions S_n that are independent of θ; denote them by $Z_n(\phi)$. The equation reduces to

$$(n+1)nZ_n + \frac{1}{\sin\phi}\frac{d}{d\phi}\left(\sin\phi\frac{dZ_n}{d\phi}\right) = 0. \tag{6.19}$$

Finally, we put $x = \cos\phi$ and $P_n(x) = Z_n(\phi)$. The reader is asked (in Exercise 6.19) to check that the equation ends up as

$$(1-x^2)P''(x) - 2xP'(x) + n(n+1)P_n(x) = 0. \tag{6.20}$$

This is the Legendre equation. □

The Laguerre polynomials are solutions of the following singular Sturm–Liouville problem. Take $I = [0, \infty[$, $p(x) = x\,e^{-x}$, $q(x) = 0$ and $w(x) = e^{-x}$. The differential equation is

$$\frac{d}{dx}\big(x\,e^{-x}\,u'(x)\big) + \lambda\,e^{-x}\,u(x) = 0 \iff$$
$$x\,u''(x) + (1-x)\,u'(x) + \lambda\,u(x) = 0, \quad x \ge 0,$$

and the "boundary conditions" are that $u(0)$ shall exist (of course) and that $u(x)/x^m$ shall tend to 0 as $x \to \infty$ for some number m. (The latter condition can be phrased thus: $u(x)$ is majorized by some power of x, as $x \to \infty$, or $u(x)$ "increases at most like a polynomial".) The eigenvalues of this problem are $\lambda = n = 0, 1, 2, \ldots$, and the Laguerre polynomials L_n are eigenfunctions.

The Hermite polynomials come from the following singular Sturm–Liouville problem. On $I = \mathbf{R}$ one studies the equation

$$\frac{d}{dx}\big(e^{-x^2}\,u'(x)\big) + \lambda\,e^{-x^2}\,u(x) = 0$$

with the "boundary condition" that the solutions are to satisfy $u(x)/x^m \to 0$ as $|x| \to \infty$ for some $m > 0$. Eigenvalues are the numbers $\lambda = 2n$, $n = 0, 1, 2, \ldots$ and the Hermite polynomials H_n are eigenfunctions.

Exercise

6.19 Check that the change of variable $x = \cos\phi$ does transform the equation (6.19) into (6.20).

Summary of Chapter 6

The Method of Separation of Variables

Given a linear partial differential equation of order 2, with independent variables (x, y) and unknown function $u(x, y)$, together with boundary and/or initial conditions

1. If necessary (and possible), homogenize the equation and as many as possible of the other conditions.
2. Look for solutions of the homogeneous sub-problem having the particular form $u(x,y) = X(x)Y(y)$. This normally leads to a Sturm–Liouville problem, and the result should be a sequence of solutions $u_n = u_n(x,y) = X_n(x)Y_n(y)$, $n = 1, 2, 3, \ldots$.
3. The homogeneous problem has the "general" solution $u = \sum c_n u_n$, where the c_n are constants.
4. Adapt the constants c_n to make the solutions satisfy also the non-homogeneous conditions.
5. If you began by homogenizing the problem, don't forget to re-adapt the solution to suit the original problem.

Definition
Assume p, q real, $w > 0$ on an interval $I = [a, b]$. Then the following is a regular Sturm–Liouville problem on I:

$$\begin{cases} (pu')' + qu + \lambda wu = 0 \\ A_0 u(a) + A_1 u'(a) = 0, \; B_0 u(b) + B_1 u'(b) = 0 \end{cases}$$

With the Sturm–Liouville problem we associate the operator A, defined for functions u that satisfy the boundary conditions by the formula

$$Au = -\frac{1}{w}\big((pu')' + qu\big).$$

Theorem
(The Sturm–Liouville theorem) The operator A, belonging to the Sturm–Liouville problem, has infinitely many eigenvalues, which can be arranged in an increasing sequence:

$$\lambda_1 < \lambda_2 < \lambda_3 < \cdots, \quad \text{where } \lambda_n \to \infty \text{ as } n \to \infty.$$

The eigenspace of each eigenvalue has dimension 1, and if φ_n is an eigenvector corresponding to λ_n, then $\{\varphi_n\}_{n=1}^{\infty}$ is a complete orthogonal system in $L^2(I, w)$.

Formulae for orthogonal polynomials are found on page 254 f.

Historical notes

JACQUES CHARLES FRANÇOIS STURM (1803–55) and JOSEPH LIOUVILLE (1809–82) both worked in Paris. Sturm was chiefly concerned with differential equations; Liouville also did remarkable work in the field of analytic functions and the theory of numbers.

Problems for Chapter 6

6.20 Using separation of variables, solve the problem $u_t = u_{xx}$, $0 < x < 1, t > 0$; $u(0,t) = 1$, $u(1,t) = 3$, $u(x,0) = 2x + 1 - \sin 2\pi x$.

6.21 Find a solution of the problem $\begin{cases} u_{xx} = u_{tt} + 2u_t, & 0 < x < \pi,\ t > 0; \\ u(0,t) = u(\pi,t) = 0,\ t > 0; \\ u(x,0) = 0,\ u_t(x,0) = \sin^3 x,\ 0 < x < \pi. \end{cases}$

6.22 Find a solution of the differential equation $u_{xx} = u_t + u$, $0 < x < \pi$, $t > 0$, that satisfies the boundary conditions $u(0,t) = u(\pi,t) = 0$, $t > 0$, and $u(x,0) = x(\pi - x)$, $0 < x < \pi$.

6.23 Find a function $u(x,t)$ such that

$$\begin{cases} u_t = 4u_{xx}, & 0 < x < 4, \quad t > 0; \\ u(0,t) = 10, & u(4,t) = 50, \quad t > 0; \\ u(x,0) = 30, & 0 < x < 4. \end{cases}$$

6.24 Find a solution of the following problem:

$$\begin{cases} u_{xx} = u_{tt}, & 0 < x < \pi,\ t > 0; \\ u(0,t) = u(\pi,t) = 0, & t > 0; \\ u(x,0) = x(\pi - x), & 0 < x < \pi; \\ u_t(x,0) = \sin 2x, & 0 < x < \pi. \end{cases}$$

6.25 Determine a solution of the boundary value problem

$$\begin{cases} u_{xx} + u_{yy} = x, & 0 < x < 1,\ 0 < y < 1; \\ u(x,0) = u(x,1) = 0, & 0 < x < 1; \\ u(0,y) = u(1,y) = 0, & 0 < y < 1. \end{cases}$$

6.26 Find a solution in the form of a series to the Dirichlet problem

$$\begin{cases} u_{xx} + u_{yy} = 0, & x^2 + y^2 < 1; \\ u(x,y) = |x|, & x^2 + y^2 = 1. \end{cases}$$

6.27 Solve the following problem for the two-dimensional Laplace equation

$$\begin{cases} u_{xx} + u_{yy} = 0, & 0 < x < \pi,\ 0 < y < \pi; \\ u_x(0,y) = u_x(\pi,y) = 0, & 0 < y < \pi; \\ u(x,0) = \sin^2 x, \quad u(x,\pi) = 0, & 0 < x < \pi. \end{cases}$$

6.28 Find a function $u(x,t)$ such that

$$u_t = u_{xx} + \cos x, \quad 0 < x < \pi,\ t > 0;$$
$$u_x(0,t) = u_x(\pi,t) = 0,\ t > 0; \quad u(x,0) = \cos^2 x + 2\cos^4 x, \quad 0 < x < \pi.$$

6.29 Solve the following problem for a modified wave equation:

$$u_{xx} = u_{tt} + 2u_t, \ 0 < x < \pi, \ t > 0;$$
$$u(0,t) = u(\pi,t) = 0, \ t > 0,$$
$$u(x,0) = \sin x + \sin 3x, \ u_t(x,0) = 0, \ 0 < x < \pi.$$

6.30 Find $u = u(x,t)$ that satisfies the equation $u_{xx} = u_t + tu$, $0 < x < \pi$, $t > 0$, with boundary conditions $u(0,t) = u(\pi,t) = 0$ for $t > 0$ and initial condition $u(x,0) = \sin 2x$, $0 < x < \pi$.

6.31 Find a bounded solution of the problem for a vibrating beam:

$$\begin{cases} u_{tt} + u_{xxxx} = 0, & 0 < x < \pi, \ t > 0; \\ u(0,t) = u(\pi,t) = u_{xx}(0,t) = u_{xx}(\pi,t) = 0, & t > 0; \\ u(x,0) = x(\pi - x), \quad u_t(x,0) = 0, & 0 < x < \pi. \end{cases}$$

6.32 A (very much) simplified model of a nuclear reactor is given by the problem

$$\begin{cases} u_t = Au_{xx} + Bu, & 0 < x < l, \ t > 0; \\ u(0,t) = u(l,t) = 0, & t > 0. \end{cases}$$

Here, u is the concentration of neutrons, while A and B are positive constants. The term Au_{xx} describes the scattering of neutrons by diffusion, and the term Bu the creation of neutrons through fission. Prove that there is a critical value L of the length l such that if $l > L$, then there exist unbounded solutions; whereas if $l < L$, then all solutions are bounded.

6.33 In order to get good tone quality from a piano, it is desirable to have vibrations rich in overtones. An exception is the *seventh* partial, which results in musical dissonance, and should thus be kept low. Under certain idealizations the vibrations of a piano string are described by

$$\begin{cases} u_{tt} = u_{xx}, & 0 < x < \pi, \ t > 0; \\ u(0,t) = u(\pi,t) = 0, & t > 0; \\ u(x,0) = 0, \quad u_t(x,0) = \begin{cases} 1/h, & \text{for } a < x < a+h, \\ 0 & \text{otherwise.} \end{cases} \end{cases}$$

Here a describes the point of impact of the hammer; and h, the width of the hammer, is a small number.
(a) In the form of a series, compute the limit of $u(x,t)$ as $h \searrow 0$.
(b) Where should the point a be located so as to eliminate the seventh partial tone? There are a number of possible answers. Which would you choose? Explain why!

7

Fourier transforms

7.1 Introduction

Suppose that f is piecewise continuous on $[-P, P]$ (and periodic with period $2P$). For the "complex" Fourier series of f we have

$$f(t) \sim \sum_{n=-\infty}^{\infty} c_n \exp\left(in \frac{\pi}{P} t\right), \tag{7.1}$$

where

$$c_n = \frac{1}{2P} \int_{-P}^{P} f(t) \exp\left(-in \frac{\pi}{P} t\right) dt. \tag{7.2}$$

One might say that f is represented by a (formal) sum of oscillations with *frequencies* $n\pi/P$ and *complex amplitudes* c_n.

Now imagine that $P \to \infty$, and we want to find a corresponding representation of functions defined on the whole real axis (without being periodic). We define, provisionally,

$$\widehat{f}(P, \omega) = \int_{-P}^{P} f(t)\, e^{-i\omega t}\, dt, \quad \omega \in \mathbf{R}, \tag{7.3}$$

so that $c_n = \dfrac{1}{2P} \widehat{f}(P, n\pi/P)$. The formula (7.1) is translated into

$$f(t) \sim \frac{1}{2P} \sum_{n=-\infty}^{\infty} \widehat{f}(P, \omega_n)\, e^{i\omega_n t} = \frac{1}{2\pi} \sum_{n=-\infty}^{\infty} \widehat{f}(P, \omega_n)\, e^{i\omega_n t} \cdot \frac{\pi}{P}, \quad \omega_n = \frac{n\pi}{P}. \tag{7.4}$$

Because of $\Delta\omega_n = \omega_{n+1} - \omega_n = \dfrac{\pi}{P}$, this last sum looks rather like a Riemann sum. Now we let $P \to \infty$ in (7.3) and define

$$\widehat{f}(\omega) = \lim_{P\to\infty} \widehat{f}(P,\omega) = \int_{-\infty}^{\infty} f(t)\, e^{-i\omega t}\, dt, \quad \omega \in \mathbf{R} \tag{7.5}$$

(at this point we disregard all details concerning convergence). If (7.4) had contained $\widehat{f}(\omega_n)$ instead of $\widehat{f}(P,\omega_n)$, the limiting process $P \to \infty$ would have resulted in

$$f(t) \sim \frac{1}{2\pi} \int_{-\infty}^{\infty} \widehat{f}(\omega)\, e^{i\omega t}\, d\omega. \tag{7.6}$$

The formula couple (7.5) + (7.6) actually will prove to be the desired counterpart of Fourier series for functions defined on all of $\mathbf{R}$. Our strategy will be the following. Placing suitable conditions on f, we let (7.5) define a new function $\widehat{f}$, called the Fourier transform of f. We then investigate the properties of $\widehat{f}$ and show that the formula (7.6) with a suitable interpretation (and under certain additional conditions on f) constitutes a means of recovering f from $\widehat{f}$.

Loosely speaking this means that while a function defined on a finite interval (such as $(-P, P)$) can be constructed as a sum of harmonic oscillations with discrete frequencies $\{\omega_n = n\pi/P : n \in \mathbf{Z}\}$, a function on the infinite interval $]-\infty, \infty[$ demands a *continuous* frequency spectrum $\{\omega : \omega \in \mathbf{R}\}$, and the *sum* is replaced by an *integral.*

7.2 Definition of the Fourier transform

Assume that f is a function on $\mathbf{R}$, such that the (improper) integral

$$\int_{-\infty}^{\infty} |f(t)|\, dt = \int_{\mathbf{R}} |f(t)|\, dt \tag{7.7}$$

is convergent; using the notation introduced in Chapter 5, this is the same as saying that $f \in L^1(\mathbf{R})$. In practice we shall only encounter functions that are piecewise continuous, i.e., they are continuous apart from possibly a finite number of finite jumps in every finite sub-interval of $\mathbf{R}$. For such an f, the following integral converges absolutely, and for every real ω its value is some complex number:

$$\widehat{f}(\omega) = \int_{\mathbf{R}} f(t) e^{-i\omega t}\, dt. \tag{7.8}$$

Definition 7.1 *The function* $\widehat{f}$, *defined by* (7.8), *is called the* Fourier transform *or* Fourier integral *of* f.

Common notations, besides $\widehat{f}$, are $\mathcal{F}[f]$ and F ("capital letter = the transform of lower-case letter"). In this connection, it is useful to work on two distinct real axes: one where f is defined, and the variable is called such things as t, x, y; and one where the transforms live and the variable is ω, ξ, λ, etc. We denote the former axis by $\mathbf{R}$ and the latter by $\widehat{\mathbf{R}}$.

Example 7.1. If $f(t) = e^{-|t|}$, $t \in \mathbf{R}$, then $f \in L^1(\mathbf{R})$, and

$$\begin{aligned}\widehat{f}(\omega) &= \int_{\mathbf{R}} e^{-|t|} e^{-i\omega t}\,dt = \int_0^\infty e^{-(1+i\omega)t}\,dt + \int_{-\infty}^0 e^{(1-i\omega)t}\,dt \\ &= \frac{1}{1+i\omega} + \frac{1}{1-i\omega} = \frac{2}{1+\omega^2},\end{aligned}$$

which can be summarized in the formula

$$\mathcal{F}[e^{-|t|}](\omega) = \frac{2}{1+\omega^2}.$$

□

Example 7.2. Let $f(t) = 1$ for $|t| < 1$, $= 0$ for $|t| > 1$ (i.e., $f(t) = H(t+1) - H(t-1)$, where H is the Heaviside function as in Sec. 2.6). Then clearly $f \in L^1(\mathbf{R})$, and

$$\widehat{f}(\omega) = \int_{-1}^1 e^{-i\omega t}\,dt = \left[\frac{e^{-i\omega t}}{-i\omega}\right]_{-1}^1 = \frac{2}{\omega} \cdot \frac{e^{i\omega} - e^{-i\omega}}{2i} = \frac{2\sin\omega}{\omega}, \quad \omega \neq 0.$$

For $\omega = 0$ one has $e^{-i\omega t} = 1$, so that $\widehat{f}(0) = 2 = \lim_{\omega \to 0} \widehat{f}(\omega)$. □

The fact noticed at the end of the last example is not accidental. It is a case of (b) in the following theorem.

Theorem 7.1 *If $f \in L^1(\mathbf{R})$, the following holds for the Fourier transform $\widehat{f}$:*

(a) *$\widehat{f}$ is bounded; more precisely, $|\widehat{f}(\omega)| \le \int_{\mathbf{R}} |f(t)|\,dt$.*

(b) *$\widehat{f}$ is continuous on $\widehat{\mathbf{R}}$.*

(c) $\lim_{\omega \to \pm\infty} \widehat{f}(\omega) = 0$.

Proof. (a) follows immediately from the estimate $\left|\int_I \varphi(t)\,dt\right| \le \int_I |\varphi(t)|\,dt$, which holds for any interval I and any Riemann-integrable function φ (even if it is complex-valued).

(b) is more complicated, and we leave the proof as an exercise (see Exercise 7.3).

(c) is a case of the Riemann–Lebesgue Lemma (Theorem 2.2, page 25).
□

When dealing with Fourier series on the interval $(-\pi, \pi)$, we have made use of special formulae in the case when the functions happen to be even or odd. Something similar can be done for Fourier transforms. For example, if f is *even*, so that $f(-t) = f(t)$, we have

$$\widehat{f}(\omega) = \frac{1}{2\pi}\int_{-\pi}^{\pi} f(t)(\cos\omega t - i\sin\omega t)\,dt = 2\int_0^{\infty} f(t)\cos\omega t\,dt, \tag{7.9}$$

from which is seen that $\widehat{f}$ is real if f is real, and that $\widehat{f}$ is even (because $\cos(-\omega t) = \cos\omega t$). Similarly, for an *odd* function g, $g(-t) = -g(t)$, it holds that

$$\widehat{g}(\omega) = -2i\int_0^{\infty} g(t)\sin\omega t\,dt; \tag{7.10}$$

if g is real, then $\widehat{g}$ is purely imaginary, and furthermore $\widehat{g}$ is odd.

Integrals such as those in (7.9) and (7.10) are sometimes called cosine and sine transforms.

Exercises

7.1 Compute the Fourier transforms of the following functions, if they exist:
(a) $f(t) = t$ if $|t| < 1$, $= 0$ otherwise.
(b) $f(t) = 1 - |t|$ if $|t| < 1$, $= 0$ otherwise.
(c) $f(t) = \sin t$.
(d) $f(t) = 1/(t - i)$.
(e) $f(t) = (\sin t)(H(t + \pi) - H(t - \pi))$ (H is the Heaviside function).
(f) $f(t) = (\cos\pi t)\big(H(t + \frac{1}{2}) - H(t - \frac{1}{2})\big)$.

7.2 Find the Fourier transforms of $f(t) = e^{-t}H(t)$ and $g(t) = e^{t}(1 - H(t))$.

7.3 A proof of the assertion (b) in Theorem 7.1 can be accomplished along the following lines:
(i) Prove that $|\widehat{f}(\omega + h) - \widehat{f}(\omega)| \leq 2\int_{\mathbf{R}} |f(t)|\big|\sin\big(\frac{1}{2}ht\big)\big|\,dt$.
(ii) Approximate f by a function g which is zero outside some bounded interval, as in the last step of the proof of Theorem 2.2, and use that $|\sin t\,| \leq |t|$. The proof even gives the result that $\widehat{f}$ is *uniformly* continuous on $\widehat{\mathbf{R}}$.

7.3 Properties

In this section we mention some properties of Fourier transforms that are useful when applying them to, say, differential equations.

Theorem 7.2 *The mapping $\mathcal{F} : f \mapsto \widehat{f}$ is a linear map from the space $L^1(\mathbf{R})$ to the space $C_0(\widehat{\mathbf{R}})$ of those continuous functions defined on $\widehat{\mathbf{R}}$ that tend to* 0 *at* $\pm\infty$.

Proof. The fact that $\widehat{f} \in C_0(\widehat{\mathbf{R}})$ is the content of (b) and (c) in Theorem 7.1. The linearity of $\mathcal{F}$ means just that

$$\mathcal{F}[f + g] = \mathcal{F}[f] + \mathcal{F}[g], \qquad \mathcal{F}[\lambda f] = \lambda\mathcal{F}[f]$$

when $f, g \in L^1(\mathbf{R})$ and λ is a scalar (i.e., a complex number). This is an immediate consequence of the definition. □

Theorem 7.3 *Suppose that $f \in L^1(\mathbf{R})$ and let a be a real number. Then the translated function $f_a(t) = f(t-a)$ and the function $e^{iat}f(t)$ also belong to $L^1(\mathbf{R})$, and*

$$\widehat{f_a}(\omega) = \mathcal{F}[f(t-a)](\omega) = e^{-ia\omega}\widehat{f}(\omega), \tag{7.11}$$

$$\mathcal{F}[e^{iat}f(t)](\omega) = \widehat{f}(\omega - a). \tag{7.12}$$

Proof. For the first formula, start with $\widehat{f_a}(\omega) = \int_{\mathbf{R}} f(t-a)\, e^{-i\omega t}\, dt$. The change of variable $t-a = y$ gives the result. The proof of the second formula is maybe even simpler. □

These results are often called the *delay rule* and the *damping rule* for Fourier transforms. Notice the pleasant mathematical symmetry of the formulae. A similar symmetry holds for the next set of formulae.

Theorem 7.4 *Suppose that f is differentiable and that both f and f' belong to $L^1(\mathbf{R})$. Then*

$$\widehat{(Df)}(\omega) = \widehat{f'}(\omega) = \mathcal{F}[f'](\omega) = i\omega\widehat{f}(\omega). \tag{7.13}$$

If both $f(t)$ and $tf(t)$ belong to $L^1(\mathbf{R})$, then $\widehat{f}$ is differentiable, and

$$\mathcal{F}[tf(t)](\omega) = i\widehat{f}\,'(\omega) = iD\widehat{f}(\omega). \tag{7.14}$$

The proof of (7.13) relies, in principle, on integration by parts:

$$\widehat{f'}(\omega) = \int_{-\infty}^{\infty} f'(t)e^{-i\omega t}\,dt = \left[f(t)e^{-i\omega t}\right]_{-\infty}^{\infty} - \int_{-\infty}^{\infty} f(t)(-i\omega)e^{-i\omega t}\,dt,$$

and one has to prove that the integrated part is zero. We omit the details, which are somewhat technical; even though $f \in L^1(\mathbf{R})$, it does not necessarily have to tend to zero in a simple way as the variable tends to $\pm\infty$. The second formula can be proved using some theorem on differentiation under the integral sign. Indeed, if this operation is permissible, we will have

$$\begin{aligned} iD\widehat{f}(\omega) &= i\frac{d}{d\omega}\left(\int_{\mathbf{R}} f(t)e^{-i\omega t}\,dt\right) = i\int_{\mathbf{R}} \frac{\partial}{\partial\omega}\left(f(t)e^{-i\omega t}\right)dt \\ &= i\int_{\mathbf{R}} f(t)(-it)e^{-i\omega t}\,dt = \int_{\mathbf{R}} tf(t)e^{-i\omega t}\,dt. \end{aligned}$$

We shall immediately use Theorem 7.4 to find the Fourier transform of the function $f(t) = e^{-t^2/2}$.

Differentiating, we get $f'(t) = -te^{-t^2/2}$, and we see that

$$f'(t) + tf(t) = 0.$$

It is easy to see that the assumptions for Theorem 7.4, both formulae, are fulfilled. Transformation gives $i\omega\widehat{f}(\omega) + i\widehat{f}'(\omega) = 0$ or, after division by i,

$$\widehat{f}'(\omega) + \omega\widehat{f}(\omega) = 0.$$

Thus, $\widehat{f}$ satisfies the same differential equation as f. The general solution of this equation is easily determined, for instance, using an integrating factor: $y' + ty = 0$ gives $y = Ce^{-t^2/2}$, $C = y(0)$. Thus, $\widehat{f}(\omega) = Ce^{-\omega^2/2}$, where

$$C = \widehat{f}(0) = \int_{\mathbf{R}} f(t)e^{-i0t}\,dt = \int_{\mathbf{R}} e^{-t^2/2}\,dt = \sqrt{2\pi}.$$

Summarizing, we have found that

$$\mathcal{F}\left[e^{-\frac{1}{2}t^2}\right](\omega) = \sqrt{2\pi}\,e^{-\frac{1}{2}\omega^2}.$$

Theorem 7.4 implies that Fourier transformation converts differentiation into an algebraic operation. This hints at the possibility of using Fourier transformation for solving differential equations, in a way that is analogous to the use of the Laplace transform. The usefulness of this idea is, however (at our present standpoint), somewhat limited, because the Fourier integral has problems with its own convergence. For example, the common homogeneous ordinary linear differential equations with constant coefficients cannot be treated at all: all solutions of this sort of equation consist of linear combinations and products of functions of the types $\cos at$, $\sin at$, e^{bt}, and polynomials in the variable t. The only function of these types that belongs to $L^1(\mathbf{R})$ is the function that is identically zero.

There are, however, categories of problems that can be treated. Later in this chapter, we shall attack some problems for the heat equation and the Laplace equation with Fourier transforms. Also, the introduction of *distributions* has widened the range of functions that have Fourier transforms. We shall have a glimpse of this in Sec. 7.11, and a fuller treatment is found in Chapter 8.

Exercises

7.4 Assume that a is a real number $\neq 0$ and that $f \in L^1(\mathbf{R})$. Let $g(t) = f(at)$. Express $\widehat{g}$ in terms of $\widehat{f}$.

7.5 Find the Fourier transform of (a) $f(t) = e^{-|t|}\cos t$, (b) $g(t) = e^{-|t|}\sin t$.

7.6 If f is defined as in Exercise 7.1 (b), page 168, then $f'(t) = 0$ for $|t| > 1$, $f'(t) = 1$ for $-1 < t < 0$, $f'(t) = -1$ for $0 < t < 1$. Compute $\mathcal{F}[f']$ in two ways, on the one hand using Theorem 7.4 and on the other hand by direct computation.
Remark. The fact that $f'(t)$ fails to exist at some points evidently does not destroy the validity of (7.13). But f must be continuous.

7.7 Find $\widehat{f}(\omega)$ if $f(t) =$ (a) $t\,e^{-t^2/2}$, (b) $e^{-(t^2+2t)}$.
Hint for (b): complete the square, combine formula (7.11), the example following Theorem 7.4 and Exercise 7.4. Another way of solving the problem is indicated in the remark below.

7.8 Suppose that $f(t)$ has Fourier transform $\widehat{f}(\omega) = e^{-\omega^4}$. Determine the transforms of $f(2t)$, $f(2t+1)$, $f(2t+1)\,e^{it}$.

7.9 Does there exist an $f \in L^1(\mathbf{R})$ such that $\widehat{f}(\omega) = 1 - \cos\omega$?

7.10 Suppose that f has the Fourier transform $\widehat{f}$. Find the transforms of $f(t)\cos at$ and $f(t)\cos^2 at$ (a real $\neq 0$).

Remark on Exercise 7.7: Problems such as 7.7 (b) and 7.8 can also be solved by writing out the Fourier integral, then rewriting it and changing variables so as to reshape the integral into a recognizable transform. For example,

$$\mathcal{F}\left[e^{-(t^2+2t)}\right](\omega) = \int_{\mathbf{R}} e^{-(t^2+2t+1)+1}\, e^{-i\omega t}\, dt = e \int_{\mathbf{R}} e^{-(t+1)^2}\, e^{-i\omega t}\, dt$$

$$\left\{\begin{array}{l} t+1 = y/\sqrt{2}, \\ t = \dfrac{y}{\sqrt{2}} - 1 \\ dt = dy/\sqrt{2} \end{array}\right\} = \frac{e}{\sqrt{2}} \int_{\mathbf{R}} e^{-y^2/2}\, e^{-i\omega(y/\sqrt{2}-1)}\, dy$$

$$= \frac{e^{1+i\omega}}{\sqrt{2}} \int_{\mathbf{R}} e^{-y^2/2}\, e^{-iy\omega/\sqrt{2}}\, dy.$$

The last integral is the Fourier transform of $e^{-t^2/2}$, computed at the point $\omega/\sqrt{2}$, which means $\sqrt{2\pi}\exp\left(-\frac{1}{2}\left(\frac{\omega}{\sqrt{2}}\right)^2\right) = \sqrt{2\pi}\,e^{-\omega^2/4}$. The answer to the problem is thus

$$\mathcal{F}\left[e^{-(t^2+2t)}\right](\omega) = \frac{e^{1+i\omega}}{\sqrt{2}} \cdot \sqrt{2\pi}\,e^{-\omega^2/4} = \sqrt{\pi}\,e^{1+i\omega-\frac{1}{4}\omega^2}.$$

7.4 The inversion theorem

We now formulate the result that constitutes our promised precise version of the formula (7.6) on page 166.

Theorem 7.5 (Inversion theorem) *Suppose that $f \in L^1(\mathbf{R})$, that f is continuous except for a finite number of finite jumps in any finite interval, and that $f(t) = \frac{1}{2}(f(t+) + f(t-))$ for all t. Then*

$$f(t_0) = \lim_{A\to\infty} \frac{1}{2\pi} \int_{-A}^{A} \widehat{f}(\omega)\, e^{i\omega t_0}\, d\omega \tag{7.15}$$

for every t_0 where f has (generalized) left and right derivatives. In particular, if f is piecewise smooth (i.e., continuous and with a piecewise continuous derivative), then the formula holds for all $t_0 \in \mathbf{R}$.

The result, and also the proof, is very similar to the convergence theorem for Fourier series. Just as for these series, the convergence properties depend essentially on the local behavior of $f(t)$ for t near t_0. To accomplish the proof we need an auxiliary lemma.

Lemma 7.1

$$\int_0^\infty \frac{\sin Au}{u}\,du = \frac{\pi}{2} \qquad \textit{for } A > 0.$$

It is easy to check, by the change of variable $Au = t$, that the integral is independent of A (if $A > 0$), so one can just as well assume that $A = 1$. The integral is not absolutely convergent, and $\int_0^\infty$ in this case stands for the limit of $\int_0^X$ as $X \to \infty$. There is no quite simple way to compute it. One method is using the calculus of residues, and textbooks on complex analysis usually contain precisely this integral as an example of that technique.

Another attempt could build on the idea that $1/u = \int_0^\infty e^{-ux}\,dx$, which might be substituted into the integral:

$$\int_0^\infty \frac{\sin u}{u}\,du = \int_0^\infty \int_0^\infty e^{-ux} \sin u\,dx\,du = \int_0^\infty \left(\int_0^\infty e^{-ux} \sin u\,du\right) dx$$
$$= \int_0^\infty \frac{dx}{1+x^2} = \frac{\pi}{2}.$$

There is, however, a difficulty here: the double integral is not absolutely convergent (the integrand is too large when x is close to 0), which makes it hard to justify the change of order of integration. However, we will not delve deeper into this problem.

Proof of Theorem 7.5. Put

$$s(t_0, A) = \frac{1}{2\pi} \int_{-A}^{A} \widehat{f}(\omega)\, e^{it_0\omega}\,d\omega$$

and rewrite this expression by inserting the definition of $\widehat{f}(\omega)$:

$$s(t_0, A) = \frac{1}{2\pi} \int_{-A}^{A} \left(\int_{-\infty}^{\infty} f(t)\, e^{-i\omega t}\,dt\right) e^{i\omega t_0}\,d\omega$$
$$= \frac{1}{2\pi} \int_{-\infty}^{\infty} \int_{-A}^{A} f(t)\, e^{i\omega(t_0-t)}\,d\omega\,dt = \frac{1}{2\pi} \int_{-\infty}^{\infty} f(t) \left[\frac{e^{i\omega(t_0-t)}}{i(t_0-t)}\right]_{\omega=-A}^{\omega=A} dt$$
$$= \frac{1}{\pi} \int_{-\infty}^{\infty} f(t)\, \frac{\sin A(t_0-t)}{t_0-t}\,dt = \frac{1}{\pi} \int_{-\infty}^{\infty} f(t_0-u)\, \frac{\sin Au}{u}\,du.$$

Switching the order of integration is permitted, because the improper double integral is absolutely convergent over the strip $(t, \omega) \in \mathbf{R} \times [-A, A]$, and in the last step we have put $t_0 - t = u$. We are now in a situation very much the same as in the proof of the convergence of Fourier series; but there is a complication inasmuch as the interval of integration is unbounded. Using the lemma we can write

$$\frac{2}{\pi}\int_0^\infty f(t_0-u)\,\frac{\sin Au}{u}\,du - f(t_0-) = \frac{2}{\pi}\int_0^\infty \bigl(f(t_0-u)-f(t_0-)\bigr)\,\frac{\sin Au}{u}\,du. \tag{7.16}$$

Now let $\varepsilon > 0$ be given. Since we have assumed that $f \in L^1(\mathbf{R})$, there exists a number X such that

$$\frac{2}{\pi}\int_X^\infty |f(t_0-u)|\,du < \varepsilon.$$

Changing the variable, we find that

$$\int_X^\infty \frac{\sin Au}{u}\,du = \int_{AX}^\infty \frac{\sin t}{t}\,dt \to 0 \quad \text{as } A \to \infty. \tag{7.17}$$

The last integral in (7.16) can be split into three terms:

$$\frac{2}{\pi}\int_0^X \frac{f(t_0-u)-f(t_0-)}{u}\cdot \sin Au\,du + \frac{2}{\pi}\int_X^\infty f(t_0-u)\,\frac{\sin Au}{u}\,du$$

$$-\frac{2}{\pi}\,f(t_0-)\int_X^\infty \frac{\sin Au}{u}\,du = I_1 + I_2 - I_3.$$

The term I_3 tends to zero as $A \to \infty$ because of (7.17). The term I_2 can be estimated:

$$|I_2| = \left|\frac{2}{\pi}\int_X^\infty f(t_0-u)\,\frac{\sin Au}{u}\,du\right| \le \frac{2}{\pi}\int_X^\infty |f(t_0-u)|\,du \le \varepsilon.$$

In the term I_1 we have the function $u \mapsto g(u) = (f(t_0-u) - f(t_0))/(-u)$. This is continuous except for jumps in the interval $(0, X)$, and it has the finite limit $g(0+) = f'_L(t_0)$ as $u \searrow 0$; this means that g is bounded and thus integrable on the interval. By the Riemann–Lebesgue lemma, we conclude that $I_1 \to 0$ as $A \to \infty$. All this together gives, since ε can be taken as small as we wish,

$$\frac{2}{\pi}\int_0^\infty f(t_0-u)\,\frac{\sin Au}{u}\,du \to f(t_0-) \quad \text{as } A \to \infty.$$

A parallel argument implies that the corresponding integral over $(-\infty, 0)$ tends to $f(t_0+)$. Taking the mean value of these two results, we have completed the proof of the theorem. □

Remark. If $\int_{\widehat{\mathbf{R}}} |\widehat{f}(\omega)|\,d\omega$ is convergent, i.e., $\widehat{f} \in L^1(\widehat{\mathbf{R}})$, then (7.15) can be written as the absolutely convergent integral

$$f(t_0) = \frac{1}{2\pi}\int_{-\infty}^\infty \widehat{f}(\omega)\,e^{i\omega t_0}\,d\omega,$$

but in general one has to make do with the symmetric limit in (7.15). □

Example 7.3. For $f(t) = e^{-|t|}$ we have $\widehat{f}(\omega) = \dfrac{2}{1+\omega^2}$. Since f is piecewise smooth it follows that

$$e^{-|t|} = \frac{1}{\pi} \lim_{A\to\infty} \int_{-A}^{A} \frac{e^{i\omega t}}{1+\omega^2}\, d\omega.$$

In this case $\widehat{f}$ happens to be absolutely integrable, and we can write simply

$$e^{-|t|} = \frac{1}{\pi} \int_{-\infty}^{\infty} \frac{e^{i\omega t}}{1+\omega^2}\, d\omega.$$

We can switch letters in this formula — t and ω are exchanged for each other — and then we also change the sign of ω, and we get (after multiplication by π) the formula

$$\pi\, e^{-|\omega|} = \int_{\mathbf{R}} \frac{e^{-i\omega t}}{1+t^2}\, dt.$$

In this way we have found the Fourier transform of $1/(1+t^2)$, which is rather difficult to reach by other methods:

$$\mathcal{F}\left[\frac{1}{1+t^2}\right](\omega) = \pi\, e^{-|\omega|}.$$

□

Example 7.4. For the function f in Example 7.2 (page 167) we have $\widehat{f}(\omega) = \dfrac{2\sin\omega}{\omega}$. In this case, the inversion integral is not absolutely convergent. The theorem here says that

$$\lim_{A\to\infty} \frac{1}{\pi} \int_{-A}^{A} \frac{\sin\omega}{\omega}\, e^{i\omega t}\, d\omega = \begin{cases} 1 \text{ as } |t| < 1, \\ \frac{1}{2} \text{ as } t = \pm 1, \\ 0 \text{ as } |t| > 1. \end{cases}$$

□

Example 7.5. When using a table of Fourier transforms (such as on page 252 f. of this book), one can make use of the evident symmetry properties of the transform itself and the inversion formula in order to transform functions that are found "on the wrong side of the table." We have actually seen an instance of this idea in Example 7.3 above. As a further example, suppose that a table contains an item like this:

$f(t)$	$\widehat{f}(\omega)$
$te^{-|t|}$	$\dfrac{-4i\omega}{(1+\omega^2)^2}$

From this one can find the transform of the function

$$g(t) = \frac{-4it}{(1+t^2)^2}$$

by performing two steps:

1. Switch sides in the table, switching variables at the same time:

$$\frac{-4it}{(1+t^2)^2} \qquad \omega e^{-|\omega|}$$

2. Multiply the right-hand side by 2π and change the sign of the variable there:

$$\frac{-4it}{(1+t^2)^2} \qquad -2\pi\omega e^{-|-\omega|}$$

This is now a true entry in the table. It may be an aesthetic gain to divide it by $-4i$ to get

$$\frac{t}{(1+t^2)^2} \qquad -\tfrac{1}{2}i\pi\omega e^{-|\omega|}$$

□

Example 7.6. When working with even or odd functions, the Fourier transform can be rewritten as a so-called cosine or sine integral, respectively (see p. 168). In these cases, the inversion formula can also be rewritten so as to contain a cosine or a sine, instead of a complex exponential. Indeed, if g is even, one gets the following couple of formulae:

$$\widehat{g}(\omega) = 2\int_0^\infty g(t)\cos\omega t\,dt, \qquad g(t) = \frac{1}{\pi}\int_0^\infty \widehat{g}(\omega)\cos t\omega\,d\omega,$$

and if h is odd, it looks like this:

$$\widehat{h}(\omega) = -2i\int_0^\infty h(t)\sin\omega t\,dt, \qquad h(t) = \frac{i}{\pi}\int_0^\infty \widehat{h}(\omega)\sin t\omega\,d\omega.$$

(The reader should check this.) In applied literature, one often meets these "cosine" and "sine" transforms with slightly modified definitions. □

Remark. In the literature one can find many variations of the definition of the Fourier transform. We have chosen the conventions illustrated by the formula couple

$$\widehat{f}(\omega) = \int_{-\infty}^{\infty} f(t)e^{-i\omega t}\,dt, \qquad f(t) \sim \frac{1}{2\pi}\int_{-\infty}^{\infty} \widehat{f}(\omega)e^{i\omega t}\,d\omega.$$

Other common conventions are described by

$$\widehat{f}(\omega) = \frac{1}{2\pi}\int_{-\infty}^{\infty} f(t)e^{-i\omega t}\,dt, \qquad f(t) \sim \int_{-\infty}^{\infty} \widehat{f}(\omega)e^{i\omega t}\,d\omega,$$

$$\widehat{f}(\omega) = \frac{1}{\sqrt{2\pi}}\int_{-\infty}^{\infty} f(t)e^{-i\omega t}\,dt, \qquad f(t) \sim \frac{1}{\sqrt{2\pi}}\int_{-\infty}^{\infty} \widehat{f}(\omega)e^{i\omega t}\,d\omega,$$

$$\widehat{f}(\omega) = \int_{-\infty}^{\infty} f(t)e^{-2\pi i\omega t}\,dt, \qquad f(t) \sim \int_{-\infty}^{\infty} \widehat{f}(\omega)e^{2\pi i t\omega}\,d\omega.$$

It also happens that the minus sign in the exponent is moved from one integral to the other. As soon as Fourier transformation is encountered in real life, one must check what definition is actually being used. This is true also for tables and handbooks. □

Exercises

7.11 Find the Fourier transforms of the following functions:
(a) $\dfrac{1}{t^2+2t+2}$, (b) $\dfrac{1}{t^2+6t+13}$, (c) $\dfrac{t}{(1+t^2)^2}$.

7.12 Find the Fourier transform of $\dfrac{1-\cos t}{t^2}$.

7.13 Find a function $f(x)$, defined for $x > 0$, such that

$$\int_0^{\infty} f(y)\cos xy\,dy = \frac{1}{1+x^2}.$$

(Hint: extend f to an even function and take a look at Example 7.3.)

7.14 Assume that f is differentiable and has the Fourier transform

$$\widehat{f}(\omega) = \frac{1+i\omega}{1+\omega^6}.$$

Compute $f'(0)$. (Note that you do not have to find a formula for $f(t)$.)

7.15 Suppose that $f \in L^1(\mathbf{R})$ and that $\widehat{f}$ has a finite number of zeroes. Prove that there cannot exist a function $g \in L^1(\mathbf{R})$ and a number a such that $g(t+a) - g(t) = f(t)$ for $-\infty < t < \infty$.

7.16 A consequence of Theorem 7.5 is that if $\widehat{f}(\omega) = 0$ for all ω, then it must hold that $f(t) = 0$ for all t where f is continuous. Using this, formulate and prove a *uniqueness theorem* for Fourier transforms.

7.5 The convolution theorem

Let f and g be two functions belonging to $L^1(\mathbf{R})$. The *convolution* $f * g$ of them is now defined to be the function defined on $\mathbf{R}$ by the formula

$$(f*g)(t) = f*g(t) = \int_{\mathbf{R}} f(t-y)\,g(y)\,dy = \int_{\mathbf{R}} f(y)\,g(t-y)\,dy.$$

It can be proved, using deeper insights in the theory of integration, that this integral is actually convergent and that the new function also belongs to $L^1(\mathbf{R})$. We content ourselves here with accepting these facts as true. The act of forming $f * g$ is also phrased as *convolving* f and g.

Theorem 7.6 (Convolution theorem)

$$\mathcal{F}[f * g] = \mathcal{F}[f]\, \mathcal{F}[g].$$

Formally, the proof runs like this:

$$\begin{aligned}
\mathcal{F}[f * g](\omega) &= \int_{\mathbf{R}} e^{-i\omega t} \left(\int_{\mathbf{R}} f(t-y)\, g(y)\, dy \right) dt \\
&= \iint_{\mathbf{R}^2} e^{-i\omega(t-y+y)}\, f(t-y)\, g(y)\, dt\, dy \\
&= \int_{\mathbf{R}} e^{-i\omega y}\, g(y)\, dy \int_{\mathbf{R}} e^{-i\omega(t-y)}\, f(t-y)\, dt \\
&= \int_{\mathbf{R}} e^{-i\omega y}\, g(y)\, dy \int_{\mathbf{R}} e^{-i\omega t}\, f(t)\, dt = \widehat{g}(\omega)\, \widehat{f}(\omega).
\end{aligned}$$

The legitimacy of changing the order of integration is taken for granted.

Example 7.7. What function f has the Fourier transform

$$\widehat{f}(\omega) = \frac{1}{(1+\omega^2)^2}\, ?$$

Solution. We start from the formula

$$\widehat{g}(\omega) = \frac{2}{1+\omega^2} \quad \text{if} \quad g(t) = e^{-|t|}.$$

By the convolution theorem we get

$$\mathcal{F}[g * g](\omega) = \left(\widehat{g}(\omega)\right)^2 = \frac{4}{(1+\omega^2)^2} = 4\widehat{f}(\omega).$$

Clearly, $f = \frac{1}{4}(g * g)$, and we thus have to convolve g with itself. For $t > 0$ we get

$$\begin{aligned}
4f(t) = g * g(t) &= \int_{-\infty}^{\infty} e^{-|t-y|}\, e^{-|y|}\, dy = \int_{-\infty}^{0} + \int_0^t + \int_t^{\infty} \\
&= \int_{-\infty}^{0} e^{-(t-y)}\, e^{y}\, dy + \int_0^t e^{-(t-y)}\, e^{-y}\, dy + \int_t^{\infty} e^{t-y}\, e^{-y}\, dy \\
&= e^{-t} \int_{-\infty}^{0} e^{2y}\, dy + e^{-t} \int_0^t dt + e^{t} \int_t^{\infty} e^{-2y}\, dy = (1+t)\, e^{-t}.
\end{aligned}$$

(Check the computations for yourself!) Since $\widehat{f}$ is an even function, f is also even, and so we must have

$$f(t) = \tfrac{1}{4}(1 + |t|)\, e^{-|t|}.$$

□

Example 7.8. If $f(t) = 1/(1+t^2)$, find $f * f$.

Solution. Put $g = f * f$. Computing the convolution directly is toilsome. Instead, we make use of Theorem 7.6. Let us start from the fact that $\widehat{f}(\omega) = \pi\, e^{-|\omega|}$ (Example 7.3, p. 174), which means that

$$\int_{\mathbf{R}} \frac{e^{-i\omega t}}{1+t^2}\, dt = \pi\, e^{-|\omega|}. \tag{7.18}$$

Theorem 7.6 gives $\widehat{g}(\omega) = \big(\widehat{f}(\omega)\big)^2 = \pi^2 e^{-|2\omega|}$. In (7.18) we now exchange ω for 2ω, multiply by π and make the change of variable $2t = y$:

$$\widehat{g}(\omega) = \pi^2\, e^{-2|\omega|} = \int_{\mathbf{R}} \frac{\pi e^{-it\cdot 2\omega}}{1+t^2}\, dt = \int_{\mathbf{R}} \frac{\pi e^{-iy\omega}}{1+\left(\dfrac{y}{2}\right)^2} \frac{dy}{2} = \int_{\mathbf{R}} \frac{2\pi e^{-i\omega t}}{4+t^2}\, dt.$$

We find that $g(t) = \dfrac{2\pi}{4+t^2}$. Thus, we have proved the formula

$$\int_{-\infty}^{\infty} \frac{dy}{(1+y^2)(1+(t-y)^2)} = \frac{2\pi}{4+t^2}, \qquad t \in \mathbf{R}.$$

□

Just as in Example 7.7, convolution can be employed to find inverse Fourier transforms. Other applications occur in the solution of certain partial differential equations, whose solutions are given in the form of convolution integrals; and convolutions occur frequently in probability theory.

Example 7.9. Prove the formula

$$\int_{-1}^{1} \frac{\sin(t-y)}{t-y}\, dy = \int_{-1}^{1} \frac{\sin y}{y}\, e^{iyt}\, dy, \quad t \in \mathbf{R}. \tag{7.19}$$

Solution. Let $f(t) = H(t+1) - H(t-1)$, and $g(t) = (\sin t)/t$. From the table of Fourier transforms we recognize that $\widehat{f}(\omega) = g(\omega)/\pi$, and by the inversion formula we have $\widehat{g}(\omega) = \frac{1}{2} f(\omega)$. The right-hand member of (7.19) can be written like this:

$$\begin{aligned}\int_{-1}^{1} \frac{\sin y}{y}\, e^{iyt}\, dy &= \int_{\mathbf{R}} f(y) g(y) e^{iyt}\, dy \\ &= \int_{\widehat{\mathbf{R}}} 2g(\omega)\cdot \pi f(\omega)\, e^{i\omega t}\, d\omega = 2\pi \int_{\widehat{\mathbf{R}}} \widehat{g}(\omega)\, \widehat{f}(\omega)\, e^{i\omega t}\, d\omega.\end{aligned}$$

The last formula consists of the inversion formula for the function whose Fourier transform is $2\pi \widehat{f}\,\widehat{g}$, and this function must be the convolution of f and g. But that convolution is just the left-hand member of (7.19). □

Example 7.10. Because of the formal symmetry between the Fourier transformation and the inversion formula, one can expect that there exists a formula involvning the convolution of transforms. Indeed, if $\widehat{f}$ and $\widehat{g}$ are sufficiently nice, to ensure that the necessary integrals converge, it is true that the Fourier transform of the product fg is the convolution of the transforms (modified by a factor of $1/(2\pi)$):

$$\widehat{fg}(\omega) = \frac{1}{2\pi} \int_{\widehat{\mathbf{R}}} \widehat{f}(\omega - \alpha)\, \widehat{g}(\alpha)\, d\alpha = \frac{1}{2\pi}\, \widehat{f} * \widehat{g}(\omega).$$

□

Exercises

7.17 Let f_a be defined for a positive number a as the function

$$f_a(t) = \frac{1}{\pi} \frac{a}{a^2 + t^2}, \qquad t \in \mathbf{R}.$$

Compute the convolution $f_{a_1} * f_{a_2}$. Generalize to more than two convolution factors.

7.18 Find a solution of the integral equation

$$\int_{-\infty}^{\infty} f(t-y)\, e^{-|y|}\, dy = \tfrac{4}{3}\, e^{-|t|} - \tfrac{2}{3}\, e^{-2|t|}.$$

7.19 Determine some f such that $\displaystyle\int_{-\infty}^{\infty} f(t-y)\, e^{-y^2/2}\, dy = e^{-t^2/4}$.

7.20 Find a function f such that $\int_{-1}^{1} f(t-y)\, dy = e^{-|t-1|} - e^{-|t+1|}$, $t \in \mathbf{R}$.

7.21 Compute the integral

$$\int_{-\infty}^{\infty} \frac{\sin[5(t-u)]\,\sin(6u)}{u(t-u)}\, du, \qquad t \in \mathbf{R}.$$

7.22 Let $f \in L^1(\mathbf{R})$ be such that f' is continuous and $f' \in L^1(\mathbf{R})$. Find a function $g \in L^1(\mathbf{R})$ such that

$$g(t) = \int_{-\infty}^{t} e^{u-t}\, g(u)\, du + f'(t), \qquad t \in \mathbf{R}.$$

7.6 Plancherel's formula

We shall now indicate an intuitive deduction of a formula that corresponds to the Parseval formula for Fourier series. If these series are written in the "complex" version, we have

$$\sum_{n=-\infty}^{\infty} |c_n|^2 = \frac{1}{2\pi}\int_{-\pi}^{\pi} |f(t)|^2\,dt, \quad \text{where} \quad c_n = \frac{1}{2\pi}\int_{-\pi}^{\pi} f(t)\,e^{-int}\,dt.$$

A simple change of variables yields the corresponding formula on the interval $(-P, P)$: put

$$c_n = \frac{1}{2P}\int_{-P}^{P} f(t)\,e^{-in\pi t/P}\,dt,$$

and we will have

$$\sum_{n=-\infty}^{\infty} |c_n|^2 = \frac{1}{2P}\int_{-P}^{P} |f(t)|^2\,dt. \tag{7.20}$$

Just as on page 165 we introduce the "truncated" Fourier transform

$$\widehat{f}(P,\omega) = \int_{-P}^{P} f(t)\,e^{-i\omega t}\,dt,$$

so that $c_n = \dfrac{1}{2P}\widehat{f}(P, n\pi/P)$, and (7.20) takes the form

$$\frac{1}{4P^2}\sum_{n=-\infty}^{\infty} \left|\widehat{f}\left(P, \frac{n\pi}{P}\right)\right|^2 = \frac{1}{2P}\int_{-P}^{P} |f(t)|^2\,dt$$

or

$$\int_{-P}^{P} |f(t)|^2\,dt = \frac{1}{2\pi}\sum_{n=-\infty}^{\infty} \left|\widehat{f}\left(P, \frac{n\pi}{P}\right)\right|^2 \cdot \frac{\pi}{P}.$$

In the same way as on page 165 we can consider the right-hand member to be almost a Riemann sum, and if we let $P \to \infty$ we ought to obtain

$$\int_{-\infty}^{\infty} |f(t)|^2\,dt = \frac{1}{2\pi}\int_{-\infty}^{\infty} |\widehat{f}(\omega)|^2\,d\omega. \tag{7.21}$$

In fact, this formula is actually true as soon as one knows that one of the integrals is convergent — if so, the other one will automatically converge as well. A correct and consistent theory of these matters cannot be achieved without having access to the integration theory of Lebesgue. The formula (7.21) is known as the PLANCHEREL formula (also sometimes as the Parseval formula).

Example 7.11. The Plancherel formula enables us to compute certain integrals. If $f(t) = 1$ for $|t| < 1$ and $= 0$ otherwise, then (see Example 7.2 p. 167) $\widehat{f}(\omega) = \dfrac{2\sin\omega}{\omega}$. Plancherel now gives

$$\int_{-1}^{1} 1\,dt = \frac{1}{2\pi}\int_{-\infty}^{\infty} \frac{4\sin^2\omega}{\omega^2}\,d\omega,$$

or, after rewriting,

$$\int_{-\infty}^{\infty} \frac{\sin^2 t}{t^2}\,dt = \pi.$$

This integral is not very easy to compute using other methods. □

Just as in Chapter 5, we can denote by $L^2(\mathbf{R})$ the set of functions f defined on $\mathbf{R}$ such that the integral $\int_R |f(t)|^2\,dt$ is convergent. If f and g are both in $L^2(\mathbf{R})$, it can be seen (just as in Sec. 5.3) that we can define an inner product by the integral

$$\langle f, g\rangle = \int_{\mathbf{R}} f(x)\,\overline{g(x)}\,dx.$$

Introducing the L^2 norm in the usual way, $\|f\| = \sqrt{\langle f, f\rangle}$, Plancherel's formula can be written in the compact form

$$\|f\|^2 = \frac{1}{2\pi}\,\|\widehat{f}\|^2.$$

There are a number of variants of the Plancherel formula. One is related to the formula for inner products in an ON basis and looks like this:

$$\int_{-\infty}^{\infty} f(t)\,\overline{g(t)}\,dt = \frac{1}{2\pi}\int_{-\infty}^{\infty} f(\omega)\,\overline{g(\omega)}\,d\omega.$$

This can be obtained from the ordinary Plancherel formula using the identity

$$\langle f, g\rangle = \tfrac{1}{4}\left(\|f+g\|^2 + i\|f+ig\|^2 - \|f-g\|^2 - i\|f-ig\|^2\right),$$

which is easily proved (it is Exercise 5.6 on page 110).

The following formula is another variation. Let $f \in L^1(\mathbf{R})$ and $g \in L^1(\widehat{\mathbf{R}})$. (Thus, g is defined on the "wrong" real line.) Then it holds that

$$\int_{-\infty}^{\infty} f(t)\,\widehat{g}(t)\,dt = \int_{-\infty}^{\infty} \widehat{f}(\omega)\,g(\omega)\,d\omega. \tag{7.22}$$

This is easily proved by considering the double integral

$$\iint_{\mathbf{R}\times\widehat{\mathbf{R}}} f(t)\,g(\omega)\,e^{-it\omega}\,dt\,d\omega,$$

and computing this in two different ways. The computation is completely legitimate, because the double integral is easily seen to be absolutely convergent. The resulting formula (7.22) plays a central role in Chapter 8.

Example 7.12. As an application of the last formula, we give a new proof of the formula

$$\int_{-1}^{1} \frac{\sin(t-y)}{t-y}\,dy = \int_{-1}^{1} \frac{\sin y}{y}\,e^{iyt}\,dy, \quad t \in \mathbf{R}. \tag{7.23}$$

(see Example 7.9 above). Let $f(y) = H(y+1) - H(y-1)$ and $g(\omega) = e^{it\omega} f(\omega)$. Then

$$\widehat{f}(\omega) = \frac{2\sin\omega}{\omega} \quad \text{and} \quad \widehat{g}(y) = \frac{2\sin(y-t)}{y-t} = \frac{2\sin(t-y)}{t-y}.$$

The identity $\int f\widehat{g} = \int \widehat{f} g$ then gives the formula (after some preening). □

Exercises

7.23 Compute the integral

$$\int_{-\infty}^{\infty} \frac{t^2}{(1+t^2)^2}\,dt$$

by studying the odd function f defined by $f(t) = e^{-t}$ for $t > 0$.

7.24 Using the results of Exercise 7.5, compute the integrals

$$\int_0^{\infty} \frac{t^2}{(t^4+4)^2}\,dt \quad \text{and} \quad \int_0^{\infty} \frac{(t^2+2)^2}{(t^4+4)^2}\,dt.$$

7.7 Application 1

We consider the following problem for the heat equation:

$$\begin{array}{llll} \text{(E)} & u_{xx} = u_t, & t > 0, & x \in \mathbf{R}, \\ \text{(I)} & u(x,0) = f(x), & & x \in \mathbf{R}. \end{array}$$

The solution $u(x,t)$ represents the temperature at the point x of an infinite rod, isolated from its surroundings, if the temperature at time $t = 0$ is given by the function $f(x)$.

Initially, we adopt the extra assumptions that $f \in L^1(\mathbf{R})$ and that for every fixed $t > 0$ the function $x \mapsto u(x,t)$ also belongs to $L^1(\mathbf{R})$. Then the Fourier transforms

$$\widehat{f}(\omega) = \int_{\mathbf{R}} f(x)\,e^{-i\omega x}\,dx,$$

$$U(\omega,t) = \mathcal{F}_x[u(x,t)](\omega) = \int_{\mathbf{R}} u(x,t)\,e^{-i\omega x}\,dx$$

exist for all $t \geq 0$. (The subscript x on $\mathcal{F}$ signifies that the transform is taken with respect to the variable x.) We also assume that we can treat the differentiations in a formal manner, by which we mean for one thing that the rule (7.13) of Theorem 7.4 can be used twice, for another that

$$\mathcal{F}_x[u_t] = \mathcal{F}_x\left[\frac{\partial u}{\partial t}\right] = \frac{\partial}{\partial t}\mathcal{F}_x[u] = \frac{\partial U}{\partial t}.$$

In this case, (E) and (I) are transformed into

$$\begin{aligned} &(\hat{\mathrm{E}}) \qquad -\omega^2 U = \frac{\partial U}{\partial t}, \quad t > 0, \qquad \omega \in \widehat{\mathbf{R}}, \\ &(\hat{\mathrm{I}}) \qquad U(\omega, 0) = \widehat{f}(\omega) \qquad\qquad\qquad \omega \in \widehat{\mathbf{R}}. \end{aligned}$$

($\hat{\mathrm{E}}$) can be solved like a common ordinary differential equation (think for a moment of ω as a constant): we get $U = C\exp(-\omega^2 t)$, where the constant of integration C need not be the same for different values of ω. Indeed, adapting to the initial condition ($\hat{\mathrm{I}}$) we find that we should have $C = \widehat{f}(\omega)$, so that

$$U(\omega, t) = \widehat{f}(\omega)\, e^{-\omega^2 t}.$$

For recovering $u(x,t)$ we notice that U is a product of two Fourier transforms. By performing a suitable change of variables in the formula

$$\mathcal{F}[\exp(-\tfrac{1}{2}x^2)](\omega) = \sqrt{2\pi}\exp(-\tfrac{1}{2}\omega^2)$$

we can find that

$$\mathcal{F}_x\left[\frac{1}{\sqrt{4\pi t}}\exp\left(-\frac{x^2}{4t}\right)\right](\omega) = e^{-\omega^2 t}.$$

Let $E(x,t)$ be the expression that is being transformed here. Then we have

$$U(\omega, t) = \mathcal{F}_x[E(x,t)](\omega) \cdot \mathcal{F}[f](\omega).$$

By the convolution theorem,

$$u(x,t) = E(x,t) * f(x) = \frac{1}{\sqrt{4\pi t}}\int_{-\infty}^{\infty} e^{-y^2/(4t)}\, f(x-y)\, dy, \qquad t > 0.$$

This integral formula has been deduced by formal calculations with Fourier transforms, but in its final appearance it does not contain any such transforms. In fact, it works in far more general situations than those indicated by our assumptions. Indeed, it is sufficient to assume that f is a continuous and bounded function on $\mathbf{R}$. Then the integral in the formula exists for all $x \in \mathbf{R}$ and $t > 0$ (show it!) and satisfies the equation $u_{xx} = u_t$, and in addition it holds that $\lim_{t \searrow 0} u(x,t) = f(x)$. The last assertion follows from the fact that $E(x,t)$ is a positive summation kernel in the variable x, indexed by the variable t tending to 0 from above.

The solution obtained also has a nice *statistical* interpretation. It is the convolution of the initial values by the density function of a normal probability distribution with expected value zero and a variance growing with time. Loosely speaking, this can be said to mean that the temperature is "smeared out" in a very regular way along the axis.

Remark. As an example of this, let the initial temperature be given by $u(x,0) = 1$ for $|x| < 1$ and 0 otherwise. The solution will be

$$u(x,t) = \frac{1}{\sqrt{4\pi t}} \int_{-1}^{1} \exp(-(x-y)^2/(4t))\,dy = \frac{1}{\sqrt{4\pi t}} \int_{x-1}^{x+1} e^{-y^2/(4t)}\,dy.$$

It is easy to see that the value of this integral is positive for *all* (x,t) with $t > 0$. This is really a cause of concern: it means that points at arbitrary distance far away on the rod will, immediately after the initial moment, be aware of the fact that the temperature near the origin was positive when t was 0. The information from the vicinity of the origin thus travels with infinite speed along the rod, which is in conflict with the wellknown statement from the theory of relativity: nothing can travel faster than light!

This indicates that the mathematical model that gives rise to the heat equation must be physically incorrect. What is wrong? Well, for one thing, in this model, matter is considered to be a perfectly homogeneous medium, which is a macroscopic approximation that does not hold at all on a small scale: in reality, matter is something discrete, consisting of atoms and subatomic particles. Yet another thing is that in the model heat itself is considered to be a homogeneous, flowing substance: in reality, heat is a macroscopic "summary" of the movements of all the particles of matter.

Yet another example of the strange behaviour of the heat equation is the following, taken from THOMAS KÖRNER's book *Fourier Analysis.* Define a function h by letting

$$h(t) = \exp\left(-\frac{1}{2t^2}\right), \quad t > 0; \qquad h(t) = 0, \quad t \le 0.$$

This function belongs to $C^\infty(\mathbf{R})$, which is not very hard to prove. If we go on to define $g(t) = h(t-1)\,h(2-t)$, then g is a C^∞-function which is positive for $1 < t < 2$ and zero elsewhere. Finally let

$$u(x,t) = \sum_{m=0}^{\infty} \frac{g^{(m)}(t)}{(2m)!}\,x^{2m}.$$

Now we have a function $u : \mathbf{R}^2 \to \mathbf{R}$ with the following properties:

(a)	$u_t(x,t) = u_{xx}(x,t)$	for all $(x,t) \in \mathbf{R}^2$,
(b)	$u(x,t) = 0$	for all $t \notin [1,2]$, $x \in \mathbf{R}$,
(c)	$u(x,t) > 0$	for all $t \in]1,2[$, $x \in \mathbf{R}$.

The proof of these assertions can be found in Körner's book, Sec. 67.

Thus, here we have an infinite rod, which has at time 0 the temperature 0 everywhere. This state of affairs remains until time reaches 1: then suddenly the whole rod acquires positive temperature, which rises and then again falls back to zero at time 2. Körner calls this "the great blast of heat from infinity."

We see again (cf. page 4) that the "natural" initial value problem for the heat equation behaves very badly as concerns uniqueness: indeed there are heaps of solutions. I cannot abstain from quoting Körner's rounding-off comment on this example:

> To the applied mathematician ... [this example] is simply an embarrassment reminding her of the defects of a model which allows an unbounded speed of propagation. To the numerical analyst it is just a mild warning that the heat equation may present problems which the wave equation does not. But the pure mathematician looks at it with the same simple pleasure with which a child looks at a rose which has just been produced from the mouth of a respectable uncle by a passing magician.

□

Exercises

7.25 Show that the function $E(x,t) = \dfrac{1}{\sqrt{4\pi t}} \exp\left(-\dfrac{x^2}{4t}\right)$ is a solution of the heat equation in the region $t > 0$. What are the initial values as $t \searrow 0$?

7.26 For an infinite rod the units of length x and time t are chosen so that the heat equation takes the form $u_{xx} = u_t$. The temperature at time $t = 0$ is given by the function $e^{-x^2} + e^{-x^2/2}$. Determine the function that describes the temperature at every moment $t > 0$.

7.27 A semi-infinite rod, materialized as the interval $[0, \infty[$, has at time $t = 0$ the temperature e^{x^2} for $0 < x < 1$, 0 for $x > 1$. When $t > 0$, the end point (i.e., the point $x = 0$) is kept at a constant temperature of 0. Determine the temperature for every x at time $t = \frac{1}{4}$.
Hint: define boundary values $f(x)$ for $x < 0$ by $f(x) = -f(-x)$, to make f an odd function. Then solve the problem as if the rod were doubly infinite. Show that this actually gives a solution with the correct boundary values for $x > 0$.

7.28 Find, in the form of an integral, a solution u of $u_{xx} = u_t$ for $t > 0$, such that $u(x, 0) = 1$ if $|x| < 1$, $= 0$ if $|x| > 1$.

7.8 Application 2

We shall treat the following problem for the Laplace differential equation:

$$\begin{cases} u_{xx} + u_{yy} = 0, & x \in \mathbf{R},\ y > 0, \\ u(x,0) = f(x), & x \in \mathbf{R}, \\ \quad u \text{ bounded for } y > 0. \end{cases}$$

Under the additional assumption that, for every fixed y, the function $x \mapsto u(x,y)$ is of class $L^1(\mathbf{R})$, we can Fourier transform the problem with respect to x. Let $U(\omega, y)$ denote this Fourier transform:

$$U(\omega, y) = \mathcal{F}_x[u(x,y)](\omega) = \int_{\mathbf{R}} u(x,y)\, e^{-i\omega x}\, dx.$$

Let us also assume that differentiation with respect to y commutes with Fourier transformation:

$$\mathcal{F}_x[u_{yy}] = \frac{\partial^2}{\partial y^2}\mathcal{F}_x[u].$$

Then the Laplace equation is transformed into

$$-\omega^2 U(\omega, y) + \frac{\partial^2}{\partial y^2} U(\omega, y) = 0.$$

Now we temporarily regard ω as a constant and solve this differential equation with the independent variable y. The general solution is

$$U(\omega, y) = A(\omega) e^{-y\omega} + B(\omega) e^{y\omega}.$$

For U to be bounded for $y > 0$ one must have $A(\omega) = 0$ for $\omega < 0$ and $B(\omega) = 0$ for $\omega > 0$, which means that we can write $U(\omega, y) = C(\omega) e^{-y|\omega|}$, where $C(\omega) = A(\omega) + B(\omega)$. For $y = 0$ we get $U(\omega, 0) = \widehat{f}(\omega) = C(\omega)$, which implies that

$$U(\omega, y) = \widehat{f}(\omega)\, e^{-y|\omega|}. \tag{7.24}$$

By inversion of this Fourier transform one can obtain the desired function u. Using the convolution theorem, we can also establish a solution formula in the form of an integral. Since

$$e^{-y|\omega|} \text{ is the Fourier transform of } \frac{y}{\pi}\,\frac{1}{y^2 + x^2} = P_y(x),$$

it holds that

$$u(x,y) = (P_y * f)(x) = \frac{y}{\pi}\int_{-\infty}^{\infty} \frac{f(t)}{(x-t)^2 + y^2}\, dt. \tag{7.25}$$

This formula is commonly called the *Poisson integral formula* for the half-plane $y > 0$. Indeed, this formula holds under more general conditions than our derivation demands (for instance, it is sufficient to assume that f is continuous and bounded.) The boundary values are right, because the functions $\{P_y\}$ constitute a positive summation kernel, as $y \searrow 0$.

In practice, it may sometimes be easier to invert the Fourier transform (7.24), in other cases it is better to use the integral formula (7.25).

Exercise

7.29 In the unbounded plane sheet $\{(x,y) : y \geq 0\}$ there is a stationary and bounded temperature distribution u. It is known that $u(x,0) = 1/(x^2+1)$. Determine $u(x,y)$ for all $y > 0$.

7.9 Application 3: The sampling theorem

Here we give an important theorem with applications in the technology of sound recording. Let c be a positive number. Assuming that a signal $f(t)$ is built up using angular frequencies ω satisfying $|\omega| \leq c$, it is possible to reconstruct the entire signal by sampling it at discrete time intervals at distance π/c. More precisely, we shall prove the following

Theorem 7.7 (Shannon's sampling theorem) *Suppose that f is continuous on $\mathbf{R}$, that $f \in L^1(\mathbf{R})$ and that $\widehat{f}(\omega) = 0$ for $|\omega| > c$. Then*

$$f(t) = \sum_{n \in \mathbf{Z}} f\Big(\frac{n\pi}{c}\Big) \frac{\sin(ct - n\pi)}{ct - n\pi},$$

where the sum is uniformly convergent on $\mathbf{R}$.

Proof. By the Fourier inversion formula, we have

$$f(t) = \frac{1}{2\pi} \int_{-c}^{c} \widehat{f}(\omega)\, e^{it\omega}\, d\omega. \tag{7.26}$$

We shall rewrite this integral. We introduce a function g as follows:

$$g(\omega) = \frac{c}{\pi}\, \widehat{f}(\omega), \qquad |\omega| < c.$$

This can be considered as a restriction to the interval $(-c, c)$ of a $2c$-periodic function with Fourier series

$$g(\omega) \sim \sum_{n \in \mathbf{Z}} c_n(g) e^{i(n\pi/c)\omega},$$

where

$$c_n(g) = \frac{1}{2c} \int_{-c}^{c} g(\omega)\, e^{-i(n\pi/c)\omega}\, d\omega = \frac{1}{2\pi} \int_{-c}^{c} \widehat{f}(\omega)\, e^{-i(n\pi/c)\omega}\, d\omega = f\Big(-\frac{n\pi}{c}\Big).$$

We also consider the function h given by

$$h(\omega) = e^{-it\omega}, \qquad |\omega| < c.$$

In the same way as for g, we have

$$h(\omega) \sim \sum_{n \in \mathbf{Z}} c_n(h) e^{i(n\pi/c)\omega},$$

with

$$c_n(h) = \frac{1}{2c} \int_{-c}^{c} e^{-it\omega}\, e^{-i(n\pi/c)\omega}\, d\omega = \frac{1}{2c} \left[\frac{e^{-it\omega - i(n\pi/c)\omega}}{-it - i\dfrac{n\pi}{c}} \right]_{\omega=-c}^{\omega=c}$$

$$= \frac{\sin(ct + \pi n)}{ct + \pi n}.$$

We now go back to (7.26) and rewrite it, using the polarized Parseval formula for functions with period $2c$:

$$f(t) = \frac{1}{2c}\int_{-c}^{c} \frac{c}{\pi}\widehat{f}(\omega)\,\overline{e^{-i\omega t}}\,d\omega = \frac{1}{2c}\int_{-c}^{c} g(\omega)\,\overline{h(\omega)}\,d\omega$$

$$= \sum_{n\in\mathbf{Z}} c_n(g)\overline{c_n(h)} = \sum_{n\in\mathbf{Z}} f\Big(-\frac{n\pi}{c}\Big)\overline{\frac{\sin(ct+\pi n)}{ct+\pi n}} = \sum_{n\in\mathbf{Z}} f\Big(\frac{n\pi}{c}\Big)\frac{\sin(ct-\pi n)}{ct-\pi n}.$$

The convergence of the series is clear, since both g and h are L^2 functions. Indeed, the convergence of symmetric partial sums $s_N = \sum_{-N}^{N}$ is uniform in t, because estimates of the remainder are uniform. The theorem is proved. □

Remark. The theorem explains why CD recordings and DAT tapes are possible. The human ear cannot hear sounds with a frequency above, say, 20 kHz. The sound signal can thus be considered to have its frequency spectrum totally within this range. If it is sampled at sufficiently small intervals, and if the sampling is precise enough, it is then possible to recover the sound from the digitalized sample record. Ordinary CD recorders use a sampling frequency of 44.1 kHz. □

7.10 *Connection with the Laplace transform

In this section we return to the Laplace transform. We also assume that the reader has some knowledge of complex analysis, in particular the theory of residues. We shall demonstrate how the Laplace transform can be considered as a special case of the Fourier transform.

Assume that $f(t)$ is defined and piecewise continuous for $t \in \mathbf{R}$ and that $f(t) = 0$ for $t < 0$. Also suppose that there exist constants t_0, M, and k so that $|f(t)| \le Me^{kt}$ for all $t > t_0$: we say that f grows (at most) exponentially. Let $s = \sigma + i\omega$ be a complex variable, where σ and ω are real. The Laplace transform of f is defined as the function

$$\widetilde{f}(s) = \int_0^{\infty} f(t)e^{-st}\,dt = \int_{-\infty}^{\infty} f(t)e^{-\sigma t}\,e^{-i\omega t}\,dt. \tag{7.27}$$

The integral converges absolutely as soon as $\sigma > k$, where k is the number introduced just above. The integral then defines a function $\widetilde{f}$, which is analytic at least in the half-plane $\sigma > k$. This can be seen by noting that the functions

$$F_n(s) = \int_0^{n} f(t)\,e^{-st}\,dt$$

are analytic and that $F_n \to \widetilde{f}$ uniformly in every interior half-plane $\sigma \ge \sigma_0$, where $\sigma_0 > k$. The details are omitted here.

What is now of interest is the fact that the formula (7.27) shows that the Laplace transform of f can be seen as the Fourier transform of the function $f(t)e^{-\sigma t}$. If we assume that f is such that the Fourier inversion formula can be applied, we can then write

$$f(t)e^{-\sigma t} = \frac{1}{2\pi} \lim_{A\to\infty} \int_{-A}^{A} \widetilde{f}(\sigma + i\omega)\, e^{it\omega}\, d\omega.$$

If this equality is multiplied by $e^{\sigma t}$ and we reintroduce $\sigma + i\omega = s$, we get

$$f(t) = \frac{1}{2\pi i} \lim_{A\to\infty} \int_{\sigma - iA}^{\sigma + iA} \widetilde{f}(s) e^{ts}\, ds, \tag{7.28}$$

where the notation $i\, d\omega = ds$ serves to indicate that the integral is a contour integral in the complex plane. The contour is a vertical line in the half-plane $\sigma > k$.

Since we have assumed that $f(t) = 0$ for $t < 0$, the integral in (7.28) will always be zero for negative values of t. For t positive, it can sometimes be calculated using residues and a half-circular contour to the left of the vertical line. We demonstrate this by an example.

Example 7.13. Find the function whose Laplace transform is $\widetilde{f}(s) = 1/(s^2 + 1)$.

Solution. We want to compute

$$\frac{1}{2\pi i} \lim_{A\to\infty} \int_{\sigma - iA}^{\sigma + iA} \frac{e^{ts}}{s^2 + 1}\, ds.$$

The integrand has simple poles at $s = \pm i$ and is analytic in the rest of the s-plane. We can choose $\sigma = 1$, say, and make a closed contour by adjoining C_A, the left-hand half of the circle $|s - 1| = A$. Taking account of the factor $1/(2\pi i)$ in the formula, the integral over the closed contour is the sum of the residues:

$$\operatorname*{Res}_{s=i} \frac{e^{ts}}{s^2+1} = \lim_{s\to i} \frac{e^{ts}}{s+i} = \frac{e^{it}}{2i}, \qquad \operatorname*{Res}_{s=-i} \frac{e^{ts}}{s^2+1} = \lim_{s\to -i} \frac{e^{ts}}{s-i} = \frac{e^{-it}}{-2i},$$
$$\sum \text{Res} = \frac{e^{it} - e^{-it}}{2i} = \sin t.$$

The integral along the circular arc can be estimated, using the fact that $|s| = |s - 1 + 1| \geq |s - 1| - 1 = A - 1$ and $\sigma = \operatorname{Re} s \leq 1$:

$$\left| \frac{1}{2\pi i} \int_{C_A} \frac{e^{st}}{s^2+1}\, ds \right| \leq \frac{1}{2\pi} \int_{C_A} \frac{e^{\sigma t}}{|s|^2 - 1}\, |ds| \leq \frac{e^t}{2\pi} \int_{C_A} \frac{|ds|}{(A-1)^2 - 1}$$
$$= \frac{e^t}{2\pi} \cdot \frac{A\pi}{(A-1)^2 - 1} \to 0 \quad \text{as } A \to \infty.$$

The conclusion is that the desired limit, which is $f(t)$ for $t > 0$, is $\sin t$. □

Exercises

7.30 Check that the integral in Example 7.13 is zero for $t < 0$. What is its value for $t = 0$?

7.31 Find the inverse Laplace transform of $s/(s^2+1)$ by the method of Example 7.13. What is the value of the integral for $t = 0$?

7.32 Example 7.13 can be generalized thus: Suppose that $\widetilde{f}(s)$ is analytic in the entire s-plane except for a finite number of isolated singularities, that $\widetilde{f}$ is analytic in $\operatorname{Re} s > k$, and that there is an estimate of the form $|\widetilde{f}(s)| \leq M/|s|^\alpha$ for $|s| \geq R$, where $\alpha > 0$. Then, for $t > 0$, $f(t) =$ the sum of all the residues for the function $e^{ts}\widetilde{f}(s)$. Prove this (or "check" it)!

7.11 *Distributions and Fourier transforms

We shall now see what happens if we try to obtain Fourier transforms of simple distributions such as those considered in Sec. 2.6–7. At this point, we treat only expressions such as $\delta_a^{(n)}(t)$. The more revolutionary aspects of the theory are postponed to Chapter 8.

It is rather obvious what the indicated transforms should be. We recall that $\delta_a^{(n)}(t)$ was defined so as to have the following effect on a sufficiently smooth function φ:

$$\int \delta_a^{(n)}(t)\varphi(t)\,dt = (-1)^n\,\varphi^{(n)}(a).$$

In particular, if we take $\varphi(t) = e^{-i\omega t}$, we should have

$$\widehat{\delta_a^{(n)}}(\omega) = \int \delta_a^{(n)}(t)e^{-i\omega t}\,dt = (-1)^n(-i\omega)^n e^{-i\omega a} = (i\omega)^n e^{-ia\omega}.$$

As a special case, we have that $\widehat{\delta}(\omega) = 1$ for all ω. A physical interpretation of this is that the δ "function" is composed out of *all* frequencies with equal amplitudes (or, rather, equal amplitude density). This kind of signal is sometimes given the name "white noise." If the situation is interpreted literally, it means that δ has *infinite energy*, and thus it cannot be realized in physical reality. However, it can be treated as a formalism that turns out to be useful.

The *convolution* of δ and a (smooth) function φ should reasonably work like this:

$$\delta * \varphi(t) = \int_{\mathbf{R}} \delta(t-u)\varphi(u)\,du = \varphi(t).$$

Thus, δ acts as an algebraic identity with respect to the convolution operation. Let us also accept that the convolution operation is commutative and associative, even when one of the objects involved is a δ.

If the functions involved are sufficiently nice, a convolution can be differentiated past the integral sign. The result is the simple formula

$$(f * g)' = f' * g = f * g'.$$

The applications in Sec. 7.7–8 can then be attacked along the following lines. Let us first study the heat equation, and assume that the initial values are δ:

$$u_{xx} = u_t, \quad x \in \mathbf{R},\ t > 0; \qquad u(x,0) = \delta(x), \quad x \in \mathbf{R}.$$

Fourier transformation gives, just as in Sec. 7.7,

$$-\omega^2 U = \frac{dU}{dt}, \quad \omega \in \widehat{\mathbf{R}},\ t > 0; \qquad U(\omega,0) = 1, \quad \omega \in \widehat{\mathbf{R}}.$$

Solving this differential equation we get

$$U(\omega,t) = e^{-\omega^2 t},$$

and inverting this Fourier transform gives

$$u(x,t) = E(x,t) = \frac{1}{\sqrt{4\pi t}} \exp\left(-\frac{x^2}{4t}\right).$$

Now assume that f is some (continuous) function, not necessarily in $L^1(\mathbf{R})$, and consider the general initial value problem:

$$u_{xx} = u_t, \quad x \in \mathbf{R},\ t > 0; \qquad u(x,0) = f(x), \quad x \in \mathbf{R}.$$

If we put $u = E * f$, we will now have a solution of this problem. Indeed, since E satisfies the heat equation, it is clear that

$$u_{xx} - u_t = \frac{\partial^2}{\partial x^2} E * f - \frac{\partial}{\partial t} E * f = E_{xx} * f - E_t * f = (E_{xx} - E_t) * f = 0 * f = 0.$$

And since $E \to \delta$ as $t \searrow 0$, it should also follow that the boundary values are right:

$$u(x,t) = E(x,t) \underset{x}{*} f(x) \to \delta * f(x) = f(x).$$

We also give an example where we use Fourier transformation to solve an ordinary differential equation.

Example 7.14. Find a solution of the equation $y''(t) - y(t) = \delta(t)$.

Solution. Fourier transformation, and using the rule for the transform of a derivative, gives

$$(i\omega)^2 \widehat{y}(\omega) - \widehat{y}(\omega) = 1.$$

If we solve for $\widehat{y}$, we find $\widehat{y}(\omega) = -1/(1+\omega^2)$, which is a well-known transform: apparently $y(t) = -\frac{1}{2}\, e^{-|t|}$. And, indeed, if we look at this function, we see that $y''(t) = y(t)$ for all $t \neq 0$; and at $t = 0$, the first derivative has an upward jump of one unit, which confirms that we have found a solution.

The reader should recognize this type of equation. Its homogeneous counterpart has the solutions $y_H = Ae^t + Be^{-t}$, and thus the given equation has the general solution $y = -\frac{1}{2}\, e^{-|t|} + y_H$. We seem to have lost all these solutions except for one. This depends on the fact that all the others actually cannot be Fourier transformed at all, not even as distributions. But as a means of finding a particular solution, the method obviously works in this case. □

Exercise

7.33 Find a solution of the equation $y''(t) + 3y'(t) + 2y(t) = \delta(t)$.

Summary of Chapter 7

Definition

If $f \in L^1(\mathbf{R})$, the Fourier transform of f is the function $\mathcal{F}[f] = \widehat{f}$ given by

$$\widehat{f}(\omega) = \int_{\mathbf{R}} f(t) e^{-i\omega t}\, dt, \quad \omega \in \widehat{\mathbf{R}}.$$

Theorem

If $f \in L^1(\mathbf{R})$, then $\widehat{f}$ is a continuous function on $\widehat{\mathbf{R}}$ that tends to 0 as $|\omega| \to \infty$.

Theorem

(Inversion theorem) Suppose that $f \in L^1(\mathbf{R})$, that f is continuous except for a finite number of finite jumps in any finite interval, and that $f(t) = \frac{1}{2}(f(t+) + f(t-))$ for all t. Then

$$f(t_0) = \lim_{A \to \infty} \frac{1}{2\pi} \int_{-A}^{A} \widehat{f}(\omega)\, e^{i\omega t_0}\, d\omega$$

for every t_0 where f has (generalized) left and right derivatives. In particular, if f is piecewise smooth (i.e., continuous and with a piecewise continuous derivative), then the formula holds for all $t_0 \in \mathbf{R}$.

A collection of formulae for the Fourier transform begins on page 252.

Historical notes

The Fourier transform, or Fourier integral, first appears in the works of FOURIER himself. Its development has run parallel to that of Fourier series ever since. Recent developments in signal processing have triggered the result known as the sampling theorem, which is attributed to CLAUDE SHANNON (1916–2001). He was the founder of information theory.

Problems for Chapter 7

7.34 Compute the Fourier transform of f, defined by

$$f(x) = \begin{cases} 2 - |x|, & |x| < 2, \\ 0, & |x| > 2. \end{cases}$$

Use the result to compute $\displaystyle\int_{-\infty}^{\infty} \left(\frac{\sin t}{t}\right)^4 dt.$

7.35 The function f is continuous on $\mathbf{R}$, and both f and f' belong to $L^1(\mathbf{R})$. Compute the Fourier transform of f, if $2f(x) - f(x+1) + f'(x) = \exp(-|x|)$, $x \in \mathbf{R}$.

7.36 Find the Fourier transform of $\dfrac{1}{x^2 + 4x + 13}$.

7.37 Find the Fourier transforms of (a) $\dfrac{1}{1+9x^2}$, (b) $\dfrac{e^{ix}}{1+9x^2}$, (c) $\dfrac{\sin x}{1+9x^2}$.

7.38 Find the Fourier transform of $f(x) = \dfrac{x}{(1+x^2)^2}$.

7.39 A function f is defined by $f(x) = 2x$ if $|x| < 1$, $f(x) = 0$ otherwise.
(a) Compute the Fourier transform $\widehat{f}$ of f.
(b) For all $x \in \mathbf{R}$, determine $\displaystyle\lim_{a \to \infty} \int_{-a}^{a} \widehat{f}(t)\, e^{itx}\, dt$.
(c) Compute $\displaystyle\int_{-\infty}^{\infty} \left(\frac{\sin t}{t^2} - \frac{\cos t}{t}\right)^2 dt.$

7.40 Let $f_N(t) = \pi N^2$ for $0 < t < 1/N$, $f(t) = 0$ for $t > 1/N$, and define $f_N(t)$ for $t < 0$ through the condition that f_N be an odd function. Compute the Fourier transform of f_N, and then find $\displaystyle\lim_{N \to \infty} \widehat{f_N}(\omega)$.

7.41 Find the Fourier transform of f, when $f(t) = \sin t$ for $|t| < \pi$, $= 0$ for $|t| \geq \pi$, and use the result to compute

$$\int_{-\infty}^{\infty} \frac{\sin^2 \pi t}{(t^2 - 1)^2}\, dt.$$

7.42 Show that $\displaystyle\int_{-\infty}^{\infty} \left(\frac{\sin x}{x}\right)^2 dx = \pi$ by transforming the function

$f(x) = \begin{cases} 1, & |x| \leq 1, \\ 0, & |x| > 1 \end{cases}$ and using the Plancherel formula.

7.43 Let $f(t) = 1 - t^2$ for $|t| < 1$, $= 0$ otherwise. Find $\widehat{f}$ and use the result to find the values of the integrals

$$\int_{-\infty}^{\infty} \frac{\sin t - t\cos t}{t^3}\, dt \quad \text{and} \quad \int_{-\infty}^{\infty} \frac{(\sin t - t\cos t)^2}{t^6}\, dt.$$

7.44 Compute the integral

$$F(\omega) = \int_{-\infty}^{\infty} \frac{\sin \alpha}{\alpha(1 + (\omega - \alpha)^2)}\, d\alpha, \quad -\infty < \omega < \infty.$$

7.45 Solve the integral equation

$$\int_{-\infty}^{\infty} \frac{f(y)}{(x-y)^2+1}\,dy = \frac{a}{a^2+x^2}, \quad x \in \mathbf{R}$$

using Fourier transforms. Conditions on a?

7.46 (a) Let $f(x) = e^{-|x|}$. Find $f * f(x) = \displaystyle\int_{-\infty}^{\infty} f(y)\, f(x-y)\,dy$.

(b) Using Fourier transforms, solve the differential equation

$$y''(x) - y(x) = e^{-|x|} \quad x \in \mathbf{R}.$$

7.47 Let the signal $f(t) = A_1 \cos(\omega_1 t + \theta_1) + A_2 \cos(\omega_2 t + \theta_2)$ be given. Compute

$$r_{xx}(t) = \lim_{T\to\infty} \frac{1}{T} \int_{-T/2}^{T/2} \overline{f(u)}\, f(t+u)\,du$$

(the auto-correlation function, ACF). Try to find the frequency spectrum $P_{xx}(\omega)$ of r_{xx} (this is called the energy spectrum of f; the connection between P_{xx} and r_{xx} is called the Wiener–Khinchin relations). To determine P_{xx} correctly, you should know something about distributions (e.g., the Dirac measure δ).

7.48 Suppose that a certain linear system transforms an incoming signal f into an outgoing signal y that is a solution of

$$y''(t) + ay'(t) + by(t) = f(t),$$

where a and b are constants. Show that if the roots of the characteristic equation $r^2 + ar + b = 0$ both have their real parts < 0, then the system is *causal;* i.e., the value $y(t)$ at any time t depends only on the values of $f(u)$ for $u \le t$.

7.49 Find all functions $f \in L^1(\mathbf{R})$ that satisfy the integral equation

$$\int_{-\infty}^{\infty} f(t-y)\, e^{-|y|}\,dy = e^{-t^2/2}, \quad t \in \mathbf{R}.$$

7.50 Find the Fourier transform of the function $f(x) = x\, e^{-|x|}$.

7.51 Determine the Fourier transform of

$$f(x) = \begin{cases} e^{-x}, & x > 0, \\ -e^{x}, & x < 0, \end{cases}$$

and then compute the integral $\displaystyle\int_{-\infty}^{\infty} \frac{x^2}{(1+x^2)^2}\,dx$.

7.52 Find a solution of the integral equation

$$\int_{-\infty}^{\infty} f(x-y)\, e^{-|y|}\,dy = (1+|x|)\, e^{-|x|}.$$

7.53 Find a solution of the integral equation

$$f(x) = e^{-|x|} + \tfrac{1}{2}\,e^{x} \int_{x}^{\infty} e^{-y} f(y)\,dy, \quad -\infty < x < \infty.$$

7.54 Using Fourier transformation, find a solution of the integral equation

$$e^{-x^2} = \int_{-\infty}^{\infty} e^{-4(x-y)^2} f(y)\,dy.$$

7.55 Compute $\displaystyle\int_{-\infty}^{\infty} \frac{\sin t}{t} \cdot \frac{e^{itx}}{1+t^2}\,dt$ for $-1 < x < 1$. Be careful about the details!

8

Distributions

8.1 History

In star-marked sections in the previous chapters we have sketched how it is possible to extend the notion of function to include things such as "instantaneous pulses" and similar phenomena. The present chapter will present a more coherent introduction to these *distributions*. The presentation is biased in the way that it centers on the kind of distribution theory that appears to be natural in connection with Fourier theory, and it is not very far-reaching. A complete study of the theory of distributions is beyond the intended scope of this book.

The reader should be able to study this chapter without having read the starred sections in the former chapters. This means that there is a certain duplication of examples, etc. This applies in particular to this introductory section, where a number of the following examples are repetitions of things that are also found in Sec. 2.6.

We are going to indicate a number of more-or-less puzzling difficulties that had vexed mathematicians for a long time. Various ways of going around the problems were suggested, until at last time was ripe, in the 1930s and 1940s, for the modern theory that we shall touch upon in this chapter.

Example 8.1. Already in Sec. 1.3 (on the wave equation) we saw difficulties in the usual demand that solutions of a differential equation of order n shall actually have (maybe even continuous) derivatives of order n. Quite natural solutions, such as those of Exercise 6.7, get disqualified for reasons that seem more of a "bureaucratic" than physical nature. This indicates that it would be a good thing to widen the notion of differentiability in one way or another. □

Example 8.2. Ever since the days of NEWTON, physicists have been dealing with situations where some physical entity assumes a very large magnitude during a very short period of time; often this is idealized so that the value is infinite at one point in time. A simple example is an elastic collision of two bodies, where the forces are thought of as infinite at the moment of impact. Nevertheless, a *finite* and well-defined amount of impulse is transferred in the collision. How is this to be treated mathematically? □

Example 8.3. A situation that is mathematically analogous to the previous one is found in the theory of electricity. An electron is considered (at least in classical quantum theory) to be a *point charge.* This means that there is a certain finite amount of electric charge localized at one point in space. The charge density is infinite at this point, but the charge itself has an exact, finite value. What mathematical object describes this? □

Example 8.4. In Sec. 2.4 we studied positive summation kernels. These consist of sequences of non-negative functions with integral equal to 1, that concentrate toward a fixed point, as a parameter tends to infinity. Can we invent a mathematical object that can be interpreted as the limit of such a sequence? □

Example 8.5. There is also a sort of inverted problem, compared with the ones in Examples 8.2–3 above. Suppose that we want to measure a physical quantity $f(t)$, that depends on time t. Is it really possible to determine $f(t)$ at a particular point in time? Every measurement takes some time to perform. A speedometer, for example, must be constructed so that it deals with time intervals of positive length, and the value indicated by it is necessarily some kind of mean value of the speed attained during the latest period of time. HEISENBERG's undecidedness principle actually tells us that certain types of measurement cannot be exact *at all*; the best we can hope for is to get some mean value. □

Example 8.6. In Sec. 7.7 and 7.8 we solved a couple of problems for partial differential equations using Fourier transformation. In order to be able to use this method we had to impose rather restrictive conditions on the solutions — they had to be integrable in a certain way, and differentiability past an integral sign had to be explicitly assumed. But in both of these cases, the result of the calculations was a formula that was actually valid in far more general situations than those demanded by the method. Is there something going on behind the stage, that we could drag out in clear view and enable us to do our Fourier transformations without hesitations and bad conscience? □

The problems in Examples 8.2 and 8.3 above have been treated by many physicists ever since the later years of the nineteenth century by using the following trick. Let us assume that the independent variable is t. Introduce a "function" $\delta(t)$ with the following properties:

$$\begin{aligned}
&(1) \quad \delta(t) \geq 0 \text{ for } -\infty < t < \infty, \\
&(2) \quad \delta(t) = 0 \text{ for } t \neq 0, \\
&(3) \quad \int_{-\infty}^{\infty} \delta(t)\, dt = 1.
\end{aligned}$$

Regrettably, there is no ordinary real-(or complex)-valued function that satisfies these conditions. Condition 2 irrevocably implies that the integral in condition 3 must be zero. Nevertheless, using formal calculations involving the object δ, it was possible to arrive at results that were both physically meaningful and "correct." A name that is commonly associated with this is P. Dirac, but he was not the only person (nor even the first one) to reason in this way. He has, however, given his name to the object δ: it is often called the *Dirac delta function* (or the Dirac measure, or the Dirac distribution).

One way of making legitimate the formal δ calculus is to follow the idea that is hinted at in Example 8.4. If δ occurs in a formula, it is replaced by a positive summation kernel K_N; upon this one then does one's calculations, and finally one passes to the limit. In a certain sense (which will be made precise at the end of Sec. 8.4), it is true that $\delta = \lim_{N\to\infty} K_N$.

The problems presented by Example 8.5 can be tackled in a very unconventional way. We simply give up the requirement (or aspiration) that the "function" f actually does possess any values $f(t)$ at all at precise points t. Instead, it is thought to have values at "fuzzy points"; i.e., it is possible to account for (weighted) means of f over intervals of positive length, these means being real or complex numbers.

Our strategy now is the following. First we make precise what we mean by a fuzzy point (in Sec. 8.2). Then, in Sec. 8.3, we define our generalization of the notion of a function, which will be called a distribution. After that, we build the machinery that will lead to the solution of the dilemmas presented as Examples 8.1–6 (including numbers 1 and 6, which have not been more closely examined in the present discussion).

The whole exposition is very sketchy. For a more complete theory we refer to more penetrating literature, such as the monumental standard work by Lars Hörmander, *The Analysis of Linear Partial Differential Operators*, Volume 1.

8.2 Fuzzy points – test functions

Here we introduce so-called test functions, which shall serve, among other things, as the "fuzzy points" mentioned at the end of last section. We do it in one dimension, but the whole theory is easily redone in an arbitrary finite dimension.

Thus, we consider a real x-axis denoted by $\mathbf{R}$, as usual. A *test function* is an infinitely differentiable complex-valued function, $\varphi : \mathbf{R} \to \mathbf{C}$, $\varphi \in C^\infty(\mathbf{R})$. In these connections one normally uses a more concise notation for the last-mentioned set: we write $\mathcal{E} = C^\infty(\mathbf{R})$.

We shall define two important subsets of $\mathcal{E}$. First we define the *support* of a function φ:

$$\text{support of } \varphi = \operatorname{supp} \varphi = \overline{\{x \in \mathbf{R} : \varphi(x) \neq 0\}}.$$

Thus, a point x belongs to $\operatorname{supp}\varphi$ if every neighborhood of x contains points where $\varphi(x) \neq 0$; the fact that x *is not* in the support of φ means that $\varphi(y) = 0$ for all y in some neighbourhood of x. The support is always a closed set.

Example 8.7. If $\varphi(x) = 1 - x^2$ for $|x| < 1$ and $\varphi(x) = 0$ for $|x| \geq 1$, then $\operatorname{supp}\varphi = [-1, 1]$. □

Example 8.8. If φ is defined by $\varphi(x) = \sin x$ for $|x| < \pi$ and 0 elsewhere, then $\operatorname{supp}\varphi = [-\pi, \pi]$. Although $\varphi(0) = 0$, the point $x = 0$ belongs to the support, because every neighborhood of this point contains points where $\varphi(x) \neq 0$. □

Saying that $\operatorname{supp}\varphi$ is *compact* means that $\varphi(x) = 0$ outside of some compact interval. The set of test functions on $\mathbf{R}$ with compact support is denoted by $\mathcal{D}$. The fact that such functions exist may not appear obvious to some readers, but indeed there are a wealth of them. If the reader is not prepared to accept this fact, some examples are constructed at the end of this section.

We shall also introduce a class of test functions situated between $\mathcal{D}$ and $\mathcal{E}$, which is the most important class in our exposition of the theory.

Definition 8.1 *We say that a function φ belongs to the* Schwartz class $\mathcal{S}$ *if φ has derivatives of all orders and these satisfy inequalities of the form*

$$(1 + |x|)^n |\varphi^{(k)}(x)| \leq C_{n,k}, \quad x \in \mathbf{R},$$

where $C_{n,k}$ are constants, for all integers n and k that are ≥ 0.

The import of the definition is that φ and all its derivatives tend to zero as $x \to \pm\infty$ faster than the inverted value of every polynomial. In particular, $\mathcal{D} \subset \mathcal{S}$, but, in addition to this, $\mathcal{S}$ contains the function $\varphi(x) = e^{-x^2}$ (and lots of others). The class $\mathcal{S}$ is named after Laurent Schwartz, who founded the theory of distributions in the 1940s.

In connection with $\mathcal{S}$, there is sometimes reason to talk about *moderately increasing* or *tempered* functions. These are functions χ that increase at most as fast as some polynomial, i.e., there are constants C and m such that $|\chi(x)| \le C(1+|x|)^m$ for all x. If χ belongs to $\mathcal{E}$, if χ and all its derivatives are tempered, and if $\varphi \in \mathcal{S}$, then the product $\chi\varphi$ will also belong to $\mathcal{S}$. Such a function χ is called a *multiplicator* (more precisely, a multiplicator on the space $\mathcal{S}$). The set of these multiplicators has no generally accepted notation, but we shall occasionally use the letter $\mathcal{M}$ for it.

The sets $\mathcal{D}$, $\mathcal{S}$ and $\mathcal{E}$ are vector spaces of infinite dimension. We want to be able to speak about convergence of sequences of elements in these spaces. In Chapter 5 we saw examples of how convergence could be defined by referring to various *norms*. The kind of convergence that we want now is more complicated to describe. Since our interest will be centered on the space $\mathcal{S}$, we content ourselves with defining convergence in this space.

Definition 8.2 *A sequence $\{\varphi_j\}_{j=1}^{\infty} \subset \mathcal{S}$ is said to converge in $\mathcal{S}$ to a function $\psi \in \mathcal{S}$, if for all integers $n \ge 0$ and $k \ge 0$ it holds that*

$$\lim_{j\to\infty} \max_{x\in\mathbf{R}} (1+|x|)^n |\varphi_j^{(k)}(x) - \psi^{(k)}(x)| = 0.$$

Thus, the functions φ_j and all their derivatives are to converge toward ψ and the respective derivatives of ψ in such a manner that the convergence is uniform even after multiplication by arbitrary polynomials in the variable x. This is quite a restrictive notion of convergence, but it turns out to be "correct" for our future needs. We write

$$\varphi_j \xrightarrow{\mathcal{S}} \psi \quad \text{as } j \to \infty.$$

Remark. The spaces $\mathcal{E}$, $\mathcal{S}$, and $\mathcal{D}$ contain many more functions than those that justify the name "fuzzy points." It is, however, desirable to be able to work with *linear spaces* of test functions: and if such a space contains "fuzzy points" around all points x of the real axis, it will automatically contain a great number of functions that are not particularly "localized" (since a linear space is closed under addition and multiplication by scalars). For this reason, it is just as well to define test functions in the more generous way that we have done. □

The fact that there actually exist test functions with compact support may not be obvious to everybody. We shall give a few examples to show that this is actually the case. First we prove a lemma (which is often taken for granted by students without really thinking).

Lemma 8.1 *Suppose that f is continuous in $[a,b]$ and differentiable in $]a,b[$, and suppose that the derivative $f'(x)$ has a limit A, as $x \searrow a$. Then f has a right-hand derivative for $x = a$, and this derivative has the value A.*

Proof. The mean-value theorem of Lagrange can be used on the subinterval $[a, a+h]$, where $h > 0$:

$$f(a+h) - f(a) = f'(\xi)\cdot h \qquad \Longleftrightarrow \qquad \frac{f(a+h)-f(a)}{h} = f'(\xi),$$

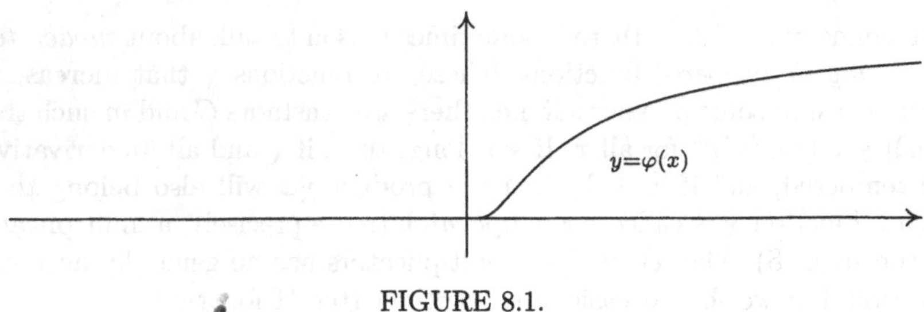

FIGURE 8.1.

where $a < \xi < a+h$. If we let $h \searrow 0$, the right-hand member of the last equation tends to A, by the assumption; thus, the left-hand member also has the limit A, and the limit of the left-hand member is, by definition, the right-hand derivative of f at a. □

Example 8.9. First define a function φ by putting

$$\varphi(x) = \begin{cases} 0, & x \leq 0; \\ e^{-1/x}, & x > 0. \end{cases}$$

The substitution $1/x = t$ and letting $t \to +\infty$, corresponding to $x \searrow 0$, shows that φ is continuous also at $x = 0$. For $x > 0$, a few differentiations give

$$\varphi'(x) = e^{-1/x} \cdot \frac{1}{x^2}, \; \varphi''(x) = e^{-1/x}\left(\frac{1}{x^4} - \frac{2}{x^3}\right), \; \varphi'''(x) = e^{-1/x}\left(\frac{1}{x^6} - \frac{6}{x^5} + \frac{6}{x^4}\right).$$

From this, one should realize that all the derivatives will have the form $e^{-1/x}$ multiplied by a polynomial in the variable $1/x$. The limit as $x \searrow 0$ will in all cases be 0. According to the lemma, φ has then right-hand derivatives of all orders equal to 0 at the origin. The left-hand derivatives are also 0, trivially. This means that φ is indefinitely differentiable everywhere and its support is the interval $[0, \infty[$. See Figure 8.1! Now define $\psi(x) = \varphi(x)\,\varphi(1-x)$. This is a C^∞ function with support $[0, 1]$. □

For the ψ of the example we thus have $\psi \in \mathcal{D}$. Other elements of $\mathcal{D}$ can be constructed by translations, dilatations, multiplication by arbitrary functions in $\mathcal{E}$, addition, etc. In fact, $\mathcal{D}$ is quite a rich space.

The function ψ, after division by the number $B = \int_0^1 \psi(x)\,dx$, can be interpreted as a "fuzzy point," localized around $x = \frac{1}{2}$. If we translate the localization to the origin by forming $\omega(x) = \psi(x - \frac{1}{2})/B$ and then re-scale according to the model $n\omega(nx)$, we do indeed obtain a positive summation kernel as in Sec. 2.4, which becomes less and less fuzzy as n increases.

Example 8.10. With ψ and B as before, let

$$\Psi(x) = \frac{1}{B}\int_{-\infty}^{x} \psi(y)\,dy.$$

This gives a function in $\mathcal{E}$, having the value 0 for $x \leq 0$ and 1 for $x \geq 1$. Then put $\Omega(x) = \Psi(x) - \Psi(x-1)$. Then $\Omega \in \mathcal{D}$, with support $[0, 2]$. Furthermore, Ω

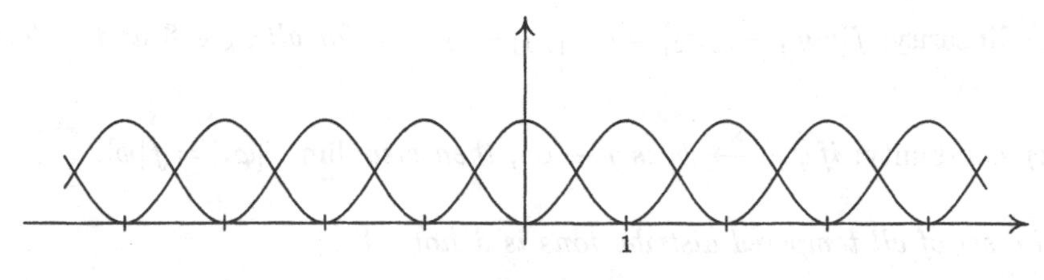

FIGURE 8.2.

has the following property:

$$\Phi(x) \equiv \sum_{n=-\infty}^{\infty} \Omega(x-n) = 1 \quad \text{for all } x \in \mathbf{R}.$$

This can be shown in the following way (compare Figure 8.2). For any fixed x, at most two terms in the sum are different from zero, and so the series is very much convergent. Furthermore, it is easy to see that $\Phi(x+1) = \Phi(x)$, so that Φ has period 1. Thus we can restrict our study to the interval $1 \le x < 2$. For these x, the sum reduces to the terms $\Omega(x) + \Omega(x-1)$, and we get

$$\begin{aligned}\Phi(x) &= \Omega(x) + \Omega(x-1) = \Psi(x) - \Psi(x-1) + \Psi(x-1) - \Psi(x-2) \\ &= \Psi(x) - \Psi(x-2) = 1 - 0 = 1.\end{aligned}$$

□

Example 8.10 shows that the function that is identically 1 for all real x can be decomposed as a sum of infinitely differentiable functions with compact supports. Such a representation of 1 is called a *partition of unity.*

Exercise

8.1 Show that if $\varphi \in \mathcal{S}$, then also $\varphi' \in \mathcal{S}$. Is the converse true: i.e., must an antiderivative of a test function in $\mathcal{S}$ also be in the same set?

8.3 Distributions

Distributions are mappings that assign to every test function in some space a complex number. If f is a distribution, we denote this value for a test function φ by writing $f[\varphi]$, using square brackets. (In literature one often sees the notation $\langle f, \varphi \rangle$, but we shall avoid this, since it does not completely share the properties of the inner product in previous chapters.) Depending on which space of test functions one chooses, one gets different classes of distributions. In connection with Fourier analysis, it turns out that the space $\mathcal{S}$ is the most natural one. The distributions belonging to these test functions are called *tempered distributions.*

Definition 8.3 *A tempered distribution f is a mapping $f : \mathcal{S} \to \mathbf{C}$ having the properties*

(1) linearity: $f[c_1\varphi_1 + c_2\varphi_2] = c_1 f[\varphi_1] + c_2 f[\varphi_2]$ *for all* $\varphi_k \in \mathcal{S}$ *and scalars* c_k;

(2) continuity: *if* $\varphi_j \xrightarrow{\mathcal{S}} \psi$ *as* $j \to \infty$, *then also* $\lim_{j\to\infty} f[\varphi_j] = f[\psi]$.

The set of all tempered distributions is denoted $\mathcal{S}'$.

We give some examples.

Example 8.11. Let f be a continuous function on $\mathbf{R}$ such that there are constants M and m such that $|f(x)| \le M(1+|x|)^m$ for all x. Then we can define a tempered distribution T_f by letting

$$T_f[\varphi] = \int_{\mathbf{R}} f(x)\,\varphi(x)\,dx.$$

It is clear that T_f is a linear mapping; the fact that it is continuous follows from the fact that convergence in $\mathcal{S}$ implies uniform convergence even after multiplication by expressions such as $(1+|x|)^{m+2}$:

$$|T_f[\varphi_j] - T_f[\psi]| = \left|\int_{\mathbf{R}} f(x)(\varphi_j(x) - \psi(x))\,dx\right|$$

$$\le \int_{\mathbf{R}} M(1+|x|)^m|\varphi_j(x) - \psi(x)|\,dx$$

$$= \int_{\mathbf{R}} M(1+|x|)^{m+2}|\varphi_j(x) - \psi(x)| \cdot \frac{dx}{(1+|x|)^2}$$

$$\le M \max_{x\in\mathbf{R}} \{(1+|x|)^{m+2}|\varphi_j(x) - \psi(x)|\} \cdot \int_{\mathbf{R}} \frac{dx}{(1+|x|)^2} \to 0$$

as $j \to \infty$. It is customary to identify the distribution T_f with the function f and write $f[\varphi]$ instead of $T_f[\varphi]$. In this way, every continuous, moderately increasing function can be considered to be a distribution, and it is in this way that distributions can be seen as generalizations of ordinary functions. □

It is not necessary that the f in Example 8.11 be continuous. It is sufficient that it is locally integrable, i.e., that $\int_a^b |f(x)|\,dx$ exists and is finite for all compact intervals $[a,b]$; on the other hand, it cannot be allowed to grow too fast as $|x| \to \infty$. An important distribution of this type is exhibited in the next example:

Example 8.12. Let H be the *Heaviside function*, defined by

$$H(x) = 0 \text{ if } x < 0, \quad H(x) = 1 \text{ if } x > 0.$$

(The value of $H(0)$ is immaterial.) This defines a distribution by the formula

$$H[\varphi] = \int_{\mathbf{R}} H(x)\,\varphi(x)\,dx = \int_0^\infty \varphi(x)\,dx. \qquad □$$

Example 8.13. Define $\delta \in \mathcal{S}'$ by $\delta[\varphi] = \varphi(0)$. (The reader is asked to check linearity and continuity.) This distribution is called the Dirac distribution (Dirac function, Dirac measure), and this is the object announced in Examples 8.2 and 8.3. □

If f is an arbitrary distribution, as a rule there exists no "value at the point x" to be denoted by $f(x)$. In spite of this, one often writes the effect of f on a test function φ as an integral:

$$f[\varphi] = \int_{\mathbf{R}} f(x)\,\varphi(x)\,dx.$$

This is indeed a very symbolic notation, but, nevertheless, it turns out to be quite useful. The whole theory develops in such a way that those suggestions that are invoked by the integral symbolism will be "correct." As an example, one writes

$$\varphi(0) = \delta[\varphi] = \int_{\mathbf{R}} \delta(x)\,\varphi(x)\,dx.$$

If we symbolically translate the δ function and write $\delta(x-a)$, the ordinary rules for changing variables yield

$$\int_{\mathbf{R}} \delta(x-a)\,\varphi(x)\,dx = \int_{\mathbf{R}} \delta(y)\,\varphi(y+a)\,dy = \big[\varphi(y+a)\big]_{y=0} = \varphi(a).$$

One often writes $\delta_a(x) = \delta(x-a)$, so that δ_a is the distribution given by $\delta_a[\varphi] = \varphi(a)$.

We include another few very mixed examples.

Example 8.14. The mapping f that is defined by

$$f[\varphi] = 3\varphi(2) - 4\varphi'(2) + 7\varphi^{(8)}(\pi) + \int_2^3 \varphi''(x)\cos 7x\,dx$$

is a tempered distribution. □

Example 8.15. The function $1/x$ cannot play the role of f in Example 8.11, since the integral will be divergent at the origin, However, we can define $f[\varphi]$ using a *symmetric limit:*

$$f[\varphi] = \lim_{\varepsilon \searrow 0} \int_{|x|>\varepsilon} \frac{\varphi(x)}{x}\,dx = \lim_{\varepsilon \searrow 0} \left(\int_{-\infty}^{-\varepsilon} + \int_{\varepsilon}^{\infty} \right) \frac{\varphi(x)}{x}\,dx$$

$$= \lim_{\varepsilon \searrow 0} \int_{\varepsilon}^{\infty} \frac{\varphi(x) - \varphi(-x)}{x}\,dx.$$

The integral converges because the integrand in the last version has a finite limit as $x \searrow 0$, namely, $2\varphi'(0)$. It is not difficult to prove that the formula actually defines a tempered distribution. This distribution is commonly called P.V.$1/x$, where P.V. stands for *principal value*, which means the symmetric limit in the formula. □

Example 8.16. If $f(x) = e^{x^2}$, then f does *not* describe a tempered distribution according to Example 8.11. Indeed, if we choose the test function $\varphi(x) = e^{-x^2}$, which clearly belongs to $\mathcal{S}$, the integral will diverge. The function f increases too fast. □

Remark. If, instead of $\mathcal{S}$, one starts with the test function sets $\mathcal{D}$ or $\mathcal{E}$ (with suitable definitions of convergence, which are omitted here), one obtains other classes of distributions. Starting from $\mathcal{D}$, consisting of test functions with compact support, one gets the class $\mathcal{D}'$ of, simply, distributions. These comprise the tempered distributions as a subset, but also lots of others. For example, the function f in Example 8.16 defines such a general distribution.

If one starts with $\mathcal{E}$ — i.e., all C^∞ functions are included among the test functions — one gets a more restricted set $\mathcal{E}'$, consisting of all distributions with compact support, which is defined in the next section. □

Exercises

8.2 Show that $f(x) = \ln|x|$ defines a distribution according to Example 8.11. (What has to be proved is essentially that f is locally integrable.)

8.3 Which of the following formulae define elements of $\mathcal{S}'$?
(a) $f[\varphi] = \int_{\mathbf{R}} (2x^2+3)\varphi''(x)\,dx$, (b) $f[\varphi] = \int_{\mathbf{R}} e^x\varphi(x)\,dx$,
(c) $f[\varphi] = (\varphi(0))^2$.

8.4 Properties

We are going to introduce some fundamental notions describing various properties of tempered distributions. (In the interest of brevity, we shall mostly omit the word "tempered.")

Two distributions f and g are *equal* (or globally equal), if $f[\varphi] = g[\varphi]$ for all $\varphi \in \mathcal{S}$.

For $f, g \in \mathcal{S}'$, the *sum* $f+g$ is defined by $(f+g)[\varphi] = f[\varphi] + g[\varphi]$ for all $\varphi \in \mathcal{S}$. If c is a scalar (a complex number), then cf is the distribution that is described by $(cf)[\varphi] = c \cdot f[\varphi]$. With these operations, $\mathcal{S}'$ itself becomes a linear space.

A distribution f is *zero on an open interval* $I =]a,b[$, if $f[\varphi] = 0$ for all $\varphi \in \mathcal{S}$ that have $\operatorname{supp}\varphi \subset I$. Two distributions f and g are *equal on an open interval* I if $f - g$ is zero on I. For example, $\delta = 0$ on the interval $]0, \infty[$.

If f and g are ordinary functions, equality on I means that $f(x) = g(x)$ for all $x \in I$ except possibly for a set of measure zero (a zero set, cf. page 89). Just as in Chapter 5, we consider such functions to be equivalent (or, loosely speaking, to be the same).

The *support* $\operatorname{supp} f$ of a distribution f should be the smallest closed set where, loosely speaking, it really does matter what values are taken by the test functions; more precisely, a point x does *not* belong to the support if there is a neighborhood (an open interval) around x where f is zero. The support is always a closed set.

Example 8.17. For a distribution that is a continuous function, $f[\varphi] = \int_{\mathbf{R}} f\varphi\,dx$, the support of f as a distribution coincides with the support of f as a function. If f is discontinuous, the expression "f is zero" in the definition above should be changed to "f is zero except possibly for a set of measure zero". □

Example 8.18. With the notation of Sec. 8.3, $\operatorname{supp} H = [0, \infty[$, $\operatorname{supp} \delta = \{0\}$, $\operatorname{supp} \delta_a = \{a\}$. The support of the distribution in Example 8.14 is the set $[2,3] \cup \{\pi\}$. □

If χ is a multiplicator function and $f \in \mathcal{S}'$, we can define the product $\chi f \in \mathcal{S}'$ by putting $(\chi f)[\varphi] = f[\chi\varphi]$ for all $\varphi \in \mathcal{S}$. (Check that this is reasonable if f is an ordinary function, and that χf actually turns out to be a tempered distribution!)

Example 8.19. What is the product $\chi\delta$? According to the definition we have

$$(\chi\delta)[\varphi] = \delta[\chi\varphi] = \chi(0)\,\varphi(0) = \chi(0) \cdot \delta[\varphi].$$

This result is often written in the form

$$\chi(x)\delta(x) = \chi(0)\delta(x).$$

In the same way one sees that in general

$$\chi(x)\delta(x-a) = \chi(a)\delta(x-a). \tag{8.1}$$

□

Example 8.20. Let f be P.V.$1/x$, and $\chi(x) = x$. What is χf? Indeed,

$$\begin{aligned} \chi f[\varphi] = f[\chi\varphi] &= \lim_{\varepsilon \searrow 0} \int_{|x|>\varepsilon} x \cdot \frac{\varphi(x)}{x}\,dx = \lim_{\varepsilon \searrow 0} \int_{|x|>\varepsilon} \varphi(x)\,dx \\ &= \int_{-\infty}^{\infty} \varphi(x)\,dx = \int_{\mathbf{R}} 1 \cdot \varphi(x)\,dx. \end{aligned}$$

We see that χf can be identified with the function which is identically 1, which we write simply as $x \cdot \text{P.V.}1/x = 1$. □

Now, at last, we arrive at the promised generalization of the notion of a derivative. The starting point is the formula for integration by parts:

$$\int_a^b f'(x)\,\varphi(x)\,dx = f(b)\varphi(b) - f(a)\varphi(a) - \int_a^b f(x)\,\varphi'(x)\,dx.$$

If f is a moderately increasing function of class C^1, and $\varphi \in \mathcal{S}$, we can let $a \to -\infty$ and $b \to \infty$. The contributions at a and b will both tend to zero, and we are left with

$$\int_{\mathbf{R}} f'(x)\,\varphi(x)\,dx = -\int_{\mathbf{R}} f(x)\,\varphi'(x)\,dx.$$

This inspires the following definition:

Definition 8.4 *If $f \in \mathcal{S}'$, a new tempered distribution f' is defined by*

$$f'[\varphi] = -f[\varphi'] \quad \textit{for all } \varphi \in \mathcal{S}.$$

We call f' the derivative of f.

It is not hard to check that actually $f' \in \mathcal{S}'$; here we profit from the rigorous conditions that we have placed upon our test functions! Let us investigate some common cases. If $f \in C^1(\mathbf{R})$, the new derivative will coincide with the old one. But now, functions that did not have a derivative in the traditional sense will find themselves to have one, this being a distribution rather than an ordinary function. Also, the definition of the derivative can be iterated any number of times, and we find that all $f \in \mathcal{S}'$ suddenly are endowed with derivatives of all orders!

Example 8.21. Find the derivative of the Heaviside function H! By definition, it should emerge from the following calculation:

$$\begin{aligned} H'[\varphi] &= -H[\varphi'] = -\int_0^\infty \varphi'(x)\,dx = -\big[\varphi(x)\big]_{x=0}^{\infty} = -(0 - \varphi(0)) \\ &= \varphi(0) = \delta[\varphi]. \end{aligned}$$

We see that $H' = \delta$, the derivative of the Heaviside function is the Dirac "function." The latter is commonly illustrated graphically by a "spike" (see Figure 8.3). In general, it can be seen that the derivative of a jump of size c at a point $x = a$ is given by $c\delta_a$ (or $c\delta(x-a)$). □

Example 8.22. Find the derivatives of δ:

$$\delta'[\varphi] = -\delta[\varphi'] = -\varphi'(0), \quad \delta''[\varphi] = -\delta'[\varphi'] = \delta[\varphi''] = \varphi''(0).$$

In general, $\delta^{(n)}[\varphi] = (-1)^n\,\varphi^{(n)}(0)$. □

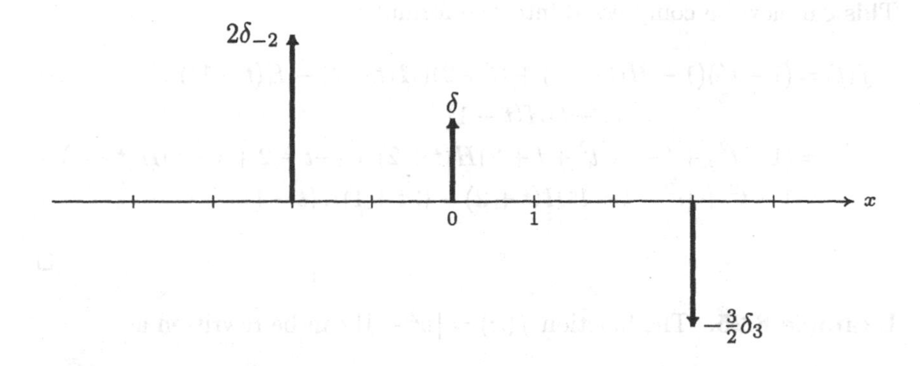

FIGURE 8.3.

Example 8.23. What is the derivative of χf, where χ belongs to $\mathcal{M}$ and $f \in \mathcal{S}'$? On the one hand,

$$(\chi f)'[\varphi] = -(\chi f)[\varphi'] = -f[\chi\varphi'],$$

and on the other,

$$\begin{aligned}(\chi' f + \chi f')[\varphi] &= (\chi' f)[\varphi] + (\chi f')[\varphi] = f[\chi'\varphi] + f'[\chi\varphi] \\ &= f[\chi'\varphi] - f[(\chi\varphi)'] = f[\chi'\varphi] - f[\chi'\varphi + \chi\varphi'] = -f[\chi\varphi'].\end{aligned}$$

(Give some thought to what motivates each individual equality sign in these calculations!) One can see that the distributions $(\chi f)'$ and $\chi' f + \chi f'$ have the same effect on arbitrary test functions; thus we have proved the rule

$$(\chi f)' = \chi' f + \chi f' \quad \text{if } \chi \text{ is moderately increasing and } f \in \mathcal{S}'.$$

□

In calculations involving functions that are defined by different formulae in different intervals, it is practical to make use of translated Heaviside functions. If $a < b$, the expression $H(t-a) - H(t-b)$ is equal to 1 for $a < t < b$ and equal to 0 outside the interval $[a, b]$. It might be called a "window" that lights up the interval (a, b) (we do not in these situations care much about whether an interval is open or closed). For unbounded intervals we can also find "windows": the function $H(t-a)$ lights up the interval (a, ∞), and the expression $1 - H(t-b)$ the interval $(-\infty, b)$.

Example 8.24. Consider the function $f : \mathbf{R} \to \mathbf{R}$ that is given by

$$f(t) = \begin{cases} 1 - t^2 & \text{for } t < -2, \\ t + 2 & \text{for } -2 < t < 1, \\ 1 - t & \text{for } t > 1. \end{cases}$$

This can now be compressed into one formula:

$$\begin{aligned}f(t) &= (1-t^2)(1-H(t+2)) + (t+2)(H(t+2)-H(t-1)) \\ &\qquad +(1-t)H(t-1) \\ &= (1-t^2) + (-1+t^2+t+2)H(t+2) + (-t-2+1-t)H(t-1) \\ &= 1-t^2 + (t^2+t+1)H(t+2) - (2t+1)H(t-1).\end{aligned}$$

□

Example 8.25. The function $f(x) = |x^2-1|$ can be rewritten as

$$\begin{aligned}f(x) &= (x^2-1)H(x-1) + (1-x^2)(H(x+1)-H(x-1)) \\ &\qquad +(x^2-1)(1-H(x+1)) \\ &= (x^2-1)\big(2H(x-1) - 2H(x+1) + 1\big).\end{aligned}$$

This formula can be differentiated, using the rule for differentiating a product:

$$\begin{aligned}f'(x) &= 2x\big(2H(x-1) - 2H(x+1) + 1\big) + (x^2-1)\big(2\delta(x-1) - 2\delta(x+1)\big) \\ &= 2x\big(2H(x-1) - 2H(x+1) + 1\big).\end{aligned}$$

In the last step, we used the formula (8.1). One more differentiation gives

$$\begin{aligned}f''(x) &= 2\big(2H(x-1) - 2H(x+1) + 1\big) + 2x\big(2\delta(x-1) - 2\delta(x+1)\big) \\ &= 2\big(2H(x-1) - 2H(x+1) + 1\big) + 4\delta(x-1) + 4\delta(x+1).\end{aligned}$$

The first term contains the classical second derivative of $|x^2-1|$, which exists for $x \neq \pm 1$; the two δ terms demonstrate that f' has upward jumps of size 4 for $x = \pm 1$. See Figure 8.4. □

Example 8.26. Let $a > 0$ and b be real constants. If $f \in \mathcal{S}'$, define g by $g(x) = f(ax+b)$. This means that (using symbolic integrals)

$$g[\varphi] = \int f(ax+b)\varphi(x)\,dx = \int f(y)\,\varphi\!\left(\frac{y-b}{a}\right)\frac{dy}{a} = \int f(x)\,\varphi\!\left(\frac{x-b}{a}\right)\frac{dx}{a}, \tag{8.2}$$

i.e.,

$$g[\varphi] = \frac{1}{a}\,f\left[x \mapsto \varphi\!\left(\frac{x-b}{a}\right)\right].$$

What connection holds between the derivatives of f and g? By the definition of g' we should have

$$g'[\varphi] = -g[\varphi'] = -\int f(ax+b)\varphi'(x)\,dx.$$

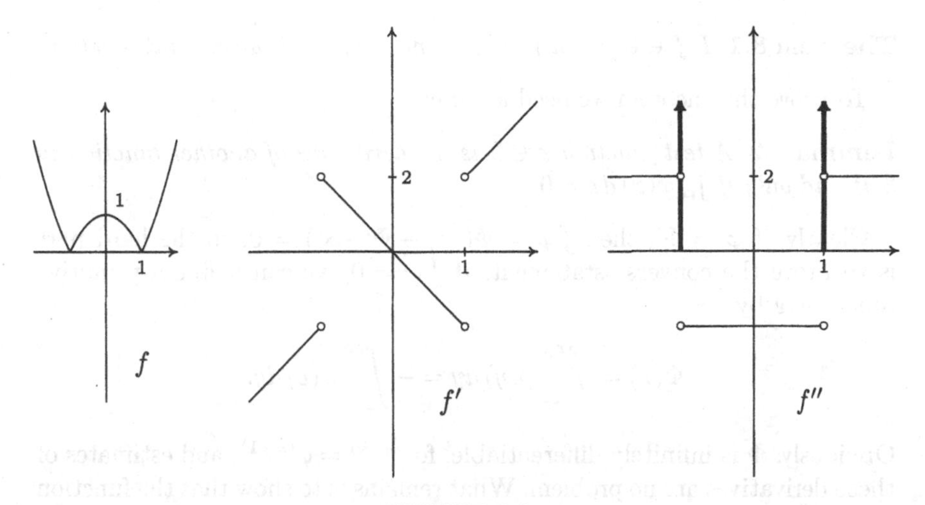

FIGURE 8.4.

What is the effect of the distribution h that is symbolically written $h(x) = af'(ax+b)$? Well, using the same change of variable as in (8.2) we get

$$h[\varphi] = a\int f'(ax+b)\varphi(x)\,dx = a\int f'(x)\,\varphi\Big(\frac{x-b}{a}\Big)\frac{dx}{a}$$

$$= \int f'(x)\,\varphi\Big(\frac{x-b}{a}\Big)\,dx,$$

and the definition of f' then yields

$$h[\varphi] = -\int f(x)\,\frac{d}{dx}\,\varphi\Big(\frac{x-b}{a}\Big)dx = -\frac{1}{a}\int f(x)\,\varphi'\Big(\frac{x-b}{a}\Big)dx,$$

which, after a change of variable, proves to be $-\int f(ax+b)\,\varphi'(x)\,dx$. If $a<0$, the computations are analogous (there will occur a couple of minus signs that cancel in the end). We have then proved that the formula $g'(x) = af'(ax+b)$ holds even for distributions. (In fact, the chain rule will hold also for more general changes of variable than these, but we will not delve deeper into this. Too "general" changes of variable may lead us out of the spaces $\mathcal{S}$ and $\mathcal{S}'$ where we have chosen to stay in this exposition.) □

Using the new notion of derivative, we have now a solution of the old trouble with the wave equation, $c^2u_{xx} = u_{tt}$. If we let f and g be two functions on $\mathbf{R}$ and form

$$u(x,t) = f(x-ct) + g(x+ct),$$

we can take derivatives of f and g in the distribution sense. Using the chain rule as in Example 8.26 it is seen immediately that u satisfies the equation.

In what follows we shall need the following result.

Theorem 8.1 *If $f \in \mathcal{S}'$, then $f' = 0$ if and only if f is a constant function.*

To prove the theorem we need a lemma.

Lemma 8.2 *A test function $\varphi \in \mathcal{S}$ is the derivative of another function in $\mathcal{S}$ if and only if $\int_{\mathbf{R}} \varphi(x)\,dx = 0$.*

Clearly, if $\varphi = \Phi'$, then $\int \varphi = \Phi(\infty) - \Phi(-\infty) = 0$, so the hard part is to prove the converse statement. If $\int \varphi = 0$, we can define a primitive function Φ by

$$\Phi(x) = \int_{-\infty}^{x} \varphi(y)\,dy = -\int_{x}^{\infty} \varphi(y)\,dy.$$

Obviously, Φ is infinitely differentiable, for $\Phi^{(k)} = \varphi^{(k-1)}$, and estimates of these derivatives are no problem. What remains is to show that the function itself tends to zero quickly enough. However, if n is an integer, we do know that $|\varphi(y)| \le C/(1+|y|)^{n+2}$ for some C, and for $x \ge 0$ we then have

$$\begin{aligned}(1+|x|)^n|\Phi(x)| &\le (1+|x|)^n \int_x^\infty \frac{C\,dy}{(1+|y|)^{n+2}} \\ &\le \frac{(1+|x|)^n}{(1+|x|)^n} \int_x^\infty \frac{C\,dy}{(1+|y|)^2} \le C_1.\end{aligned}$$

For $x < 0$, we can do the analogous thing to the other one of the integrals that define Φ. This shows that $\Phi \in \mathcal{S}'$, and the lemma is proved.

Proof of Theorem 8.1. Saying that f is the constant c means that f is identified with the differentiable function $f(x) = c$; in this case, the new derivative coincides with the traditional one, so that $f' = 0$. The converse is harder. Assume that $f' = 0$. Then

$$f[\varphi'] = -f'[\varphi] = 0 \text{ for all } \varphi \in \mathcal{S}. \tag{8.3}$$

However, we must know what f' does to an arbitrary test function. Let ψ_0 be a fixed test function such that $\int \psi_0(x)\,dx = 1$. For an arbitrary $\varphi \in \mathcal{S}$, put $A = \int \varphi(x)\,dx$. Then $\varphi - A\psi_0$ is a test function with integral 0, and by the lemma it is the derivative of some function belonging to $\mathcal{S}$. Put $c = f[\psi_0]$. Using (8.3) we find that

$$\begin{aligned}0 = f[\varphi - A\psi_0] &= f[\varphi] - Af[\psi_0] = f[\varphi] - \int \varphi(x)\,dx \cdot f[\psi_0] \\ &= f[\varphi] - c\int \varphi(x)\,dx,\end{aligned}$$

and it follows that $f[\varphi] = \int c\varphi(x)\,dx$, which means that f can be identified with the constant c. □

Let $f_j \in \mathcal{S}'$ for $j = 1, 2, 3, \ldots$. Suppose that $g \in \mathcal{S}'$ has the property that $\lim_{j\to\infty} f_j[\varphi] = g[\varphi]$ for all $\varphi \in \mathcal{S}$. Then we say that *the sequence*

$\{f_j\}$ *converges in* $\mathcal{S}'$ *to* g, which we write as $f_j \to g$ or, when precision is needed, $f_j \xrightarrow{\mathcal{S}'} g$. This sort of convergence is clearly simple and natural. An immediate consequence is that every positive summation kernel (Sec. 2.4) converges to the Dirac distribution.

Remark. A curious reader is probably missing one operation among the definitions that we have made: multiplication. It turns out, however, that it is not possible to define the product of two distributions in general. It is possible to multiply certain couples, but other products cannot be given a meaningful interpretation. In this text we have treated a simple case: the product of a so-called multiplicator function and a distribution. This case will be sufficient for our present needs. □

Exercises

8.4 Show that $x^2\delta''' = 6\delta'$.

8.5 Prove the following formula: $\chi(x)\delta'(x-a) = \chi(a)\delta'(x-a) - \chi'(a)\delta(x-a)$.

8.6 Find the third derivative of $|x^3 - x^2 - x + 1|$.

8.7 Show that the distributional derivative of $\ln|x|$ is P.V.$(1/x)$.

8.8 Show that the derivative of $f = \text{P.V.}(1/x)$ is described by the formula

$$f'[\varphi] = -\lim_{\varepsilon \searrow 0} \int_{|x| \geq \varepsilon} \frac{\varphi(x) - \varphi(0)}{x^2}\, dx.$$

8.9 Consider Exercise 2.20, page 25. Interpret this exercise as a result in distribution theory. Sketch the graph of the "kernel" occurring in the problem.

8.5 Fourier transformation

If $\varphi \in \mathcal{S}$, it is certainly true that $\varphi \in L^1(\mathbf{R})$, and in addition $\varphi^{(k)} \in L^1(\mathbf{R})$ for all $k \geq 1$. This means that φ and all its derivatives have Fourier transforms:

$$\widehat{\varphi^{(k)}}(\omega) = \int_{\mathbf{R}} \varphi^{(k)}(x)\, e^{-i\omega x}\, dx.$$

Integrating by parts and using the fact that $\varphi^{(m)}(x)$ decreases rapidly as $x \to \pm\infty$, one can see that

$$\widehat{\varphi^{(k)}}(\omega) = (-1)^k \int_{\mathbf{R}} \varphi(x)\, D_x^k\big(e^{-i\omega x}\big)\, dx = (i\omega)^k\, \widehat{\varphi}(\omega).$$

Since the transform of $\varphi^{(k)}$ is bounded (Theorem 7.1), it follows that $|\omega|^k |\widehat{\varphi}(\omega)| \leq C_k$, i.e., $\widehat{\varphi}$ decreases at least as fast as $1/|\omega|^k$ at infinity; and this is true for all $k \geq 0$. Furthermore, we can form the derivative of $\widehat{\varphi}$:

$$\frac{d^n}{d\omega^n}\, \widehat{\varphi}(\omega) = \int_{\mathbf{R}} \varphi(x)\, (-ix)^n\, e^{-i\omega x}\, dx,$$

where differentiation under the integral sign is allowed because the resulting integral converges uniformly (again we use that $\varphi(x)$ is small when $|x|$ is large). Thus, $\widehat{\varphi} \in C^\infty$. But on the last integral we can again apply the same procedure as above to see that the derivatives also decrease rapidly as $\omega \to \pm\infty$. Collecting our results, we have proved part of the following theorem.

Theorem 8.2 *The Fourier transformation $\mathcal{F}$ is a continuous bijection of the space $\mathcal{S}$ onto itself.*

In plain words: if $\varphi \in \mathcal{S}$ on the x-axis, then $\mathcal{F}(\varphi) = \widehat{\varphi} \in \mathcal{S}$ on the ω-axis, and every function $\psi(\omega)$ belonging to $\mathcal{S}$ is the Fourier transform of some function $\varphi \in \mathcal{S}$. The latter statement follows from the fact that if $\psi(\omega)$ belongs to $\mathcal{S}$, then one can form the inverse Fourier transform φ of ψ:

$$\varphi(x) = \frac{1}{2\pi} \int_{\widehat{\mathbf{R}}} \psi(\omega)\, e^{ix\omega}\, d\omega. \tag{8.4}$$

Just as in the argument before the theorem, one sees that φ will be a member of $\mathcal{S}$ on the x-axis and that the Fourier inversion formula gives that $\widehat{\varphi} = \psi$.

The fact that the Fourier transformation is a *continuous* mapping from $\mathcal{S}$ to $\mathcal{S}$ means that if $\varphi_j \xrightarrow{\mathcal{S}} \psi$, then also $\widehat{\varphi_j} \xrightarrow{\mathcal{S}} \widehat{\psi}$. We omit the details of the proof of this fact; essentially it hinges on the fact that the expressions

$$\max_{\omega \in \widehat{\mathbf{R}}} (1 + |\omega|)^n |\widehat{\varphi_j}^{(k)}(\omega) - \widehat{\psi}^{(k)}(\omega)|$$

can be estimated by the corresponding expressions for $(1+|x|)^k |\varphi_j^{(n)} - \psi^{(n)}|$. (Roughly speaking, differentiation on one side corresponds to multiplication by the variable on the other side.)

Again we stress that if $\varphi \in \mathcal{S}$, then the Fourier inversion formula (8.4) will *always* hold. There are no convergence problems for the integral, since $\widehat{\varphi}(\omega)$ tends rapidly to zero at both infinities. The definition of $\mathcal{S}$ is tailored so that the Fourier transformation shall have all these nice properties.

Now we want to define the Fourier transform of a distribution. As a preparation, we do a "classical" calculation. Assume that $f(x)$ and $g(\omega)$ are functions belonging to L^1, each on its own axis:

$$\|f\|_1 = \int_{\mathbf{R}} |f(x)|\, dx < \infty, \qquad \|g\|_1 = \int_{\widehat{\mathbf{R}}} |g(\omega)|\, d\omega < \infty.$$

The Fourier transforms $\widehat{f}$ and $\widehat{g}$ both exist:

$$\widehat{f}(\omega) = \int_{\mathbf{R}} f(x)\, e^{-i\omega x}\, dx, \qquad \widehat{g}(x) = \int_{\widehat{\mathbf{R}}} g(\omega)\, e^{-ix\omega}\, d\omega,$$

and they satisfy the inequalities $|\widehat{f}(\omega)| \le \|f\|_1$ and $|\widehat{g}(x)| \le \|g\|_1$. The function $f(x)\, g(\omega)\, e^{-ix\omega}$ will then be absolutely integrable over the plane

$\mathbf{R} \times \widehat{\mathbf{R}}$, and the integral can be computed by iteration in two ways. This gives the identity

$$\iint\limits_{\mathbf{R}\times\widehat{\mathbf{R}}} f(x)\, g(\omega)\, e^{-ix\omega}\, dx\, d\omega = \int_{\mathbf{R}} f(x)\, \widehat{g}(x)\, dx = \int_{\widehat{\mathbf{R}}} \widehat{f}(\omega)\, g(\omega)\, d\omega.$$

The equality of the two last members is our inspiration. It is, for example, true if $g = \varphi \in \mathcal{S}$ and f is any L^1 function. We extend its domain of validity in the following definition.

Definition 8.5 *The Fourier transform of $f \in \mathcal{S}'$ is the distribution $\widehat{f}$ that is defined by the formula*

$$\widehat{f}[\varphi] = f[\widehat{\varphi}] \quad \textit{for all } \varphi \in \mathcal{S}.$$

Just as in Chapter 7, we shall also write $\widehat{f} = \mathcal{F}(f)$ (but now we use ordinary brackets instead of square ones, in order to avoid confusing it with the notation $f[\varphi]$).

Remark. The equality $\int f\widehat{g} = \int \widehat{f}g$ is sometimes considered to be a variant of the polarized version of the Plancherel formula. □

We proceed to check that $\widehat{f}$ is actually a tempered distribution. It is clear that it is linear:

$$\begin{aligned}\widehat{f}[c_1\varphi_1 + c_2\varphi_2] &= f\big[\,(c_1\varphi_1 + c_2\varphi_2)^\wedge\big] = f[c_1\widehat{\varphi_1} + c_2\widehat{\varphi_2}] \\ &= c_1 f[\widehat{\varphi_1}] + c_2[\widehat{\varphi_2}] = c_1\widehat{f}[\varphi_1] + c_2\widehat{f}[\varphi_2].\end{aligned}$$

The continuity is a simple consequence of the continuity of the Fourier transformation on $\mathcal{S}$: if $\varphi_j \xrightarrow{\mathcal{S}} \psi$, then

$$\widehat{f}[\varphi_j] = f[\widehat{\varphi_j}] \to f[\widehat{\psi}] = \widehat{f}[\psi],$$

which tells us precisely that $\widehat{f}$ is continuous, and thus a distribution.

Let us compute the Fourier transforms of some distributions. We start with a few examples that are ordinary functions, but do not belong to $L^1(\mathbf{R})$.

Example 8.27. Let $f(x) = 1$ for all x. What is the Fourier transform $\widehat{f}$? We should have

$$\begin{aligned}\widehat{f}[\varphi] = f[\widehat{\varphi}] &= \int_{\mathbf{R}} f(x)\, \widehat{\varphi}(x)\, dx = \int_{\mathbf{R}} \widehat{\varphi}(x)\, dx \\ &= 2\pi \cdot \frac{1}{2\pi} \int_{\mathbf{R}} \widehat{\varphi}(x)\, e^{i0x}\, dx = 2\pi\varphi(0) = 2\pi\delta[\varphi].\end{aligned}$$

It follows that $\widehat{f} = 2\pi\,\delta$, or $\widehat{f}(\omega) = 2\pi\,\delta(\omega)$ if we want to stress the name of the independent variable. (Notice that the test functions and their transforms have reversed independent variables in this connection!) □

Example 8.28. Take $f(x) = x^n$ (n integer ≥ 0). This function defines a tempered distribution, and its transform satisfies

$$\widehat{f}[\varphi] = f[\widehat{\varphi}] = \int_{\mathbf{R}} x^n \,\widehat{\varphi}(x)\, dx.$$

By the ordinary rules for Fourier transforms we have that $(ix)^n\,\widehat{\varphi}(x)$ is the transform of the function $\varphi^{(n)}$, which means that $x^n\,\widehat{\varphi}(x)$ is the transform of $(-i)^n\,\varphi^{(n)}$. The inversion formula then gives that $\widehat{f}[\varphi] = 2\pi(-i)^n\varphi^{(n)}(0)$. In the preceding section we saw that the nth derivative of δ is described by $\delta^{(n)}[\varphi] = (-1)^n\varphi^{(n)}(0)$. Thus we must have $\widehat{f} = 2\pi i^n\delta^{(n)}$. □

Before giving further examples we present a list of rules of computation, which on the face look quite familiar, but now are in need of new proofs. To simplify the formulation, we introduce two new notations. First, if $f \in \mathcal{S}'$, then $\check{f}$ is what could be symbolically written as $f(-x)$; more precisely, we define $\check{f}[\varphi] = \int f(x)\varphi(-x)\,dx$. We say that f is *even* if $\check{f} = f$, *odd* if $\check{f} = -f$.

Secondly, if $a \in \mathbf{R}$ and $\varphi \in \mathcal{S}$, we define the translated function φ_a by $\varphi_a(x) = \varphi(x-a)$. For $f \in \mathcal{S}'$ the translated distribution f_a is $f(x-a)$, which means $f_a[\varphi] = \int f(x)\varphi(x+a)\,dx = f[\varphi_{-a}]$. (This notation is a generalization of the notation δ_a to arbitary distributions.)

Theorem 8.3 *If $f, g \in \mathcal{S}'$, then*

(a) *f is even/odd $\iff$ $\widehat{f}$ is even/odd.*

(b) $\widehat{\widehat{f}} = 2\pi\check{f}$.

(c) $\widehat{f_a} = e^{-ia\omega}\widehat{f}$.

(d) $\mathcal{F}(e^{iax}f) = \left(\widehat{f}\right)_a$.

(e) $\widehat{f'} = i\omega\widehat{f}$.

(f) $\widehat{xf} = i\left(\widehat{f}\right)'$.

Proving these formulae are excellent exercises in the definitions of the notions involved. As examples, we perform the proofs of rules (d) and (e):

(d): The effect of the left-hand member on a test function φ is rewritten:

$$\mathcal{F}(e^{iax}f)[\varphi] = (e^{iax}f)[\widehat{\varphi}] = f[e^{iax}\widehat{\varphi}] = f[\mathcal{F}(\varphi_{-a})] = \widehat{f}[\varphi_{-a}] = (\widehat{f})_a[\varphi].$$

Each equality sign corresponds to a definition or a theorem: the first one is the definition of the Fourier transform; the second one is the definition of the product of a function and a distribution; the third one is a rule for "classical" Fourier transforms; the fourth is again the definition of the Fourier transform; and the last one is the definition of the translate of a distribution.

(e) is proved similarly; the reader is asked to identify the reason for each equality sign in the following formula:

$$\widehat{f}'[\varphi] = f'[\widehat{\varphi}] = -f[(\widehat{\varphi})'] = -f[\mathcal{F}(-i\omega\varphi)] = f[\mathcal{F}(i\omega\varphi)] = \widehat{f}[i\omega\varphi]$$
$$= (i\omega\widehat{f})[\varphi].$$

□

We proceed to give more examples of transforms of distributions.

Example 8.29. Transform $f = \text{P.V.}1/x$ (Example 8.15, page 205): we have seen that $xf = 1$. Transformation gives $iD\widehat{f} = 2\pi\delta = 2\pi H'$, which can be rewritten as $iD(\widehat{f} + 2\pi iH) = 0$. By Theorem 8.1 it follows that $\widehat{f} + 2\pi iH = c =$ constant, whence $\widehat{f} = c - 2\pi iH$. To determine the constant we notice that f is odd, and thus $\widehat{f}$ must also be odd. This gives $c = \pi i$ and $\widehat{f} = \pi i(1 - 2H)$.

If we introduce the function $\operatorname{sgn} x = x/|x|$ (the sign of x), we can write the result as $\widehat{f} = -\pi i \operatorname{sgn} x$. □

Example 8.30. $H =$ the Heaviside function. In Example 8.29 we saw that

$$\mathcal{F}\big(\text{P.V.}\,\frac{1}{x}\big)(\omega) = \pi i(1 - 2H(\omega)).$$

Since both sides are odd distributions, rule (b) gives

$$\pi i\mathcal{F}(1 - 2H)(\omega) = 2\pi\left(\text{P.V.}\frac{1}{\omega}\right)^{\vee} = -2\pi\text{P.V.}\frac{1}{\omega}.$$

On the other hand, $\mathcal{F}(1 - 2H) = \widehat{1} - 2\widehat{H} = 2\pi\delta - 2\widehat{H}$. From this we can solve

$$\widehat{H} = \pi\delta(\omega) - i\text{P.V.}\frac{1}{\omega}\,.$$

□

As a finale to this section we prove the following result.

Theorem 8.4 *If $f \in \mathcal{S}'$, then $xf(x)$ is the zero distribution if and only if $f = A\delta$ for some constant A.*

Proof. Transformation of the equation $xf(x) = 0$ gives $i\widehat{f}' = 0$, and by Theorem 8.1 this means that $\widehat{f} = C$, where C is a constant. Transform back again: since $\widehat{1} = 2\pi\delta$, we find that f must be a constant times δ, and the proof is complete. □

By translation of the situation in the theorem, it is seen that the following also holds: if $f \in \mathcal{S}'$ and $(x - a)f(x) = 0$, then $f = A\delta_a$ for some constant A. This can be further generalized. If χ is a multiplicator function that has a *simple* zero at the point $x = a$, and $\chi f = 0$, then $f = A\delta_a$. The proof is built on writing $\chi(x) = (x-a)\psi(x)$, where $\psi(a) \neq 0$, and then the previous result is applied to the distribution ψf.

A different kind of generalization of Theorem 8.4 is given in Exercise 8.14.

Exercises

8.10 Determine the Fourier transform of $f(x) = e^{-x}H(x)$. What is the transform of $1/(1+ix)$? Of $x/(1-ix)$?

8.11 Let $f_1 = \text{P.V.}1/x$. Define recursively f_n for $n = 2, 3, \ldots$ by $f_{n+1} = -f_n'/n$. Prove that $x^n f_n = 1$ for $n = 1, 2, 3, \ldots$.

8.12 Find the Fourier transform of $x^3/(1+x^2)$.

8.13 Find a tempered distribution f that solves the integral equation

$$\int_0^\infty e^{-u} f(t-u)\,du = H(t), \quad -\infty < t < \infty,\ t \neq 0.$$

Check your result by substituting into the equation.

8.14 Suppose that $f \in \mathcal{S}'$ is such that $x^n f = 0$ for an integer n. Prove that f is a linear combination of $\delta^{(k)}$, $k = 0, 1, \ldots, n-1$.

8.15 Let f be a function of moderate growth on $\mathbf{R}$ such that

$$f(x) = \frac{1}{2}\int_{-1}^{1} f(x-t)\,dt \quad \text{for all } x \in \mathbf{R}.$$

(That is, $f(x)$ is always equal to its mean value over the interval of length 2 with x in its middle.) Prove that f must be a polynomial of degree at most 1. (Hint: take Fourier transforms and use the result of the preceding exercise.)

8.6 Convolution

Two test functions in $\mathcal{S}$ can always be *convolved* according to the recipe in Sec. 7.5, because the defining integral

$$f * g(x) = \int_{\mathbf{R}} f(x-y)g(y)\,dy = \int_{\mathbf{R}} f(y)g(x-y)\,dy$$

converges very nicely. But the operation of convolution can be extended to more general situations. If one of the functions is continuous and bounded, and the other one belongs to $L^1(\mathbf{R})$, it works nicely, too. Now we shall take one of them to be a distribution and the other one a test function: thus let $f \in \mathcal{S}'$ and $\varphi \in \mathcal{S}$. By the convolution of f and φ we mean the function $f*\varphi$ given by $f * \varphi(x) = f[\check{\varphi}_x] = f_y[\varphi(x-y)]$, where the subscript y indicates that the distribution acts with respect to the variable y. With a symbolic integral: $f * \varphi(x) = \int f(y)\varphi(x-y)\,dy$.

The convolution is thus an ordinary function of the variable x. One can also prove that it is infinitely differentiable. This follows from the fact that for each fixed x, the sequence of functions ψ_h, defined by

$$y \mapsto \psi_h(y) = \frac{\varphi(x+h-y) - \varphi(x-y)}{h},$$

will converge in the sense of $\mathcal{S}$, as $h \to 0$ (with the function $y \mapsto \varphi'(x-y)$ as the limit). (The verification of this statement is somewhat complicated; it involves the notion of uniform continuity.) This implies that

$$\frac{f*\varphi(x+h) - f*\varphi(x)}{h} = \frac{f_y[\varphi(x+h-y)] - f_y[\varphi(x-y)]}{h} = f[\psi_h]$$
$$\to f_y[\varphi'(x-y)] = f*\varphi'(x).$$

The reasoning can be iterated, and one finds that $f*\varphi$ has derivatives of all orders, that also satisfy

$$D^n(f*\varphi) = f*\varphi^{(n)}.$$

What is the result of Fourier transforming a convolution of this type? The reader should be prepared for the answer. A proof of this runs along these lines: let ψ be an arbitrary test function:

$$\widehat{\varphi * f}[\psi] = \varphi * f[\widehat{\psi}] = \int \varphi * f(x)\,\widehat{\psi}(x)\,dx = \iint \varphi(x-y)\,f(y)\,\widehat{\psi}(x)\,dx\,dy,$$

$$\widehat{\varphi}\,\widehat{f}[\psi] = \widehat{f}[\widehat{\varphi}\,\psi] = f[\widehat{\widehat{\varphi}\,\psi}] = f\left[\frac{1}{2\pi}\,\widehat{\widehat{\varphi}} * \widehat{\psi}\right] = f[\check{\varphi} * \widehat{\psi}]$$
$$= f_x[\textstyle\int \varphi(y-x)\widehat{\psi}(y)\,dy] = \displaystyle\iint f(x)\,\varphi(y-x)\,\widehat{\psi}(y)\,dx\,dy.$$

Since $\widehat{\varphi * f}$ and $\widehat{\varphi}\widehat{f}$ have the same effect on any test function, they are the same distribution. (The proof may seem defective inasmuch as certain integrals are "symbolic", but this can be justified.)

The definition of convolution can be extended further by going a roundabout way via the Fourier transform and the formula $f*g = \mathcal{F}^{-1}(\widehat{f}\widehat{g})$, but for this we refer to deeper texts.

The Dirac distribution δ has a special relation to the convolution operation. As soon as the convolution is defined, one has $\delta * f = f$. Indeed, if f is continuous, we have

$$\delta * f(x) = \int \delta(y) f(x-y)\,dy = [f(x-y)]_{y=0} = f(x).$$

Algebraically, this means that δ is a unit element for the operation $*$.

Exercises

8.16 Compute the convolution $\delta^{(n)} * f$, where f is a function belonging to $\mathcal{E}$.

8.17 Prove that $f*\varphi[\psi] = f[\check{\varphi} * \psi]$, whenever $f \in \mathcal{S}$ and $\varphi, \psi \in \mathcal{S}'$. (This could be taken as an alternative definition of $f*\varphi$.)

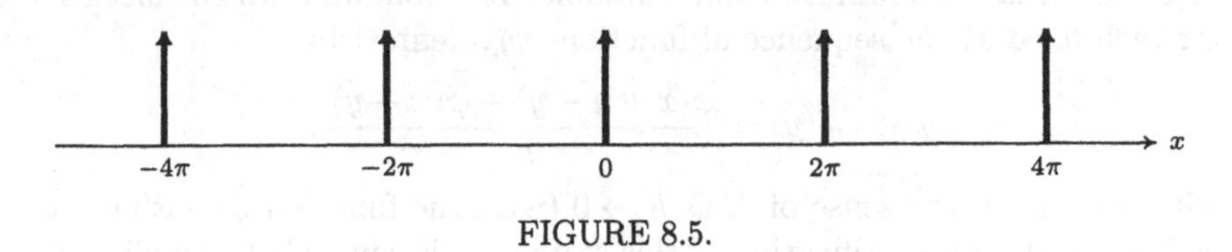

FIGURE 8.5.

8.7 Periodic distributions and Fourier series

The intention of this section is to give just a hint about how distribution theory actually can be used to unify the classical notions of Fourier series and Fourier transforms to make them special cases of one notion.

A tempered distribution f is said to be periodic with period $2P$, if $f[\varphi] = f[\varphi_{2P}]$ for all $\varphi \in \mathcal{S}$ or, symbolically, $\int f(x)\varphi(x)\,dx = \int f(x)\varphi(x-2P)\,dx$ for all $\varphi \in \mathcal{S}$.

Example 8.31. A simple example of a distribution with period 2π is the so-called *pulse train* (see Figure 8.5):

$$\text{Ш} = \sum_{n\in\mathbf{Z}} \delta_{2\pi n} \quad \text{or} \quad \text{Ш}(x) = \sum_{n\in\mathbf{Z}} \delta(x - n\cdot 2\pi).$$

That this is actually a tempered distribution hinges essentially on the fact that the sum

$$\text{Ш}[\varphi] = \sum_{n\in\mathbf{Z}} \varphi(x + n\cdot 2\pi)$$

is convergent, which follows from the estimate $|\varphi(x+n\cdot 2\pi)| \le M/(1+n^2)$, which is true for all $\varphi \in \mathcal{S}$ (with wide margins). □

Let us investigate what the Fourier transform of a periodic distribution f should look like. For simplicity, we assume that the period is 2π, so that we have $f_{2\pi} = f$. Direct transformation gives, using rule (c) in Theorem 8.3,

$$\widehat{f} = e^{-2\pi i\omega}\widehat{f} \quad \Longleftrightarrow \quad (1 - e^{-2\pi i\omega})\widehat{f} = 0.$$

Evidently, for all ω such that $e^{-2\pi i\omega} \ne 1$, $\widehat{f}$ must be zero; this holds for all $\omega \ne$ integer. At integer points $\omega = n$, the factor $1 - e^{-2\pi i\omega}$ has a simple zero, i.e., it behaves essentially like a nonzero constant times the factor $(\omega - n)$. Using Theorem 8.4 (in a "local" version) we see that $\widehat{f}$ has a point mass at $\omega = n$, i.e., a multiple of δ_n. Thus we can write

$$\widehat{f} = \sum_{n\in\mathbf{Z}} \gamma_n\delta_n \quad \text{or} \quad \widehat{f}(\omega) = \sum_{n\in\mathbf{Z}} \gamma_n\delta(\omega - n)$$

for certain constants γ_n.

In order to identify the coefficients γ_n, we consider the particular case when f is a "nice" 2π-periodic function. Then f has a nicely convergent ordinary Fourier series, so that we can write

$$f(x) = \sum_{n \in \mathbf{Z}} c_n e^{inx}, \qquad c_n = \frac{1}{2\pi} \int_{-\pi}^{\pi} f(x)\, e^{-inx}\, dx.$$

If it is permissible to form the Fourier transform term by term, we could use the fact that the transform of $e^{i\alpha x}$ is $2\pi\delta_\alpha$, and we would get

$$\widehat{f}(\omega) = \sum_{n \in \mathbf{Z}} c_n \cdot 2\pi\delta(\omega - n).$$

Thus it seems that the coefficients of the pulse train that makes up $\widehat{f}$ are nothing but the classical Fourier coefficients of f (multiplied by the factor 2π).

Formally, the inversion formula for the periodic distribution f would look like this:

$$\begin{aligned} f(x) &\sim \frac{1}{2\pi} \int_{\widehat{\mathbf{R}}} \widehat{f}(\omega) e^{ix\omega}\, d\omega = \frac{1}{2\pi} \int_{\widehat{\mathbf{R}}} \sum_{n \in \mathbf{Z}} c_n \cdot 2\pi\delta(\omega - n)\, e^{ix\omega}\, d\omega \\ &= \sum_{n \in \mathbf{Z}} c_n \delta_n[e^{ix\omega}] = \sum_{n \in \mathbf{Z}} c_n e^{inx}, \end{aligned}$$

i.e., the inversion formula is the ordinary Fourier series.

All these calculations can in fact be justified, and this is done in more complete texts. It turns out that an arbitrary 2π-periodic distribution f has a Fourier series/transform of the form $\sum c_n e^{inx}$ and $2\pi \sum c_n \delta_n$, respectively, where the coefficients satisfy an equality of the form $|c_n| \leq M(1 + |n|)^k$ for some constants M and k; and, conversely, a pulse train with such coefficients is the Fourier transform of some periodic distribution.

Exercise

8.18 Find the Fourier transform of the pulse train Ш in Example 8.31.

8.8 Fundamental solutions

Let $P(r)$ be a polynomial in the variable r, with constant coefficients. $P(r)$ is the characteristic polynomial of the *differential operator* $P(D)$, which can operate on functions or distributions:

$$\begin{aligned} P(r) &= a_n r^n + a_{n-1} r^{n-1} + \cdots + a_1 r + a_0, \\ P(D) &= a_n D^n + a_{n-1} D^{n-1} + \cdots + a_1 D + a_0, \\ P(D)y &= a_n y^{(n)} + a_{n-1} y^{(n-1)} + \cdots + a_1 y' + a_0 y. \end{aligned}$$

For example, $P(D)$ can be considered as a linear mapping $\mathcal{S}' \to \mathcal{S}'$.

If $P(D)$ operates on a convolution $f * \varphi$, where, say, $f \in \mathcal{S}'$ and $\varphi \in \mathcal{S}$, then the linearity and the rule for differentiation of a convolution imply that the result can be written

$$P(D)(f * \varphi) = (P(D)f) * \varphi = f * (P(D)\varphi).$$

Now suppose that $E \in \mathcal{S}'$ is a distribution such that $P(D)E = \delta$, and let f be an arbitrary continuous function. Then we have

$$P(D)(E * f) = (P(D)E) * f = \delta * f = f.$$

Thus, if we have found such an E, we have a recipe for finding a particular solution of the differential equation $P(D)y = f$, where the right-hand member f is an arbitrary (continuous) function. One says that E is a *fundamental solution* of the operator $P(D)$.

Example 8.32. Let $a > 0$. Let us find a fundamental solution of the familiar operator $P(D) = D^2 + a^2$, i.e., we want to find a distribution E such that $E'' + a^2 E = \delta$. Fourier transformation gives

$$(i\omega)^2 \widehat{E} + a^2 \widehat{E} = \widehat{\delta} = 1,$$

and at least one solution of this ought to be found by solving for $\widehat{E}$ like this:

$$\widehat{E} = \frac{1}{-\omega^2 + a^2} = \frac{1}{2a}\left(\frac{1}{\omega + a} - \frac{1}{\omega - a}\right) = \frac{i}{4a}\left(\frac{2}{i(\omega + a)} - \frac{2}{i(\omega - a)}\right).$$

(The two fractions in the last expression are interpreted as P.V.'s, as in Example 8.15.). We recognize that the Fourier transform of $\operatorname{sgn} t$ is $2/(i\omega)$, which gives

$$y(t) = \frac{i}{4a}\left(e^{-iat}\operatorname{sgn} t - e^{iat}\operatorname{sgn} t\right) = \frac{1}{2a}\operatorname{sgn} t \sin at = \frac{1}{2a}\sin at \cdot (2H(t) - 1).$$

Just to be safe, we check by differentiating:

$$\begin{aligned}
y'(t) &= \frac{1}{2a}(a\cos at(2H(t) - 1) + \sin at \cdot 2\delta(t)) = \tfrac{1}{2}(\cos at(2H(t) - 1)), \\
y''(t) &= \tfrac{1}{2}(-a\sin at(2H(t) - 1) + \cos at \cdot 2\delta(t)) \\
&= -\frac{a}{2}\sin at \cdot (2H(t) - 1) + \delta(t) = -a^2 y(t) + \delta(t).
\end{aligned}$$

□

Exercise

8.19 Find fundamental solutions of the operators (a) $P(D) = D - a$, (b) $P(D) = D^2 + 3D + 2$, (c) $P(D) = D^2 + 2D + 1$, (d) $P(D) = D^2 + 2D + 2$.

8.9 Back to the starting point

We round off the chapter by looking back at the first section, 8.1, where we presented a number of "problems." Most of these have now found some sort of solution.

Problem 8.1 was settled in Example 8.26: even an angular wave-shape has derivatives in the distributional sense, and these derivatives satisfy the wave equation.

Problems 8.2–4 dealt with point charges of different kinds; the solution, as we have seen throughout the chapter, is to make the δ notion legitimate by viewing it as a distribution.

Problem 8.5 may be said to have been solved by the very idea of considering distributions as linear functionals on a space of test functions.

There remains Problem 8.6. Let us first take the heat problem with the unknown function $u = u(x,t)$:

$$u_{xx} = u_t,\ x \in \mathbf{R},\ t > 0; \qquad u(x,0) = f(x),\ x \in \mathbf{R}.$$

We can now let f be a *tempered distribution* on $\mathbf{R}$. Assume that for every fixed t, the thing that is symbolically denoted by $x \mapsto u(x,t)$ is a tempered distribution, and also assume that one can reverse the order of differentiation with respect to t and Fourier transformation with respect to x. If the Fourier transform of u with respect to x is denoted by $U(\omega,t)$, as in Sec. 7.7, we get the same transformed problem as we got there:

$$-\omega^2 U = \frac{\partial U}{\partial t},\ t > 0; \qquad U(\omega,0) = \widehat{f}.$$

This differential equation can be solved just as before, and we get $U(\omega,t) = e^{-\omega^2 t} \cdot \widehat{f}$, where the right-hand member is now a product of a test function and a tempered distribution. Transforming back again we get $u(x,t) = E(x,t) * f$, which is a convolution of a test function and a tempered distribution, as in Sec. 8.6. An interesting fact is now that such a convolution is actually an ordinary function, and furthermore this function has derivatives of all orders that actually satisfy the heat equation. The initial values are the distribution f in the sense that

$$u(\cdot\,,t) \xrightarrow{\mathcal{S}'} f \quad \text{as } t \searrow 0.$$

The Dirichlet problem in the half-plane, which was considered in Sec. 7.8, can be treated in a similar way. A minor complication is the fact that the Poisson kernel

$$P_y(x) = \frac{y}{\pi}\,\frac{1}{y^2 + x^2}$$

is not a test function, but the definition of convolution in Sec. 8.6 can be extended to encompass this case as well.

Summary of Chapter 8

Definition
We say that a function φ belongs to the Schwartz class $\mathcal{S}$ if φ has derivatives of all orders and these satisfy inequalities of the form

$$(1+|x|)^n|\varphi^{(k)}(x)| \leq C_{n,k}, \quad x \in \mathbf{R},$$

where $C_{n,k}$ are constants for all integers n and k that are ≥ 0.

Definition
A sequence $\{\varphi_j\}_{j=1}^{\infty} \subset \mathcal{S}$ is said to converge in $\mathcal{S}$ to a function $\psi \in \mathcal{S}$ if for all integers $n \geq 0$ and $k \geq 0$ it holds that

$$\lim_{j\to\infty} \max_{x\in\mathbf{R}} (1+|x|)^n|\varphi_j^{(k)}(x) - \psi^{(k)}(x)| = 0.$$

Definition
A tempered distribution f is a mapping $f : \mathcal{S} \to \mathbf{C}$ having the properties

(1) linearity: $f[c_1\varphi_1 + c_2\varphi_2] = c_1 f[\varphi_1] + c_2 f[\varphi_2]$ for all $\varphi_k \in \mathcal{S}$ and scalars c_k;

(2) continuity: if $\varphi_j \xrightarrow{\mathcal{S}} \psi$ as $j \to \infty$, then also $\lim_{j\to\infty} f[\varphi_j] = f[\psi]$.

The set of all tempered distributions is denoted by $\mathcal{S}'$.

Intuitively, a distribution on $\mathbf{R}$ is a sort of generalized function; it need not have ordinary function values, but somehow it lives on the axis, and its "global" behavior is a sort of sum of its "local behavior". For example, the Dirac distribution δ is zero everywhere except at the origin, where it is in a certain sense infinitely large.

Definition
If $f \in \mathcal{S}'$, a new tempered distribution f' is defined by

$$f'[\varphi] = -f[\varphi'] \quad \text{for all } \varphi \in \mathcal{S}.$$

We call f' the derivative of f.

Theorem
If $f \in \mathcal{S}'$, then $f' = 0$ if and only if f is a constant function.

Theorem
The Fourier transformation $\mathcal{F}$ is a continuous bijection of the space $\mathcal{S}$ onto itself.

Definition
The Fourier transform of $f \in \mathcal{S}'$ is the distribution $\widehat{f}$ that is defined by the formula

$$\widehat{f}[\varphi] = f[\widehat{\varphi}] \quad \text{for all } \varphi \in \mathcal{S}.$$

If $f \in \mathcal{S}'$, then $\check{f}$ is what could be symbolically written as $f(-x)$; more precisely, $\check{f}[\varphi] = \int f(x)\varphi(-x)\,dx$. We say that f is *even* if $\check{f} = f$, *odd* if $\check{f} = -f$.

If $a \in \mathbf{R}$ and $\varphi \in \mathcal{S}$, we define the translated function φ_a by $\varphi_a(x) = \varphi(x-a)$. For $f \in \mathcal{S}'$ the translated distribution f_a is $f(x-a)$, which means $f_a[\varphi] = \int f(x)\varphi(x+a)\,dx = f[\varphi_{-a}]$.

Theorem
If $f, g \in \mathcal{S}'$, then

(a) f is even/odd $\iff$ $\widehat{f}$ is even/odd.

(b) $\widehat{\widehat{f}} = 2\pi\check{f}$.

(c) $\widehat{f_a} = e^{-ia\omega}\widehat{f}$.

(d) $\mathcal{F}(e^{iax}f) = \left(\widehat{f}\right)_a$.

(e) $\widehat{f'} = i\omega\widehat{f}$.

(f) $\widehat{xf} = i\left(\widehat{f}\right)'$.

Historical notes

The notion of a point mass was imagined already by Isaac Newton (1642–1727) in his description of the laws of gravity. Many physicists and applied mathematicians, such as Oliver Heaviside, in the nineteenth century, used this and analogous concepts, with a varying sense of bad conscience, because they had no stringent mathematical formulation. Paul Dirac (1902–84), who won the Nobel Prize in Physics in 1933, discussed the problem closely and his name was associated to the object δ.

A mathematically acceptable definition of distributions was given around the middle of the twentieth century by Laurent Schwarz (1915–2002). Simultaneously, a number of Russian mathematicians developed a similar theory in a quite different way. They talked about *generalized functions,* defined as limits of sequences of ordinary functions in a certain manner. It was soon discovered that the two definitions in fact give rise to equivalent notions.

The definitions are easily extended to several dimensions, so that one talks about distributions on $\mathbf{R}^n$. These objects have an enormous importance in the study of partial differential equations. There are also other sorts of generalizations to even more general objects, such as *hyperfunctions.* Research in the field is intensive today.

Problems for Chapter 8

If the star-marked sections in Chapters 2–7 have not been studied previously, the exercises there could now be looked up and solved. Here are another few problems.

8.20 Which of the following functions belong to $\mathcal{S}$ or/and $\mathcal{M}$: e^{-x^2}, $e^{-|x|^5}$, $\sin(x^2)$, x^n (n integer), $1/(1+x^2)$.

8.21 Decide which of these functions can be considered as distributions: e^{2x}, e^{-2x}, $e^{2x}H(x)$, $e^{-2x}H(x)$, $e^{\sin x}$, $(x^2-1)^3$.

8.22 Find the first and second derivatives (in the sense of distributions) of (a) $xH(x)$, (b) $|x|$, (c) $|x^2-x|$.

8.23 Simplify $f(x) = \psi(x)\delta''(x-a)$, where $\psi \in \mathcal{M}$.

8.24 Compute f'', where $f(x) = |\sin x|$. Draw pictures of f, f', and f''.

8.25 An electric charge q at a point a on the x-axis can be represented by $q\,\delta(x-a)$. Suppose that we have one charge n at the point $1/n$ and an opposite charge $-n$ at the point $-1/n$. The "limit" of this system, as $n \to \infty$, is called an electric dipole. Describe this limit as a distribution!

8.26 Find all tempered distributions f satisfying the differential equation $f'(x) + 2xf(x) = \delta(x-2)$.

8.27 Determine the Fourier transforms of (a) $\sin ax$, (b) $\cos bx$, (c) $x^n H(x)$ ($n = 0, 1, 2, \ldots$), (d) $\delta(x-a) + \delta(x+a)$, (e) $x\delta'(x)$.

8.28 Let $a > 0$. Find the Fourier transforms of $e^{-ax}H(x)$ and $e^{ax}(1-H(x))$ and finally of $1/(x+bi)$, where b is a real number.

8.29 Find a solution in the form of a tempered distribution of the problem $y'' + a^2y = \delta$. Then show how this distribution can be used to construct a solution of the equation $y'' + a^2y = f$, where f is an "arbitrary" function (i.e., a function belonging to some suitable class).

9
Multi-dimensional Fourier analysis

In this chapter we give a sketch of what Fourier series and integrals look like in the multivariable case. There are almost no proofs. Sections 9.1–2 tackle the problem of summing a series when the terms have no natural sequential order, which happens as soon as they are numbered in some other way than by index sequences such as $1, 2, 3, \ldots$ and similar sequences. In the next few sections we indicate what the Fourier analysis looks like. The intention behind these sections is to provide a sort of moral support for a student who comes across, say, in his physics studies, such things as Fourier transforms taken over all of 3-space – here a real mathematician is saying that these things do exist and can be used!

We have not included anything about distributions. This would have gone far beyond the intended scope of this book. But the interested reader should know that distributions in $\mathbf{R}^n$ are an extremely useful tool, and the theory of these objects is both beautiful and exciting. Those who are interested should go on to some of the books mentioned in the bibliography.

9.1 Rearranging series

In this section we presuppose that the reader is acquainted with the elements of the theory of numerical series. By this we mean notions such as convergence, absolute convergence, and comparison tests.

As is well known, if we want to compute the sum of finitely many terms, the order in which the terms are taken does not matter:

$$1+2+3+4+5 = 4+1+5+3+2.$$

This need not be the case when there are infinitely many terms, i.e., when the sum is a *series.*

Example 9.1. The series

$$\sum_{k=1}^{\infty} \frac{(-1)^{k+1}}{k} = 1 - \tfrac{1}{2} + \tfrac{1}{3} - \tfrac{1}{4} + \tfrac{1}{5} - \tfrac{1}{6} + \cdots$$

is convergent according to the so-called Leibniz test and has a certain sum $s > \frac{1}{2}$ (it can actually be shown that $s = \ln 2$). If we rearrange the terms so that each positive term is followed by two negative terms, so that we try to sum

$$1 - \tfrac{1}{2} - \tfrac{1}{4} + \tfrac{1}{3} - \tfrac{1}{6} - \tfrac{1}{8} + \tfrac{1}{5} - \tfrac{1}{10} - \tfrac{1}{12} + \tfrac{1}{7} - \tfrac{1}{14} - \tfrac{1}{16} + \cdots,$$

it is rather easy to show that the partial sums now tend to $s/2 \neq s$. □

The reason things can be as bad as in the example is the following: If we add those terms of the original series that are positive, which means that we write

$$1 + \tfrac{1}{3} + \tfrac{1}{5} + \tfrac{1}{7} + \cdots + \tfrac{1}{2k-1} + \cdots,$$

we get a *divergent* series; in like manner the negative terms also make up a divergent series:

$$-\tfrac{1}{2} - \tfrac{1}{4} - \tfrac{1}{6} - \tfrac{1}{8} - \cdots - \tfrac{1}{2k} - \cdots.$$

This means that the sum of the original series is really an expression of the type "$\infty - \infty$": there is an infinity of "positive mass" and an infinity of "negative mass." Actually, it is not very difficult to describe a process that can assign *any given number* as the sum of such a series, by taking the terms in a suitable order. It can also be made to diverge in various different ways. Se Exercise 9.1 below.

A convergent series $\sum_{k=1}^{\infty} a_k$ of this type, i.e., such that $\sum_{k=1}^{\infty} |a_k|$ is divergent ($= +\infty$, as it is often written), is said to be *conditionally* convergent. The alternative is an *absolutely convergent* series. For such a series the troubles indicated above cannot happen. We are going to prove a theorem about this. First we want to give a proper definition of a rearrangement of a series. Let $\mathbf{Z}_+$ be the positive integers: $\mathbf{Z}_+ = \{1, 2, 3, \ldots\}$.

Definition 9.1 *Let $\varphi : \mathbf{Z}_+ \to \mathbf{Z}_+$ be a bijection, i.e., a one-to-one mapping of the positive integers onto themselves. Then we say that the series $\sum_{k=1}^{\infty} a_{\varphi(k)}$ is a* rearrangement *of the series $\sum_{k=1}^{\infty} a_k$.*

Theorem 9.1 *An absolutely convergent series remains absolutely convergent and its sum does not change as a result of any rearrangement of its terms.*

Proof. Put $s = \sum_{k=1}^{\infty} a_k$. Let φ be a rearrangement bijection and put $t_n = \sum_{k=1}^{n} a_{\varphi(k)}$. The starting point is that $\sum a_k$ is absolutely convergent, i.e., that

$$\sum_{k=1}^{\infty} |a_k| = \sigma < +\infty.$$

Let $M(n) = \max_{1 \le k \le n} \varphi(k)$. Then certainly

$$\sum_{k=1}^{n} |a_{\varphi(k)}| \le \sum_{j=1}^{M(n)} |a_j| \le \sigma < +\infty,$$

and since σ does not depend on n it follows that the rearranged series is also absolutely convergent. Now let $\varepsilon > 0$ be an arbitrary positive number. Then there exists an $N_0 = N_0(\varepsilon)$ such that $\sum_{j=N_0+1}^{\infty} |a_j| < \varepsilon$. If m is large enough, the numbers $\varphi(1), \varphi(2), \varphi(3), \ldots, \varphi(m)$ will surely include all the numbers $1, 2, 3, \ldots, N_0$ (and probably a lot of other numbers). Then we can estimate: for all sufficiently large values of m it holds that

$$|t_m - s| = \left| \sum_{k=1}^{m} a_{\varphi(k)} - \sum_{j=1}^{\infty} a_k \right| = \left| \sum_{\substack{\text{some} \\ j > N_0}} a_j \right| \le \sum_{\substack{\text{some} \\ j > N_0}} |a_j| \le \sum_{j=N_0+1}^{\infty} |a_j| < \varepsilon,$$

which means that $t_m \to s$ as $m \to \infty$, and this is what we wanted to prove. □

In certain situations it is natural to consider series with terms indexed by *all integers:* we have seen Fourier series in Chapter 4, and in complex analysis one considers Laurent series. Such a series $\sum_{k \in \mathbf{Z}} a_k$ is (classically) considered as convergent if the two "halves" $\sum_{k=0}^{\infty} a_k$ and $\sum_{k=1}^{\infty} a_{-k}$ are both separately convergent. As we have seen in Chapter 4, there are also other ways to define the convergence of such a series. In the case of Fourier series it turns out to be natural to study the symmetric partial sums $\sum_{k=-n}^{n} a_k$. The example

$$\sum_{k=-\infty}^{\infty} k$$

shows that these may well converge, although the series is divergent in the classical sense.

Exercise

9.1 (a) Describe (in principle) how to rearrange a conditionally convergent series (for example the one in the introductory example above) to make it converge to the sum 4.
(b) Describe how to rearrange the same series so that it diverges to $-\infty$.
(c) The series can also be rearranged so that its partial sums make up a sequence that is dense in the interval $[-13, 2003]$. How?
(d) The partial sums can be dense on the whole real axis. How?

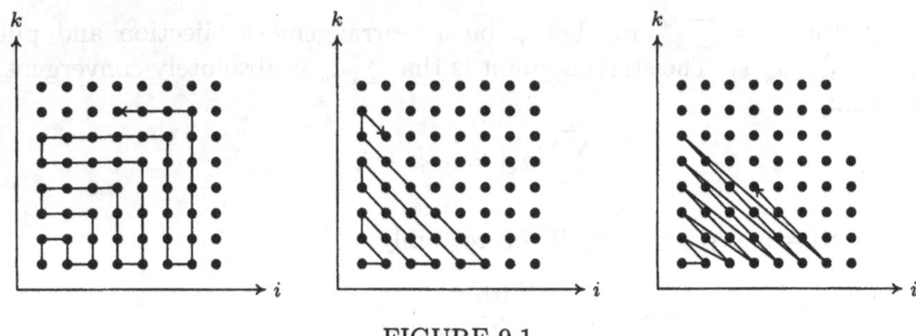

FIGURE 9.1.

9.2 Double series

Let there be given a *doubly indexed* sequence of numbers, i.e., a set of numbers a_{ij}, where both i and j are integers (positive, say). We want to add all these numbers, that is to say we want to compute what could be denoted by $\sum_{i,j=1}^{\infty} a_{ij}$. We immediately meet a difficulty: there is no "natural" order in which to take the terms, when we want to define partial sums. The problem can be illustrated as in Figure 9.1: to each *lattice point* (i, j) (i.e., point with integer coordinates) in the first quadrant there corresponds the term a_{ij}. The terms can be enumerated in different ways, as indicated in the figure. None of these orderings is really more natural than the others. As seen in the previous section, we must expect to get different end results when we try different orderings.

The mathematician's way out of this dilemma is to evade it. He restricts the choice of series that he cares to consider. Each way of enumerating the terms in the double series actually amounts to re-thinking it as an "ordinary" series with a certain ordering of the terms. Suppose that *one* enumeration results in a series that turns out to be *absolutely convergent.* By Theorem 9.1, every other enumeration will also result in a series that is absolutely convergent, and which even has the same sum as the first one. This means that the order of summation does not really matter, and it can be chosen so as to be convenient in some way or other.

We summarize: *A double series* $\sum_{i,j=1}^{\infty} a_{ij}$ *is accepted as convergent only if it is absolutely convergent.*

The following theorem, which deals with a different kind of rearrangement, can also be proved.

Theorem 9.2 *Suppose that the double series* $\sum_{i,j=1}^{\infty} a_{ij}$ *is (absolutely) convergent with sum* s. *Put*

$$V_i = \sum_{j=1}^{\infty} a_{ij}, \qquad H_j = \sum_{i=1}^{\infty} a_{ij}.$$

Then also

$$s = \sum_{i=1}^{\infty} V_i = \sum_{j=1}^{\infty} H_j, \tag{9.1}$$

where all the occurring series are also absolutely convergent.

In plain words, the formula (9.1) says that the series can be summed "first vertically and then horizontally," or the other way round; (9.1) can be written as

$$\sum_{i,j=1}^{\infty} a_{ij} = \sum_{i=1}^{\infty}\left(\sum_{j=1}^{\infty} a_{ij}\right) = \sum_{j=1}^{\infty}\left(\sum_{i=1}^{\infty} a_{ij}\right).$$

We omit the proof.

An important case of double series occurs when two simple series are multiplied.

Theorem 9.3 *Suppose that* $s = \sum_{i=1}^{\infty} a_i$ *and* $t = \sum_{j=1}^{\infty} b_j$ *are absolutely convergent series. Then the double series* $\sum_{i,j=1}^{\infty} a_i b_j$ *is also absolutely convergent and has the sum* $s \cdot t$.

Again, we omit the proof and round off with a few examples.

Example 9.2. It is well known that $\dfrac{1}{1-x} = \sum_{k=0}^{\infty} x^k$, where the series is absolutely convergent for $|x| < 1$. Then, by Theorem 9.3, we also have

$$\frac{1}{(1-x)^2} = \frac{1}{1-x} \cdot \frac{1}{1-x} = \sum_{j=0}^{\infty} x^j \sum_{k=0}^{\infty} x^k = \sum_{j,k=0}^{\infty} x^{j+k}.$$

We choose to sum this series "diagonally" (draw a picture!): $j + k = n$, $n = 0, 1, 2, \ldots$, and for fixed n we let j run from 0 to n:

$$\frac{1}{(1-x)^2} = \sum_{n=0}^{\infty}\left(\sum_{j+k=n} x^{j+k}\right) = \sum_{n=0}^{\infty}\left(\sum_{j=0}^{n} x^n\right) = \sum_{n=0}^{\infty} x^n \left(\sum_{j=0}^{n} 1\right)$$

$$= \sum_{n=0}^{\infty} (n+1)x^n \quad (= 1 + 2x + 3x^2 + 4x^3 + \cdots),$$

which must be absolutely convergent for $|x| < 1$. (The doubtful reader can check this by, say, the ratio test.) □

Compare the calculations in the example and the ordinary maneuvers when switching the order of integration in a double integral!

Example 9.3. *Convolutions* have been encountered a few times in different situations earlier in the book. (A unified discussion of this notion is found in Appendix A.) Here we take a look at the case of *convolutions of sequences.* If $a = \{a_i\}_{i=-\infty}^{\infty}$ and $b = \{b_i\}_{i=-\infty}^{\infty}$ are two sequences of numbers, a third sequence $c = \{c_i\}_{i=-\infty}^{\infty}$, the *convolution of* a *and* b, is formed by the prescription

$$c_i = \sum_{j=-\infty}^{\infty} a_{i-j}b_j = \sum_{j\in\mathbf{Z}} a_{i-j}b_j, \quad \text{all } i \in \mathbf{Z},$$

provided the series is convergent. If, say, a and b are absolutely summable, which means that the series $\sum_{-\infty}^{\infty} |a_i| = s$ and $\sum_{-\infty}^{\infty} |b_i| = t$ are convergent, this is true, as is seen by the following computations:

$$\begin{aligned}\sum_{i\in\mathbf{Z}} |c_i| &= \sum_{i\in\mathbf{Z}} \Bigl|\sum_{j\in\mathbf{Z}} a_{i-j}b_j\Bigr| \le \sum_i \sum_j |a_{i-j}b_j| = \sum_j |b_j| \sum_i |a_{i-j}| \\ &= \sum_j |b_j| \cdot s = t \cdot s < +\infty.\end{aligned}$$

Not only does the convolution c exist, it is actually absolutely summable, too! One writes $c = a * b$. □

In what follows, we shall even encounter series where the terms are numbered by indices of dimension greater than 2. For these series, the same results apply as in the case we have considered: by "convergence" one must mean absolute convergence, or else some particular order of summation must be explicitly indicated.

Exercises

9.2 For $a > 0$, compute the value of the sum $\displaystyle\sum_{k=1}^{\infty}\sum_{j=1}^{\infty} 2^{-(ak+j)}$.

9.3 Compute $\displaystyle\sum_{k,n=2}^{\infty} k^{-n}$.

9.4 Suppose that $f(x) \sim \sum a_n e^{inx}$ and $g(x) \sim \sum b_n e^{inx}$, where the Fourier series are assumed to be absolutely convergent (i.e., $\sum |a_n| < \infty$ and $\sum |b_n| < \infty$). Find the Fourier coefficients of $h = fg$.

9.3 Multi-dimensional Fourier series

We shall indicate how to treat Fourier series in several variables. Practical computation with such series in concrete cases easily leads to volumes of computation that are hardly manageable by hand. Thus, there is no reason to try to acquire much skill at such computations; but it is of great interest for applications that these series exist.

We shall study functions defined on $\mathbf{T}^d$, where d is a positive integer. $\mathbf{T}^d$ is the Cartesian product of d copies of the circle $\mathbf{T}$ and is called the *d-dimensional torus.* A typical element of $\mathbf{T}^d$ is thus a d-tuple

$$\mathbf{x} = (x_1, x_2, \ldots, x_d),$$

where $x_k \in \mathbf{T}$ for $k = 1, 2, \ldots, d$. Thus, a function $f : \mathbf{T}^d \to \mathbf{C}$ is a rule that to each $\mathbf{x} \in \mathbf{T}^d$ assigns a complex number $f(\mathbf{x}) = f(x_1, x_2, \ldots, x_d)$.

An alternative description is to consider functions $f : \mathbf{R}^d \to \mathbf{C}$ that are 2π-periodic in each argument, i.e.,

$$f(x_1 + 2\pi n_1,\, x_2 + 2\pi n_2,\, \ldots,\, x_d + 2\pi n_d) = f(x_1, x_2, \ldots, x_d)$$

for all $x_k \in \mathbf{R}$ and all integers n_k, $k = 1, 2, \ldots, d$.

Integration over $\mathbf{T}^d$ should be interpreted and described in the following way:

$$\int_{\mathbf{T}^d} f(\mathbf{x})\, d\mathbf{x} = \int_{\mathbf{T}} \int_{\mathbf{T}} \cdots \int_{\mathbf{T}} f(x_1, x_2, \ldots, x_n)\, dx_1\, dx_2 \cdots dx_d$$

$$= \int_{a_1}^{a_1+2\pi} \int_{a_2}^{a_2+2\pi} \cdots \int_{a_d}^{a_d+2\pi} f(x_1, x_2, \ldots, x_n)\, dx_1\, dx_2 \cdots dx_d,$$

where the numbers $a_1, a_2, \ldots, a_d$ can be chosen at will (because of the periodicity). The space $L^p(\mathbf{T}^d)$ consists of all (Lebesgue measurable) functions such that

$$\|f\|^p_{L^p(\mathbf{T}^d)} := \int_{\mathbf{T}^d} |f(\mathbf{x})|^p\, d\mathbf{x} < \infty.$$

Most important are, as earlier in the book, the cases $p = 1$, when the functions are also called absolutely integrable, and $p = 2$. In the latter case, there is also an *inner product*

$$\langle f, g \rangle = \int_{\mathbf{T}^d} f(\mathbf{x})\, \overline{g(\mathbf{x})}\, d\mathbf{x}.$$

Functions that agree except for a zero set are identified, just as before; what is *meant* by a zero set is, however, someting new. A zero set is something with the d-dimensional volume measure equal to zero. Thus, in $\mathbf{T}^2$ a part of a straight line is a zero set, in $\mathbf{T}^3$ a part of a plane (or some other smooth surface) is a zero set.

We shall work with d-tuples of integers, which we write as

$$\mathbf{n} = (n_1, n_2, \ldots, n_d).$$

These are the elements of the set $\mathbf{Z}^d$. For $\mathbf{x} \in \mathbf{T}^d$ and $\mathbf{n} \in \mathbf{Z}^d$ we form the "scalar product"

$$\mathbf{n} \cdot \mathbf{x} = n_1 x_1 + n_2 x_2 + \cdots n_d x_d = \sum_{k=1}^{d} n_k x_k.$$

For this product, a few natural rules hold. Some of these are the following, which the reader may want to prove (if it seems necessary):

$$(\mathbf{m} + \mathbf{n}) \cdot \mathbf{x} = \mathbf{m} \cdot \mathbf{x} + \mathbf{n} \cdot \mathbf{x}, \quad \mathbf{n} \cdot (\mathbf{x} + \mathbf{y}) = \mathbf{n} \cdot \mathbf{x} + \mathbf{n} \cdot \mathbf{y}.$$

We shall also need the functions $\varphi_{\mathbf{n}}$, defined by

$$\varphi_{\mathbf{n}}(\mathbf{x}) = e^{i\mathbf{n}\cdot\mathbf{x}} = \exp(i(n_1 x_1 + n_2 x_2 + \cdots n_d x_d)).$$

They have a nice property in the space $L^2(\mathbf{T}^d)$:

$$\begin{aligned}
\langle \varphi_{\mathbf{m}}, \varphi_{\mathbf{n}} \rangle &= \int_{\mathbf{T}^d} e^{i\mathbf{m}\cdot\mathbf{x}}\, \overline{e^{i\mathbf{n}\cdot\mathbf{x}}}\, d\mathbf{x} = \int_{\mathbf{T}^d} e^{i(\mathbf{m}-\mathbf{n})\cdot\mathbf{x}}\, d\mathbf{x} \\
&= \int_{\mathbf{T}^d} e^{i(m_1-n_1)x_1}\, e^{i(m_2-n_2)x_2} \ldots e^{i(m_d-n_d)x_d}\, dx_1\, dx_2 \cdots dx_d \\
&= \prod_{k=1}^{d} \int_{\mathbf{T}} e^{i(m_k-n_k)x_k}\, dx_k = 0 \quad \text{as soon as } m_k \neq n_k \text{ for some k.}
\end{aligned}$$

It follows that $\varphi_{\mathbf{m}}$ is orthogonal to $\varphi_{\mathbf{n}}$ as soon as the d-tuple $\mathbf{m}$ is different from the d-tuple $\mathbf{n}$. But if $\mathbf{m} = \mathbf{n}$, each of the integrals in the last product will be equal to 2π, so that

$$\|\varphi_{\mathbf{n}}\|^2 = \langle \varphi_{\mathbf{n}}, \varphi_{\mathbf{n}} \rangle = (2\pi)^d.$$

For a function $f \in L^2(\mathbf{T}^d)$, we can define the Fourier coefficients

$$\begin{aligned}
c_{\mathbf{n}} = \widehat{f}(\mathbf{n}) &= \frac{1}{(2\pi)^d} \int_{\mathbf{T}^d} f(\mathbf{x})\, e^{-i\mathbf{n}\cdot\mathbf{x}}\, d\mathbf{x} \\
&= \frac{1}{(2\pi)^d} \int \cdots \int_{\mathbf{T}^p} f(x_1, x_2, \ldots, x_d)\, e^{-i(n_1 x_1 + n_2 x_2 + \cdots + n_d x_d)}\, dx_1\, dx_2 \cdots dx_d.
\end{aligned}$$

The formal Fourier series can be written

$$f(\mathbf{x}) \sim \sum_{\mathbf{n}\in\mathbf{Z}^d} c_{\mathbf{n}}\, e^{i\mathbf{n}\cdot\mathbf{x}}$$
$$= \sum_{n_1=-\infty}^{\infty}\sum_{n_2=-\infty}^{\infty}\cdots\sum_{n_d=-\infty}^{\infty} \widehat{f}(n_1,n_2,\ldots,n_d)e^{i(n_1x_1+n_2x_2+\cdots+n_dx_d)}.$$

Because of Bessel's inequality, it will converge in the sense of L^2. Indeed, it can be proved that the system $\{\varphi_{\mathbf{n}}\}_{\mathbf{n}\in\mathbf{Z}^d}$ is complete in $L^2(\mathbf{T}^d)$.

The Fourier coefficients can also be defined if we only assume that $f \in L^1(\mathbf{T}^d)$, and we could try to imitate the theory of Chapter 4. Convergence theorems are, however, considerably more complicated to formulate and prove. For most practical purposes it is sufficient to know that the Fourier series "represents" the function, without actually knowing whether it converges pointwise in some sense or other.

Remark. A sufficient condition for pointwise convergence in the case $d = 2$ is that $f \in C^1(\mathbf{T}^2)$ and that in addition the mixed second derivative f_{xy} is continuous on $\mathbf{T}^2$. Conditions of this type can, in principle, be established for any value of d. Higher dimension requires, in general, higher regularity of the function. □

When the dimension is low, say, 2, it can be practical to write such things as (x, y) and (m, n) instead of (x_1, x_2) and (n_1, n_2). We shall do so in the following example.

Example 9.4. Let f be defined by $f(x, y) = xy$ for $|x| < \pi$, $|y| < \pi$. We compute its Fourier coefficients c_{mn}. We shall need the value of the integral

$$\alpha_k := \int_{-\pi}^{\pi} t\, e^{-ikt}\, dt = \left[t\,\frac{e^{-ikt}}{-ik}\right]_{-\pi}^{\pi} - \frac{1}{-ik}\int_{-\pi}^{\pi} e^{-ikt}\, dt$$
$$= 2\pi i\,\frac{(-1)^k}{k} \quad \text{for } k \in \mathbf{Z}\setminus\{0\}, \qquad \alpha_0 = 0.$$

We get

$$\begin{aligned} c_{mn} &= \frac{1}{4\pi^2}\int_{-\pi}^{\pi}\int_{-\pi}^{\pi} xy\, e^{-i(mx+ny)}\, dx\, dy \\ &= \frac{1}{4\pi^2}\int_{-\pi}^{\pi} x\, e^{-imx}\, dx \int_{-\pi}^{\pi} y\, e^{-iny}\, dy \\ &= \frac{\alpha_m\alpha_n}{4\pi^2} = -\frac{(-1)^m}{m}\,\frac{(-1)^n}{n} = \frac{(-1)^{m+n+1}}{mn} \text{ if } mn \neq 0, \\ c_{mn} &= 0 \text{ if } mn = 0. \end{aligned}$$

The series can be written

$$f(x,y) \sim \sum_{\substack{m,n\in\mathbf{Z}\\ mn\neq 0}} \frac{(-1)^{m+n+1}}{mn}\, e^{i(mx+ny)}.$$

□

Example 9.5. Of course, other periods than 2π can be treated, and the periods need not be the same in different directions. Define f by $f(x,y) = xy - x^2$ for $-1 < x < 1$ and $0 < y < 2\pi$, and assume the period to be 2 in the x variable and 2π in the y variable. The Fourier coefficients are given by the formula

$$c_{mn} = \frac{1}{2 \cdot 2\pi} \int_{-1}^{1} dx \int_{0}^{2\pi} (xy - x^2)\, e^{-i(m\pi x + ny)}\, dy.$$

(Effective computation of these coefficients is rather messy, because there occur a number of cases such as for one or the other of m and n being zero, etc.) □

It is also possible to study Fourier series for periodic *distributions* on $\mathbf{R}^d$, but we leave this out.

9.4 Multi-dimensional Fourier transforms

The theory of Fourier transforms is extended to $\mathbf{R}^d$ in a way that is completely analogous to our treatment of Fourier series in the last section. Notation: elements of $\mathbf{R}^d$ are $\mathbf{x} = (x_1, x_2, \ldots, x_d)$, and elements of what is often called the *dual* space $\widehat{\mathbf{R}}^d$ are written as $\boldsymbol{\omega} = (\omega_1, \omega_2, \ldots, \omega_d)$. We define a "scalar product"

$$\boldsymbol{\omega} \cdot \mathbf{x} = \omega_1 x_1 + \omega_2 x_2 + \cdots + \omega_d x_d.$$

The fact that $f : \mathbf{R}^d \to \mathbf{C}$ belongs to the class $L^1(\mathbf{R}^d)$ means that (f is Legesgue measurable and) the integral

$$\|f\|_1 := \int_{\mathbf{R}^d} |f(\mathbf{x})|\, d\mathbf{x} = \underset{\mathbf{R}^d}{\int \cdots \int} |f(x_1, x_2, \ldots, x_d)|\, dx_1\, dx_2 \cdots dx_d$$

is convergent. Then for all $\boldsymbol{\omega} \in \widehat{\mathbf{R}}^d$, the integral

$$\widehat{f}(\boldsymbol{\omega}) = \int_{\mathbf{R}^d} f(\mathbf{x})\, e^{-i\boldsymbol{\omega}\cdot\mathbf{x}}\, d\mathbf{x},$$

exists, and the function $\widehat{f} : \widehat{\mathbf{R}}^d \to \mathbf{C}$ is called the Fourier transform of f. Under suitable conditions, one can recover f, in principle, through the formula

$$f(\mathbf{x}) = \frac{1}{(2\pi)^d} \int_{\widehat{\mathbf{R}}^d} \widehat{f}(\boldsymbol{\omega})\, e^{i\mathbf{x}\cdot\boldsymbol{\omega}}\, d\boldsymbol{\omega}.$$

Sufficient conditions for this to hold pointwise are, for example, that $\widehat{f} \in L^1(\widehat{\mathbf{R}}^d)$ or that f is sufficiently regular (has continuous derivatives of sufficiently high order, and this order depends on the dimension d). But also in other cases, the Fourier transform can be said to "represent" the function in some sense, and it can be used in various kinds of calculations. The L^2 theory indicated in Section 7.6 (including Plancherel's theorem) can also be generalized to higher dimensions.

Example 9.6. An important application of the transform is found in the theory of probability, where the multi-dimensional normal distribution is used. In dimension d, a normalized version of this is described by the density function

$$f(\mathbf{x}) = \frac{1}{(2\pi)^{d/2}} \exp(-\tfrac{1}{2}(x_1^2 + x_2^2 + \cdots + x_d^2)) = \frac{1}{(2\pi)^{d/2}} \exp(-\tfrac{1}{2}|\mathbf{x}|^2).$$

It holds that $\|f\|_1 = 1$, and it is easy to compute the Fourier transform or *characteristic function*

$$\widehat{f}(\boldsymbol{\omega}) = \frac{1}{(2\pi)^{d/2}} \int_{\mathbf{R}^d} e^{-|\mathbf{x}|^2/2} e^{-i\boldsymbol{\omega}\cdot\mathbf{x}} \, d\mathbf{x} = \prod_{k=1}^{d} \frac{1}{(2\pi)^{1/2}} \int_{\mathbf{R}} e^{-x_k^2/2} e^{-i\omega_k x_k} \, dx_k$$

$$= \prod_{k=1}^{d} e^{-\omega_k^2/2} = \exp(-\tfrac{1}{2}|\boldsymbol{\omega}|^2).$$

Here we have made use of our knowledge of the ordinary one-dimensional transform of the function $e^{-x^2/2}$. □

Appendix A
The ubiquitous convolution

The operation known as *convolution* appears, in a variety of versions, throughout the theory. We shall here indicate what also makes this operation so important in applications.

In a purely mathematical setting, we find that convolutions of *number sequences* occur when we multiply polynomials and related objects such as power series. Given two polynomials

$$P(x) = a_0 + a_1 x + a_2 x + \cdots + a_m x^m, \quad Q(x) = b_0 + b_1 + b_2 x^2 + \cdots + b_n x^n,$$

we multiply these term by term to get a polynomial

$$PQ(x) = c_0 + c_1 x + c_2 x^2 + \cdots + c_{m+n} x^{m+n}.$$

It is easily seen that its coefficients are given by

$$c_0 = a_0 b_0, \quad c_1 = a_0 b_1 + a_1 b_0, \quad c_2 = a_0 b_2 + a_1 b_1 + a_2 b_0,$$

and, in general

$$c_k = a_0 b_k + a_1 b_{k-1} + a_2 b_{k-2} + \cdots + a_k b_0 = \sum_{i+j=k} a_j b_j \,.$$

This formula exhibits the characteristic property of a convolution: two "numbered objects" are combined so that the sums of indices is constant, to form a new "numbered object". If the objects are "numbered" using a continuous variable, we have to deal with integrals instead of sums.

Now we turn to applications. We study a "black box," i.e., a device that converts an *insignal* $f(t)$ into an *outsignal* $g(t)$. We shall assume that the device satisfies a few reasonable conditions, namely, the following:

(a) It is *invariant under translation of time*, which means that if we feed it a translated input $f(t-a)$, then the output is similarly translated to have the shape of $g(t-a)$. In plain words, this condition amounts to saying that the device operates in the same way whatever the clock says.

(b) It is *linear*, which means that if we input a linear combination such as $\alpha f_1(t) + \beta f_2(t)$, the output looks like $\alpha g_1(t) + \beta g_2(t)$ (with the natural interpretation of letters). This is a reasonable assumption for many (but certainly not all) black boxes.

(c) It is *continuous* in some way (which we shall not specify explicitly here), so that "small" alterations of the input generate "small" changes in the output.

(d) It is *causal*, which means that the output at any point t_0 in time cannot be influenced by the values taken by $f(t)$ for $t > t_0$.

In the first case to consider, we assume that we sample both input and output at discrete points in time. This means that the input can be represented by a *sequence* $f = \{f_t\}_{t=-\infty}^{\infty}$, and the output is another sequence $g = \{g_t\}_{t=-\infty}^{\infty}$. For a start, we assume that the input is 0 for all $t < 0$. The conditions (d) and (b) then force the output to have the same property. It will be practical to denote these sequences as $\mathbf{f} = (f_0, f_1, f_2, \ldots)$.

Let $\boldsymbol{\delta}$ be the input sequence $(1, 0, 0, 0, \ldots)$, and suppose that it results in the output $\mathbf{a} := (a_0, a_1, a_2, \ldots)$ (the impulse response). By causality and translation invariance, it is then clear that the postponed input $\boldsymbol{\delta}_n = (0, \ldots, 0, 1, 0, 0, \ldots)$ yields the output $\mathbf{a}_n = (0, \ldots, 0, a_0, a_1, a_2, \ldots)$. By linearity, then, the input

$$\mathbf{f} = (f_0, f_1, \ldots, f_n, 0, 0, \ldots) = \sum_{k=0}^{n} f_k \boldsymbol{\delta}_k$$

must produce the output

$$\sum_{k=0}^{n} f_k \mathbf{a_k}$$

$$= \left(f_0 a_0,\ f_0 a_1 + f_1 a_0,\ f_0 a_2 + f_1 a_1 + f_2 a_0,\ \ldots, \textstyle\sum_{j=0}^{n} f_j a_{n-j},\ 0, 0, \ldots\right).$$

Finally, by continuity, we find that an arbitrary input $\mathbf{f} = (f_0, f_1, f_2, \ldots)$ must produce an output $\mathbf{g} = (g_0, g_1, g_2, \ldots)$, where

$$g_n = f_0 a_n + f_1 a_{n-1} + f_2 a_{n-2} + \cdots + f_n a_0, \quad n = 0, 1, 2, \ldots.$$

We call $\mathbf{g}$ the convolution of $\mathbf{f}$ and $\mathbf{a}$, and we write $\mathbf{g} = \mathbf{f} * \mathbf{a}$.

Now we remove the condition that everything starts at time 0. We return to the notation $f = \{f_t\} = \{f_t\}_{t=-\infty}^{\infty}$. We we assume that an input f may start at any time $t = -T$, where T may also be positive. The invariance of the black box under translation of time indicates what must happen now. Let $\widetilde{f}$ and $\widetilde{g}$ be the translated sequences defined by $\widetilde{f}_t = f_{t-T}$ and $\widetilde{g}_t = g_{t-T}$ for $t \geq 0$. By property (a) we must have $\widetilde{g} = \widetilde{f} * a$, where a is the sequence **a** above, extended by zeroes for negative indices. This means that $\widetilde{g}_t = \sum_{k=0}^{t} a_k \widetilde{f}_{t-k}$ for all $t \geq 0$. We rewrite this with t replaced by $t + T$ and get

$$g_t = \widetilde{g}_{t+T} = \sum_{k=0}^{t+T} a_k \widetilde{f}_{t+T-k} = \sum_{k=0}^{t+T} a_k f_{t-k}, \quad t \geq -T.$$

If we let $f_t = 0$ for all $t < -T$, we can write this as

$$g_t = \sum_{k=0}^{\infty} a_k f_{t-k} = \sum_{k=-\infty}^{t} a_{t-k} f_t.$$

This formula defines the convolution of two sequences a and f, where now all the f_t may be different from zero — we can treat the case, theoretically possible, of an "input" that has no beginning.

It is also possible to consider the same sort of notion where the other convolution factor a also has "no beginning." This leads us to the formula for the convolution of two doubly infinite sequences $a = \{a_t\}_{-\infty}^{\infty}$ and $f = \{f_t\}_{-\infty}^{\infty}$:

$$(f * a)_t = \sum_{k=-\infty}^{\infty} a_k f_{t-k} = \sum_{k=-\infty}^{\infty} f_k a_{t-k}. \tag{A.1}$$

This case may not be physically interesting for the description of (causal) black boxes and the like, but it is interesting as a mathematical construction.

We now turn our interest to functions defined on a continuous t-axis. The analysis above, that depends on simple notions of linear algebra, cannot be imitated directly. Also, we do not attempt a completely stringent treatment, but content ourselves by a more intuitive approach.

Let $f(t)$ be the input and $g(t)$ the output, as before, and introduce a "black box function" $\varphi(t)$ that describes the device. In fact, let $\varphi(t)$ be the impulse response, i.e., it descibes the output resulting from inputting a Dirac pulse $\delta(t)$ at $t = 0$. By causality and linearity, $\varphi(t) = 0$ for $t < 0$. For any number u, by translation invariance, the input $\delta(t - u)$ must result in the output $\varphi(t - u)$. Now consider an arbitrary input $f(t)$. The properties of δ imply that

$$f(t) = \int_{-\infty}^{\infty} f(u)\delta(t - u)\, du. \tag{A.2}$$

Linearity and continuity now imply that the output due to f should be

$$g(t) = f * \varphi(t) = \int_{-\infty}^{\infty} f(u)\varphi(t - u)\, du. \tag{A.3}$$

(This conclusion could be supported in the following way: the integral in (A.2) might be approximated by "Riemann sums" $\sum f(u_k)\delta(t - u_k)$, and linearity tells us that the response to this is the corresponding sum $\sum f(u_k)\varphi(t - u_k)$, which is an approximation to (A.3). This is, however, not logically rigorous, because a Riemann sum involving δ is really rather nonsensical.)

In the considerations leading up to (A.3) we assumed that $\varphi(t) = 0$ for $t < 0$, which means that the interval of integration is actually $]-\infty, t[$. If, in addition, the input does not start before $t = 0$, the integral is to be taken over just $[0, t]$, and we recognize the variant of the convolution that appears in connection with the Laplace transform.

Just as in the discussion of sequences, these restrictions on f and φ can be totally removed for more general applications.

Appendix B
The discrete Fourier transform

This appendix is an introduction to a discrete (i.e., "non-continuous") counterpart of Fourier series. If you like, you can view it as an approximation to ordinary Fourier series, but it has considerable interest on its own. In applications during the last half-century, it has acquired great importance for treating numerical data. One then uses a further development of the elementary ideas that are presented here, called the *fast Fourier transform* (FFT).

For convenience, we study the interval $(0, 2\pi)$. In this interval we single out the points $x_k = k \cdot 2\pi/N$, $k = 0, 1, \ldots, N-1$, which make up a set $G = G_N$:

$$G = G_N = \{x_k : k = 0, 1, 2, \ldots, N-1\} = \left\{ \frac{2\pi k}{N} : k = 0, 1, 2, \ldots, N-1 \right\}.$$

Consider the set $l_N = l^2(G_N)$ of functions $f : G_N \to \mathbf{C}$ (cf. Sec. 5.3). For each k, $f(x_k)$ is thus a complex number. The set l_N is a complex vector space. It is easy to construct a natural basis in this space: let e_n be the function on G_N defined by $e_n(x_k) = 1$ for $k = n$, $= 0$ for $k \neq n$; then it is easily seen that $\{e_n\}_{n=0}^{N-1}$ makes up a linearly independent set, and for each $f \in l_N$ it holds that $f = \sum_{k=0}^{N-1} f(x_k) e_k$, which means that $\{e_n\}$ spans the space l_N. Thus, the dimension of the complex vector space is N.

In l_N we define (just as in Sec. 5.3) an *inner product* by putting

$$\langle f, g \rangle = \sum_{k=0}^{N-1} f(x_k) \, \overline{g(x_k)} \, .$$

The basis $\{e_n\}$ is orthonormal with respect to this inner product. We now proceed to construct another basis, which will prove to be orthogonal. Begin by letting $\omega = \omega_N = \exp(2\pi i/N)$ (ω is a so-called primitive Nth root of unity). Then define a function φ_n by putting $\varphi_n(x) = e^{inx}$, or, more precisely,

$$\varphi_n(x_k) = \exp\left(in \cdot \frac{k}{N} \cdot 2\pi\right) = \omega^{nk}.$$

Notice that this definition implies that $\overline{\varphi_n(x_k)} = \omega^{-nk}$.

Theorem B.1 *$\{\varphi_n\}_{n=0}^{N-1}$ is an orthogonal basis in l_N, and $\|\varphi_n\|^2 = N$.*

Proof. The number of vectors is right, so all that remains is to show that they are orthogonal. Let $0 \le m, n \le N-1$, and form

$$\langle \varphi_m, \varphi_n \rangle = \sum_{k=0}^{N-1} \omega^{mk}\omega^{-nk} = \sum_{k=0}^{N-1} \omega^{(m-n)k}.$$

If $m = n$, all the terms in the sum are equal to 1, which gives $\langle \varphi_n, \varphi_n \rangle = N$. If $m \ne n$, we have a finite geometric sum with the ratio $\omega^{m-n} \ne 1$. (Since $0 \le m, n \le N-1$, it must hold that $-N+1 \le m-n \le N-1$, and in that case $\omega^{m-n} = 1$ is possible only if $m - n = 0$.) The formula for a geometric sum then gives

$$\langle \varphi_m, \varphi_n \rangle = \frac{1 - \omega^{(m-n)N}}{1 - \omega^{m-n}}.$$

But $\omega^{(m-n)N} = \exp\left(\dfrac{2\pi i}{N} \cdot (m-n)N\right) = e^{2\pi i(m-n)} = 1$, which implies that $\langle \varphi_m, \varphi_n \rangle = 0$, and the theorem is proved. □

Since we have an orthogonal basis, we can represent an arbitrary $f \in l_N$ as $f = \sum c_n \varphi_n$, where the coefficients are given by

$$c_n = \frac{\langle f, \varphi_n \rangle}{\langle \varphi_n, \varphi_n \rangle} = \frac{1}{N} \langle f, \varphi_n \rangle.$$

It is also common to write $c_n = \widehat{f}(n)$, which results in the formula

$$\widehat{f}(n) = \frac{1}{N}\langle f, \varphi_n \rangle = \frac{1}{N} \sum_{k=0}^{N-1} f(x_k)\omega^{-kn} = \frac{1}{N} \sum_{k=0}^{N-1} f(x_k) e^{-i2\pi kn/N}.$$

With these Fourier coefficients we have thus

$$f = \sum_{n=0}^{N-1} \widehat{f}(n)\varphi_n,$$

or, written out in full,

$$\begin{aligned} f(x_k) &= \sum_{n=0}^{N-1} \widehat{f}(n)\varphi_n(x_k) = \sum_{n=0}^{N-1} \widehat{f}(n)\omega^{nk} = \sum_{n=0}^{N-1} \widehat{f}(n)e^{i2\pi kn/N} \\ &= \sum_{n=0}^{N-1} \widehat{f}(n)e^{inx_k}. \end{aligned}$$

Compare the "complex" form of an ordinary Fourier series. (We could formulate a "real" counterpart of this, too, but the formulae for that construction are messier.)

The theorem of Pythagoras or the Parseval formula for the system $\{\varphi_n\}$ looks like $\|f\|^2 = \sum_{n=0}^{N-1} |\widehat{f}(n)|^2 \|\varphi_n\|^2$, or

$$\frac{1}{N}\sum_{k=0}^{N-1} |f(x_k)|^2 = \sum_{n=0}^{N-1} |\widehat{f}(n)|^2.$$

In practical use, the computation of $\widehat{f}(n)$ can be speeded up by the use of an idea by Cooley and Tukey (1965) (the fast Fourier transform). The idea is that the right-hand member in the formula

$$\widehat{f}(n) = \frac{1}{N}\sum_{k=0}^{N-1} f(x_k)\omega^{-nk} = \frac{1}{N}\sum_{k=0}^{N-1} f(x_k)\left(\omega^{-n}\right)^k$$

can be seen as a polynomial in the variable ω^{-n}, which is swiftly computed using the method of Horner; this reduces the number of operations considerably. Further rationalizations are possible, using factorization of the number N. More about this can be found in books on numerical analysis.

Appendix C
Formulae

C.1 Laplace transforms

Take care to use the correct definition when dealing with distributions (cf. Sec 3.5, p. 57).

	$f(t)$	$\widetilde{f}(s) = F(s) = \mathcal{L}[f](s)$
General rules		
$\mathcal{L}$01.	$f(t)$	$\displaystyle\int_0^\infty e^{-st} f(t)\,dt$
$\mathcal{L}$02.	$\alpha f(t) + \beta g(t)$	$\alpha F(s) + \beta G(s)$
$\mathcal{L}$03.	$t^n f(t)$	$(-1)^n F^{(n)}(s)$
$\mathcal{L}$04.	$e^{-at} f(t)$, a constant	$F(s+a)$
$\mathcal{L}$05.	$f(t-a)\,H(t-a)$, $a > 0$	$e^{-as} F(s)$
$\mathcal{L}$06.	$f(at)$, $a > 0$	$\dfrac{1}{a} F\left(\dfrac{s}{a}\right)$
$\mathcal{L}$07.	$f'(t)$	$sF(s) - f(0)$
$\mathcal{L}$08.	$f^{(n)}(t)$	$s^n F(s) - s^{n-1} f(0) - s^{n-2} f'(0) - \cdots - f^{(n-1)}(0)$
$\mathcal{L}$09.	$\displaystyle\int_0^t f(u)\,du$	$\dfrac{F(s)}{s}$
$\mathcal{L}$10.	$\displaystyle f * g(t) = \int_0^t f(u) g(t-u)\,du$	$F(s)\,G(s)$

Laplace transforms of particular functions

$\mathcal{L}11.$	$\delta(t)$	1
$\mathcal{L}12.$	$\delta^{(n)}(t)$	s^n
$\mathcal{L}13.$	$H(t)$	$\dfrac{1}{s}$
$\mathcal{L}14.$	$t^n, \quad n = 0, 1, 2, \ldots$	$\dfrac{n!}{s^{n+1}}$
$\mathcal{L}15.$	e^{-at}	$\dfrac{1}{s+a}$
$\mathcal{L}16.$	$t^a\, e^{ct}, \quad a > -1, \; c \in \mathbf{C}$	$\dfrac{\Gamma(a+1)}{(s-c)^{a+1}}$
$\mathcal{L}17.$	$\cos bt$	$\dfrac{s}{s^2+b^2}$
$\mathcal{L}18.$	$\sin bt$	$\dfrac{b}{s^2+b^2}$
$\mathcal{L}19.$	$t \sin bt$	$\dfrac{2bs}{(s^2+b^2)^2}$
$\mathcal{L}20.$	$t \cos bt$	$\dfrac{s^2-b^2}{(s^2+b^2)^2}$
$\mathcal{L}21.$	$\dfrac{1}{2b^3}(\sin bt - bt \cos bt)$	$\dfrac{1}{(s^2+b^2)^2}$
$\mathcal{L}22.$	$\delta(t-a), \quad a \geq 0$	e^{-as}
$\mathcal{L}23.$	$\dfrac{\sin t}{t}$	$\arctan \dfrac{1}{s}$
$\mathcal{L}24.$	$\dfrac{a}{\sqrt{4\pi t^3}}\, e^{-a^2/(4t)}, \quad a > 0$	$e^{-a\sqrt{s}}$
$\mathcal{L}25.$	$l_n(t) = \dfrac{e^{t/n}}{n!} \dfrac{d^n}{dt^n}\left(t^n e^{-t}\right)$	$\dfrac{\left(s-\frac{1}{2}\right)^n}{\left(s+\frac{1}{2}\right)^{n+1}}$
$\mathcal{L}26.$	$\ln t$	$-\dfrac{\ln s + \gamma}{s}, \quad \gamma = 0.5772156\ldots$
$\mathcal{L}27.$	$\Gamma'(t) - \ln t$	$\dfrac{\ln s}{s}$
$\mathcal{L}28.$	$\dfrac{e^{bt} - e^{at}}{t}, \quad a, b \in \mathbf{R}$	$\ln \left\| \dfrac{s-a}{s-b} \right\|$
$\mathcal{L}29.$	$\operatorname{Erf}(\sqrt{t}) = \dfrac{2}{\sqrt{\pi}} \displaystyle\int_0^{\sqrt{t}} e^{-u^2}\, du$	$\dfrac{1}{s\sqrt{s+1}}$
$\mathcal{L}30.$	$\operatorname{Ei}(t) = \displaystyle\int_t^{\infty} \dfrac{e^{-u}}{u}\, du$	$\dfrac{\ln(s+1)}{s}$

$\mathcal{L}31.$	$\operatorname{Si}(t) = \int_t^\infty \frac{\sin u}{u}\,du$	$\frac{\arctan s}{s}$
$\mathcal{L}32.$	$\operatorname{Ci}(t) = \int_t^\infty \frac{\cos u}{u}\,du$	$\frac{\ln(s^2+1)}{2s}$
$\mathcal{L}33.$	$J_0(t)$ (Bessel function)	$\frac{1}{\sqrt{s^2+1}}$
$\mathcal{L}34.$	$J_0(2\sqrt{t})$	$\frac{e^{-1/s}}{s}$

$\Gamma(x) = \int_0^\infty t^{x-1}e^{-t}\,dt,\ x > 0.\quad \Gamma(x+1) = x\Gamma(x).$

$\Gamma(n+1) = n!$ for $n = 0, 1, 2, \ldots.\quad \Gamma(\frac{1}{2}) = \sqrt{\pi}.$

$$\gamma = \text{Euler's constant} = \lim_{n\to\infty}\left(\sum_{k=1}^{n}\frac{1}{k} - \ln n\right) \approx 0.5772156.$$

C.2 Z transforms

$$A(z) = Z[a](z) = \sum_{n=0}^{\infty} a_n z^{-n}, \quad |z| > \sigma = \sigma_a.$$

$$\text{Inversion:} \quad a_n = \frac{1}{2\pi i} \int_{|z|=r} A(z) z^{n-1}\, dz, \quad n \in \mathbf{N}, r > \sigma_a.$$

	a_n	$A(z)$
General rules		
$\mathcal{Z}1.$	$\lambda a_n + \mu b_n$	$\lambda A(z) + \mu B(z)$
$\mathcal{Z}2.$	$\lambda^n a_n$	$A(z/\lambda)$
$\mathcal{Z}3.$	$(k \geq 0): \quad a_{n+k}$	$z^k \left(A(z) - a(0) - \dfrac{a(1)}{z} - \cdots - \dfrac{a(k-1)}{z^{k-1}} \right)$
$\mathcal{Z}4.$	$n a_n$	$-zA'(z)$
$\mathcal{Z}5.$	$(a * b)_n = \sum_{k=0}^{n} a_k b_{n-k}$	$A(z)\, B(z)$
Particular sequences		
$\mathcal{Z}6.$	1	$\dfrac{z}{z-1}$
$\mathcal{Z}7.$	n	$\dfrac{z}{(z-1)^2}$
$\mathcal{Z}8.$	n^2	$\dfrac{z^2+z}{(z-1)^3}$
$\mathcal{Z}9.$	λ^n	$\dfrac{z}{z-\lambda}$
$\mathcal{Z}10.$	$n\lambda^n$	$\dfrac{\lambda z}{(z-\lambda)^2}$
$\mathcal{Z}11.$	$(n+1)\lambda^n$	$\dfrac{z^2}{(z-\lambda)^2}$
$\mathcal{Z}12.$	$\binom{n+m}{m} \lambda^n$	$\dfrac{z^{m+1}}{(z-\lambda)^{m+1}}$
$\mathcal{Z}13.$	$\binom{n}{m} \lambda^n$	$\dfrac{\lambda^m z}{(z-\lambda)^{m+1}}$
$\mathcal{Z}14.$	$\cos \alpha n$	$\dfrac{z^2 - z\cos\alpha}{z^2 - 2z\cos\alpha + 1}$
$\mathcal{Z}15.$	$\sin \alpha n$	$\dfrac{z \sin\alpha}{z^2 - 2z\cos\alpha + 1}$
$\mathcal{Z}16.$	$\dfrac{\lambda^n}{n!}$	$e^{\lambda/z}$

C.3 Fourier series

$$f(x) \sim \sum_{-\infty}^{\infty} c_n e^{inx} \sim \frac{a_0}{2} + \sum_{n=1}^{\infty} (a_n \cos nx + b_n \sin nx)$$

$$\text{where} \quad c_n = \frac{1}{2\pi} \int_{-\pi}^{\pi} f(x)\, e^{-inx}\, dx \quad \text{resp.} \quad \begin{cases} a_n = \frac{1}{\pi} \int_{-\pi}^{\pi} f(x) \cos nx\, dx \\ b_n = \frac{1}{\pi} \int_{-\pi}^{\pi} f(x) \sin nx\, dx \end{cases}.$$

$$\begin{cases} c_n = \frac{1}{2}(a_n - i b_n), & n \geq 0 \\ c_{-n} = \frac{1}{2}(a_n + i b_n), & n \geq 0 \end{cases} \qquad \begin{cases} a_n = c_n + c_{-n}, & n \geq 0 \\ b_n = i(c_n - c_{-n}), & n \geq 1 \end{cases}$$

$$f \text{ even} \Longrightarrow b_n = 0 \text{ and } a_n = \frac{2}{\pi} \int_0^{\pi} f(x) \cos nx\, dx.$$

$$f \text{ odd} \Longrightarrow a_n = 0 \text{ and } b_n = \frac{2}{\pi} \int_0^{\pi} f(x) \sin nx\, dx.$$

$$\text{Parseval:} \quad \begin{cases} \displaystyle \frac{1}{2\pi} \int_{-\pi}^{\pi} |f(x)|^2\, dx = \sum_{n=-\infty}^{\infty} |c_n|^2 \\ \displaystyle \frac{1}{\pi} \int_{-\pi}^{\pi} |f(x)|^2\, dx = \frac{|a_0|^2}{2} + \sum_{n=1}^{\infty} \left(|a_n|^2 + |b_n|^2\right) \end{cases}$$

$$\text{Polarized Parseval:} \quad \begin{cases} \displaystyle \frac{1}{2\pi} \int_{-\pi}^{\pi} f(x)\, \overline{g(x)}\, dx = \sum_{n=-\infty}^{\infty} c_n\, \overline{\gamma_n} \\ \displaystyle \frac{1}{\pi} \int_{-\pi}^{\pi} f(x)\, \overline{g(x)}\, dx = \frac{a_0\, \overline{\alpha_0}}{2} + \sum_{n=1}^{\infty} \left(a_n\, \overline{\alpha_n} + b_n\, \overline{\beta_n}\right) \end{cases}$$

If f has period $2P$, then $(P\Omega = \pi)$

$$f(x) \sim \sum_{-\infty}^{\infty} c_n\, e^{in\Omega x} \sim \frac{a_0}{2} + \sum_{n=1}^{\infty} (a_n \cos n\Omega x + b_n \sin n\Omega x),$$

$$\frac{1}{2P} \int_a^{a+2P} |f(x)|^2\, dx = \sum_{-\infty}^{\infty} |c_n|^2,$$

$$\frac{1}{P} \int_a^{a+2P} |f(x)|^2\, dx = \frac{|a_0|^2}{2} + \sum_{n=1}^{\infty} \left(|a_n|^2 + |b_n|^2\right),$$

$$c_n = \frac{1}{2P} \int_a^{a+2P} f(x) e^{-in\Omega x}\, dx, \qquad \begin{matrix} a_n \\ b_n \end{matrix} = \frac{1}{P} \int_a^{a+2P} f(x) \begin{matrix} \cos \\ \sin \end{matrix} n\Omega x\, dx.$$

C.4 Fourier transforms

General rules

		$f(t)$	$\widehat{f}(\omega)$
$\mathcal{F}01.$		$f(t)$	$\displaystyle\int_{-\infty}^{\infty} f(t)\, e^{-i\omega t}\, dt$
$\mathcal{F}02.$		$\displaystyle\frac{1}{2\pi}\int_{-\infty}^{\infty} \widehat{f}(\omega)\, e^{i\omega t}\, d\omega$	$\widehat{f}(\omega)$
$\mathcal{F}03.$		f even $\iff \widehat{f}$ even,	f odd $\iff \widehat{f}$ odd
$\mathcal{F}04.$	Linearity	$\alpha f(t) + \beta g(t)$	$\alpha \widehat{f}(\omega) + \beta \widehat{g}(\omega)$
$\mathcal{F}05.$	Scaling	$f(at) \quad (a \neq 0)$	$\displaystyle\frac{1}{\lvert a\rvert}\, \widehat{f}\Big(\frac{\omega}{a}\Big)$
$\mathcal{F}06.$		$f(-t)$	$\widehat{f}(-\omega)$
$\mathcal{F}07.$		$\overline{f(t)}$	$\overline{\widehat{f}(-\omega)}$
$\mathcal{F}08.$	Time translation	$f(t-T)$	$e^{-iT\omega}\, \widehat{f}(\omega)$
$\mathcal{F}09.$	Frequency translation	$e^{i\Omega t}\, f(t)$	$\widehat{f}(\omega - \Omega)$
$\mathcal{F}10.$	Symmetry	$\widehat{g}(t)$	$2\pi g(-\omega)$
$\mathcal{F}11.$	Time derivative	$\displaystyle\frac{d^n}{dt^n}\, f(t)$	$(i\omega)^n\, \widehat{f}(\omega)$
$\mathcal{F}12.$	Frequency derivative	$(-it)^n\, f(t)$	$\displaystyle\frac{d^n}{d\omega^n}\, \widehat{f}(\omega)$
$\mathcal{F}13.$	Time convolution	$\displaystyle\int_{-\infty}^{\infty} f(t-u)\, g(u)\, du$	$\widehat{f}(\omega)\, \widehat{g}(\omega)$
$\mathcal{F}14.$	Frequency convolution	$f(t)\, g(t)$	$\displaystyle\frac{1}{2\pi}\int_{-\infty}^{\infty} \widehat{f}(\omega - \alpha)\, \widehat{g}(\alpha)\, d\alpha$

Plancherel's formulae:

$$\int_{-\infty}^{\infty} f(t)\, \overline{g(t)}\, dt = \frac{1}{2\pi}\int_{-\infty}^{\infty} \widehat{f}(\omega)\, \overline{\widehat{g}(\omega)}\, d\omega,$$

$$\int_{-\infty}^{\infty} \lvert f(t)\rvert^2\, dt = \frac{1}{2\pi}\int_{-\infty}^{\infty} \lvert \widehat{f}(\omega)\rvert^2\, d\omega.$$

Fourier transforms of particular functions

	$f(t)$	$\widehat{f}(\omega)$
$\mathcal{F}15.$	$\delta(t)$	1
$\mathcal{F}16.$	$\delta^{(n)}(t)$	$(i\omega)^n$
$\mathcal{F}17.$	$f(t) = 1$ for $\lvert t\rvert < 1$, $= 0$ otherwise	$\dfrac{2\sin\omega}{\omega}$
$\mathcal{F}18.$	$f(t) = 1 - \lvert t\rvert$ for $\lvert t\rvert < 1$, $= 0$ otherwise	$\left(\dfrac{2\sin\frac{1}{2}\omega}{\omega}\right)^2$
$\mathcal{F}19.$	$e^{-t}H(t)$	$\dfrac{1}{1+i\omega}$
$\mathcal{F}20.$	$e^{t}(1-H(t))$	$\dfrac{1}{1-i\omega}$
$\mathcal{F}21.$	$e^{-\lvert t\rvert}$	$\dfrac{2}{1+\omega^2}$
$\mathcal{F}22.$	$e^{-\lvert t\rvert}\operatorname{sgn} t$	$\dfrac{-2i\omega}{1+\omega^2}$
$\mathcal{F}23.$	$\operatorname{sgn} t$	$\dfrac{2}{i\omega}$
$\mathcal{F}24.$	$H(t)$	$\dfrac{1}{i\omega} + \pi\delta(\omega)$
$\mathcal{F}25.$	1	$2\pi\delta(\omega)$
$\mathcal{F}26.$	$\dfrac{\sin\Omega t}{\pi t}$	$H(\omega+\Omega) - H(\omega-\Omega)$
$\mathcal{F}27.$	$\dfrac{1}{\sqrt{2\pi}}e^{-t^2/2}$	$e^{-\omega^2/2}$
$\mathcal{F}28.$	$\dfrac{1}{\sqrt{4\pi A}}e^{-t^2/(4A)} \quad (A>0)$	$e^{-A\omega^2}$

C.5 Orthogonal polynomials

$L^2_w(a,b)$: $\langle f,g\rangle = \int_a^b f(t)\,\overline{g(t)}\,w(t)\,dt$, $\|f\| = \sqrt{\langle f,f\rangle}$.

$\delta_{kn} = 1$ if $k=n$, $\delta_{kn} = 0$ if $k \neq n$ ("KRONECKER delta").

LEGENDRE polynomials $P_n(x)$: $(a,b) = (-1,1)$, $w(t) \equiv 1$.

P1. $P_n(t) = \dfrac{1}{2^n n!} D^n\big((t^2-1)^n\big)$.

P2. $P_0(t) = 1$, $P_1(t) = t$, $P_2(t) = \frac{1}{2}(3t^2-1)$, $P_3(t) = \frac{1}{2}(5t^3-3t)$.

P3. $(1-2tz+z^2)^{-1/2} = \displaystyle\sum_{n=0}^{\infty} P_n(t)\,z^n \quad (|z|<1,\ |t|\le 1)$.

P4. $(n+1)P_{n+1}(t) = (2n+1)t\,P_n(t) - n\,P_{n-1}(t)$.

P5. $\displaystyle\int_{-1}^{1} P_n(t)\,P_k(t)\,dt = \frac{2}{2n+1}\,\delta_{nk}$.

P6. $(1-t^2)P_n''(t) - 2t\,P_n'(t) + n(n+1)P_n(t) = 0$.

LAGUERRE polynomials $L_n(t)$: $(a,b) = (0,\infty)$, $w(t) = e^{-t}$.

L1. $L_n(t) = \dfrac{e^t}{n!} D^n\big(t^n e^{-t}\big)$.

L2. $L_0(t) = 1$, $L_1(t) = 1-t$, $L_2(t) = 1 - 2t + \frac{1}{2}t^2$.

L3. $\dfrac{1}{1-z}\exp\left(\dfrac{-tz}{1-z}\right) = \displaystyle\sum_{n=0}^{\infty} L_n(t)\,z^n \quad (|z|<1)$.

L4. $(n+1)L_{n+1}(t) = (2n+1-t)L_n(t) - nL_{n-1}(t)$.

L5. $\displaystyle\int_0^{\infty} L_k(t)\,L_n(t)\,e^{-t}\,dt = \delta_{kn}$.

L6. $tL_n''(t) + (1-t)L_n'(t) + nL_n(t) = 0$.

HERMITE polynomials $H_n(t)$: $(a,b) = (-\infty,\infty)$, $w(t) = e^{-t^2}$.

H1. $H_n(t) = (-1)^n e^{t^2} D^n\big(e^{-t^2}\big)$.

H2. $H_0(t) = 1$, $H_1(t) = 2t$, $H_2(t) = 4t^2 - 2$, $H_3(t) = 8t^3 - 12t$.

H3. $e^{2tz-z^2} = \displaystyle\sum_{n=0}^{\infty} H_n(t)\,\frac{z^n}{n!}$.

H4. $H_{n+1}(t) = 2tH_n(t) - 2nH_{n-1}(t)$.

H5. $\displaystyle\int_{-\infty}^{\infty} H_k(t)\,H_n(t)\,e^{-t^2}\,dt = n!\,2^n\sqrt{\pi}\,\delta_{kn}$.

H6. $H_n''(t) - 2tH_n'(t) + 2nH_n(t) = 0$.

CHEBYSHEV polynomials $T_n(t)$: $(a, b) = (-1, 1)$, $w(t) = 1/\sqrt{1-t^2}$.

T1. $T_n(t) = \cos(n \arccos t)$, $T_n(\cos\theta) = \cos n\theta$, $0 \le \theta \le \pi$.

T2. $T_0(t) = 1$, $T_1(t) = t$, $T_2(t) = 2t^2 - 1$, $T_3(t) = 4t^3 - 3t$.

T3. $T_n(t) = 2tT_{n-1}(t) - T_{n-2}(t)$.

T4. $\displaystyle\int_{-1}^{1} T_k(t)\, T_n(t) \frac{dt}{\sqrt{1-t^2}} = \tfrac{1}{2}\pi\delta_{kn}$ if $k > 0$ or $n > 0$;
$= \pi$ if $k = n = 0$.

T5. $(1-t^2)T_n''(t) - tT_n'(t) + n^2 T_n(t) = 0$.

Appendix D
Answers to selected exercises

Chapter 1

1.1. $u(x,t) = \frac{1}{2}(e^{-(x-ct)^2} + e^{-(x+ct)^2}) + \dfrac{1}{2c}(\arctan(x+ct) - \arctan(x-ct))$.

Chapter 2

2.1. i, $(1-i)/\sqrt{2}$, $(-\sqrt{3}+i)/2$, $\sqrt{3}-i$.

2.3. $\cos 3t = 4\cos^3 t - 3\cos t$.

2.12. $\lim a_k/k = 0$.

2.23. $\delta(2t) = \frac{1}{2}\delta(t)$; $\delta(at) = \dfrac{1}{|a|}\,\delta(t)$.

2.25. $\chi(t)\delta_a''(t) = \chi(a)\delta_a''(t) - 2\chi'(a)\delta_a'(t) + \chi''(a)\delta_a(t)$.

2.27. $f''(x) = -2H(x+1) + 2H(x-1) + 2\delta(x+1) + 2\delta(x-1)$;
$(x^2-1)f''(x) = 2f(x)$.

2.29. $y = e^{a^2-t^2}H(t-a) + Ce^{-t^2}$, where C is an arbitrary constant.

2.31. $y = \frac{1}{4}(1+x^2)H(x-1) + 1 + x^2$.

2.33. $y = (1-e^{-x^2})H(x) - e^{1-x^2}H(x-1) + (1+e)e^{-x^2}$.

2.36. $\varphi'(0)$.

2.38. $f'(t) = 2tH(t)$, $f''(t) = 2H(t)$, $f'''(t) = 2\delta(t)$, $f^{(4)}(t) = 2\delta'(t)$.

2.40. $f'''(x) = 24x(H(x+1) - H(x-1) + 8(\delta(x+1) - \delta(x-1))$.

Chapter 3

3.3. $\widetilde{f}(s) = (1 + e^{-\pi s})/(s^2+1)$.

3.4. (a) $\dfrac{4}{s^3} - \dfrac{1}{2+1}$ (b) $\dfrac{s^4+4s^2+24}{s^5}$. (c) $\dfrac{1}{s} - \dfrac{2}{s^2+4}$.

3.6. $\widetilde{f}(s) = \dfrac{2e^{-s}}{s^3}$.

3.7. $\widetilde{f}(s) = \dfrac{1}{s}\ln\left|\dfrac{s+1}{s}\right|$.

3.9. $\widetilde{f}(s) = \dfrac{\pi e^{-\pi s}}{s^2+1} + \dfrac{2s(e^{-\pi s}+1)}{(s^2+1)^2}$.

3.11. $\dfrac{2(s+1)}{((s+1)^2+1)^2}$.

3.14. $\dfrac{1-(1+s)e^{-s}}{s^2(1-e^{-s})}$.

3.17. (a) $1 - e^{-t}$. (b) $3te^t$. c) $\frac{1}{4}\big(1 - (2t+1)e^{-2t}\big)$.

3.18. (a) $e^{-t} - 1 + t$. (b) $1 - \cos bt$.

3.19. (a) $f(t) = 1 + H(t-1)$. (b) $f(t) = (e^{2(t-1)} - e^{t-1})H(t-1)$.
(c) $\dfrac{e^{-2t} - e^{-3t}}{t}$.

3.21. $y = e^t(2 - 2\cos t - t\sin t)$.

3.23. $x = 2 + \frac{1}{2}(t^2 + e^{-t} + \cos t - 3\sin t)$, $y = \frac{1}{2}(2 - e^{-t} - \cos t + 3\sin t)$.

3.25. $y(t) = 2e^t - e^{2t} + \frac{1}{2}(1 - e^{2(t-2)} - 2e^{t-2})H(t-2)$.

3.27. $y = e^t - t - 1$.

3.29. $\dfrac{1}{a-b}(e^{at} - e^{bt})$ if $a \neq b$; te^{at} if $a = b$.

3.31. $\frac{1}{2}(t\cos t + \sin t)$.

3.33. $f(t) = 3$.

3.35. $y(t) = e^{-t}\sin t$.

3.37. $f(t) = \frac{5}{2}(1 - e^{-2t}) - 4te^{-t}$.

3.39. $y = \big(1 - \frac{1}{2}(t^2+1)e^{1-t}\big)H(t-1) + \frac{1}{2}(t-2)^2e^{2-t}H(t-2)$.

3.41. (a) $E(t) = \frac{1}{2}\sin 2t\,H(t)$. (b) $E(t) = \frac{1}{2}e^{-2t}\sin 2t\,H(t)$.
(c) $E(t) = \frac{1}{2}t^2e^{-t}H(t)$.

3.43. $f(t) = (2 - e^{-t})\,H(t)$.

3.44. (a) $\dfrac{2z}{2z-1}$. (b) $\dfrac{3z}{(z-3)^2}$. (c) $\dfrac{2z^2+4z}{(z-2)^3}$. (d) $\dfrac{z}{(z-1)^{p+1}}$.

3.45. (a) $a_n = \frac{1}{3}\left(\frac{2}{3}\right)^n$. (b) $a_1 = 1$, $a_n = 0$ for all $n \neq 1$.

3.47. $a_n = 1 - \cos\dfrac{n\pi}{2}$, $b_n = 1 - \sin\dfrac{n\pi}{2}$.

3.49. $a_n = (-1)^n \binom{n}{3} = \frac{1}{6}(-1)^n(n^3 - 3n^2 + 2n)$.

3.51. $y(0) = \frac{1}{3}$, $y(1) = -\frac{5}{3}$, $y(n) = \frac{4}{3}$ for all $n \geq 2$.

3.53. $x(n) = \frac{1}{5} \cdot 2^n + \frac{4}{5} \cos \dfrac{n\pi}{2} - \frac{2}{5} \sin \dfrac{n\pi}{2}$.

3.55. (a) is stable, (b) and (c) are unstable.

3.57. $y = e^t - e^{-t} \sin t$.

3.59. $y = \frac{1}{2} \sin 2t + \left(\frac{1}{4}(t-1)^2 - \frac{1}{8} + \frac{1}{8} \cos 2(t-1)\right) H(t-1)$.

3.61. $f(t) = e^t \cos 2t$, $t > 0$.

3.63. $y(t) = \frac{1}{2}t^2 - t + 1 + \cos t + \sin t$.

3.65. $y(t) = \sin t$, $z(t) = e^{-t} - \cos t$.

3.67. $y(t) = 2(t+1) \sin t$, $z = 2e^t - 2(t+1) \cos t$.

3.69. $f(t) = 3t - 3 + 8e^{-t} + \cos\left(t\sqrt{2}\right) - \dfrac{1}{\sqrt{2}} \sin\left(t\sqrt{2}\right)$.

3.71. $y(t) = 2t - 1 + \sin 2t$.

3.73. $y(t) = \frac{1}{2} t^2$.

3.75. $y(t) = 3t + 5$.

3.77. $y(t) = 4 \sin 2t - 2 \sin t$.

Chapter 4

4.4. $f(t) \sim 1 + 2 \displaystyle\sum_{n=1}^{\infty} \frac{(-1)^{n+1}}{n} \sin nt$.

4.6. (a) $f(t) \sim \cos 2t$. (b) $g(t) \sim \frac{1}{2} + \frac{1}{2} \cos 2t$. (c) $h(t) \sim \frac{3}{4} \sin t - \frac{1}{4} \sin 3t$. *Sens moral:* If a function consists entirely of terms that can be terms in a Fourier series, then the function *is* its own Fourier series.

4.9. $f(t) \sim \dfrac{1 - e^{-\pi}}{\pi} + \dfrac{2}{\pi} \displaystyle\sum_{n=1}^{\infty} \frac{1 - (-1)^n e^{-\pi}}{1 + n^2} \cos nt$.

4.12. $f(t) \sim -\frac{7}{15} \pi^4 - 48 \displaystyle\sum_{n=1}^{\infty} \frac{(-1)^n}{n^4} \cos nt$; $\zeta(4) = \dfrac{\pi^4}{90}$.

4.16. If a has the form $n^2 + (-1)^n$ for some integer $n \neq 0$, then the problem has the solutions $y(t) = Ae^{int} + Be^{-int}$, where A and B are arbitrary constants (the solutions can also be written in "real" form as $y(t) = C_1 \cos nt + C_2 \sin nt$). If $a = 1$ there are the solutions $y(t) =$ constant. For other values of a there are no nontrivial solutions.

4.18. $f(t) \sim 1 - \frac{1}{2} \cos t + 2 \displaystyle\sum_{n=2}^{\infty} \frac{(-1)^{n+1}}{n^2 - 1} \cos nt$. Converges to $f(t)$ for all t.

(Sketch the graph of f!)

4.20. (a) $f(t) \sim \dfrac{\pi}{4} + \dfrac{2}{\pi} \displaystyle\sum_{n=1}^{\infty} \frac{\cos(2n-1)t}{(2n-1)^2} - \sum_{n=1}^{\infty} \frac{\sin nt}{n}$; (b) $\frac{1}{8} \pi^2$.

4.22. $\cos\alpha t \sim \dfrac{\sin\alpha\pi}{\alpha\pi} - \dfrac{2\alpha\sin\alpha\pi}{\pi}\sum_{n=1}^{\infty}\dfrac{(-1)^n}{n^2-\alpha^2}\cos nt$. The series converges for all t to the periodic continuation of $\cos\alpha t$. Substitute $t=\pi$, divide by $\sin\alpha\pi$, and stir around; and the formula for the cotangent will materialize.

4.24. $y(t) = \frac{1}{2} + \dfrac{1}{\pi}\sum_{n\in\mathbf{Z}\setminus\{0\}}\dfrac{\sin\frac{1}{2}n\pi}{n\left(1-\frac{1}{4}n^2+\frac{1}{2}in\right)}e^{int/2}$.

4.26. $f(x) \sim \dfrac{8}{\pi^3}\sum_{k=0}^{\infty}\dfrac{\sin(2k+1)\pi x}{(2k+1)^3}$; the particular sum is $\dfrac{\pi^3}{32}$.

4.28. $f(t) \sim -\dfrac{8}{\pi}\sum_{n=1}^{\infty}\dfrac{n(-1)^n}{4n^2-1}\sin nt$.

4.30. $f(t) \sim \dfrac{\pi}{2} + \dfrac{2}{\pi}\sum_{n\text{ odd}}\dfrac{e^{int}}{n^2}$.

4.32. (a) $f(t) \sim \frac{4}{2} - 4\sin 2t + 7\cos 3t$.

(b) $f(t) \sim \dfrac{2}{\pi} - \dfrac{4}{\pi}\sum_{n=1}^{\infty}\dfrac{\cos 2nt}{4n^2-1}$.

4.33. The same as 4.32 (b) (draw pictures, as always!).

4.35. $f(t) \sim \dfrac{b-a}{2\pi} + \dfrac{1}{2\pi i}\sum_{n\in\mathbf{Z}\setminus\{0\}}\dfrac{e^{-ina}-e^{-inb}}{n}e^{int}$. It is convergent for all t. The sum is $s(t)=1$ for $a<t<b$, $s(t)=\frac{1}{2}$ for $t=a \vee t=b$, $s(t)=0$ for all other $t\in[-\pi,\pi]$.

4.37. $f(t) \sim -\frac{1}{6} + \dfrac{1}{\pi^2}\sum_{n=1}^{\infty}\dfrac{\cos 2\pi nt}{n^2}$.

4.39. $r(t) \sim \sum_{n\in\mathbf{Z}}|c_n|^2 e^{2\pi int/T}$.

4.41. $f(x) \sim \dfrac{2}{\pi} - \dfrac{4}{\pi}\sum_{n=1}^{\infty}\dfrac{\cos nx}{4n^2-1}$; $s_1=\frac{1}{2}$, $s_2=\frac{1}{2}-\frac{1}{4}\pi$.

4.43. $y(t) = c_0 + \cos t$, where c_0 is any constant.

4.45. $f(x) \sim \dfrac{1}{\pi}\sum_{n\in\mathbf{Z}}\dfrac{(-1)^n\sin\alpha\pi}{\alpha-n}e^{inx}$.

4.48. $f(x) \sim \frac{7}{15} + \dfrac{48}{\pi^4}\sum_{n=1}^{\infty}\dfrac{(-1)^n}{n^4}\cos n\pi x$; $\zeta(4) = \dfrac{\pi^4}{90}$.

Chapter 5

5.1. $\|\mathbf{u}\| = \sqrt{19}$, $\|\mathbf{v}\| = \sqrt{11}$, $\langle\mathbf{u},\mathbf{v}\rangle = 1+8i$.

5.3. Yes.

5.5. (a) $(1,2,3)$, $(5,-4,1)$, $(1,1,-1)$. (b) 1, x, $x^2-\frac{1}{3}$.

5.7. $p(x) = 3x + \frac{1}{2}(e^2 - 7)$.

5.9. $p(x) = \dfrac{\pi}{2\pi^2 - 16}\, x$.

5.11. $\pi^{-1/4}$, $\pi^{-1/4}\sqrt{2}\cdot x$, $\pi^{-1/4}\sqrt{2}(x^2 - \frac{1}{2})$.

5.13. $c_0 = \frac{3}{8}$, $c_2 = \frac{1}{2}$, $c_1 = c_3 = 0$.

5.18. (a) $p(x) = \frac{1}{2} + \frac{45}{32}\, x - \frac{35}{32}\, x^3$. (b) $p(x) = \frac{15}{4}(3 - \pi)x^2 + \frac{3}{4}(2\pi - 5)$.

5.27. $p(x) = \dfrac{1}{3\sqrt{\pi}}(7 - 2x^2)$.

5.29. First $\frac{3}{16} + \frac{15}{16}x^2$; second $\dfrac{2}{3\pi}(1 + 4x^2)$; third $\frac{5}{32} + \frac{35}{32}x^2$.

5.31. $s_1 = \frac{1}{18}$, $s_2 = (2 + 3\pi)/36$, $s_3 = \frac{1}{144}\,\pi^2 - \frac{1}{162}$.

5.33. $\zeta(8) = \frac{1}{9450}\pi^8$.

5.35. $\|\varphi_{mn}\| = 2\pi$.

5.37. $f(x) \sim 6L_0(x) - 18L_1(x) + 18L_2(x) - 6L_3(x)$.

5.39. $\displaystyle\int_{-1}^{1} |f(x)|^2\, dx = \sum_{n=0}^{\infty}(n + \tfrac{1}{2})|c_n|^2$.

5.41. The coefficients are $\sin 1$ resp. $(2\cos 1 + \sin 1 - 2)\sqrt{3}$.

5.43. $a_0 = -\frac{3}{35}$, $a_2 = \frac{6}{7}$, $a_1 = a_3 = 0$.

5.45. $a = \dfrac{3(20 - \pi^2)}{\pi^3}$, $b = 0$, $c = \dfrac{15(\pi^2 - 12)}{\pi^3}$.

5.47. $P(x) = \dfrac{4}{15\pi}(11 - 12x^2)$.

Chapter 6

6.1. (a) $u(x,t) = \frac{3}{4}e^{-t}\sin x - \frac{1}{4}e^{-9t}\sin 3x$.

(b) $u(x,t) = \dfrac{8}{\pi}\displaystyle\sum_{k=1}^{\infty}\frac{k}{4k^2 - 9}\, e^{-4k^2 t}\sin 2kx$.

6.3. $u(x,t) = \frac{1}{2}(1 + e^{-9t}\cos 3x)$.

6.5. $u(x,t) = (2e^{-t} - 1)\sin x + e^{-4t}\sin 2x$.

6.7. The solution is $u(x,t) = \dfrac{4a}{\pi}\displaystyle\sum_{k=0}^{\infty}\frac{(-1)^k}{(2k+1)^2}\cos(2k+1)t\sin(2k+1)x$.

Only partials with odd numbers are heard, which is natural because the even partials have vibration nodes at the middle point of the string.

6.9. $u(x,t) = \dfrac{2}{\pi}\displaystyle\sum_{k=0}^{N-1}\frac{n(-1)^n}{n^2 + h}\, e^{-(n^2+h)t}\sin nx + \frac{\sinh x\sqrt{h}}{\sinh \pi\sqrt{h}}$.

6.11. $u(x,t) = \sin\dfrac{x}{2} + \dfrac{2}{\pi}\displaystyle\sum_{n=1}^{\infty}\frac{(-1)^n n}{n^2 - \frac{1}{4}}\, e^{(\frac{1}{4} - n^2)t}\sin nx$.

6.13. $u(x,y) = \frac{3}{4} + \frac{1}{4}(x^4 - 6x^2y^2 + y^4)$.

6.14. $u(r,\theta) = \frac{3}{4}\, r\sin\theta - \frac{1}{4}\, r^3\sin 3\theta$.

6.17. $\varphi_n(x) = \sin \omega_n x$, where ω_n are the positive solutions of the equation $\tan(\omega\pi) = -\omega$, $n = 1, 2, 3, \ldots$. Draw a picture: it holds that $n - \frac{1}{2} < \omega_n < n$.

6.21. $u(x,t) = \frac{3}{4} e^{-t} t \sin x - \dfrac{1}{8\sqrt{2}} e^{-t} \sin\bigl(t\sqrt{8}\bigr) \sin 3x$.

6.23. $u(x,t) = 10(x+1) + \dfrac{40}{\pi} \displaystyle\sum_{n=1}^{\infty} \frac{1}{n} e^{-n^2\pi^2 t} \sin \frac{n\pi x}{2}$.

6.25. $u(x,y) = \dfrac{2}{\pi^3} \displaystyle\sum_{n=1}^{\infty} \frac{(-1)^{n+1}}{n^3(e^{n\pi}+1)} \bigl(e^{n\pi y} + e^{n\pi - n\pi y}\bigr) \sin n\pi x + \frac{1}{6}\bigl(x^3 - x\bigr)$.

6.27. $u(x,y) = \dfrac{\pi - y}{2\pi} + \dfrac{e^{2y} - e^{4\pi - 2y}}{2(e^{4\pi} - 1)} \cos 2x$.

6.29. $u(x,t) = e^{-t}\bigl((1+t)\sin x + (\cos t\sqrt{8} + \frac{1}{\sqrt{8}} \sin t\sqrt{8}) \sin 3x\bigr)$.

6.31. $u(x,t) = \dfrac{8}{\pi} \displaystyle\sum_{k=1}^{\infty} \frac{\cos(2k-1)^2 t \sin(2k-1)x}{(2k-1)^3}$.

6.33. (a) $u(x,t) = \dfrac{2}{\pi} \displaystyle\sum_{n=1}^{\infty} \frac{\sin na}{n} \sin nt \sin nx$.

(b) a should satisfy $\sin 7a = 0$, i.e., $a = \dfrac{k\pi}{7}$ for some $k = 1, 2, \ldots, 6$. For practical reasons one prefers $k = 1$ or 6 for a grand piano. (For an upright piano some other value may be more practical.)

Chapter 7

7.1. (a) $\widehat{f}(\omega) = 2i \dfrac{\omega \cos \omega - \sin \omega}{\omega^2}$, $\omega \neq 0$; $\widehat{f}(0) = 0$.

(b) $\widehat{f}(\omega) = 2(1 - \cos\omega)/(\omega^2) = 4\sin^2(\omega/2)/(\omega^2)$, $\omega \neq 0$; $\widehat{f}(0) = 1$.

(c) and (d) $f \notin L^1(\mathbf{R})$, and $\widehat{f}$ does not exist.

7.5. (a) $\widehat{f}(\omega) = 2(2+\omega^2)/(\omega^4+4)$, (b) $\widehat{g}(\omega) = -4i\omega/(\omega^4+4)$.

7.7. (a) $-i\sqrt{2\pi}\,\omega \exp\bigl(-\frac{1}{2}\omega^2\bigr)$. (b) See the remark following the exercises.

7.9. No (because $1 - \cos\omega$ does not tend to 0 as $\omega \to \infty$).

7.11. (a) $\pi e^{i\omega - |\omega|}$, (b) $\frac{1}{2}\pi\, e^{3i\omega - 2|\omega|}$, (c) $-\frac{1}{2}\pi i\omega e^{-|\omega|}$.

7.13. $f(x) = e^{-x}$ for $x > 0$.

7.17. $f_{a_1} * f_{a_2} = f_{a_1 + a_2}$. In general, $\mathop{*}\limits_{k=1}^{n} f_{a_k} = f_{\Sigma_{k=1}^n a_k}$.

7.19. $f(t) = \dfrac{1}{\sqrt{\pi}} \exp\bigl(-\frac{1}{2}t^2\bigr)$.

7.21. $\dfrac{\pi \sin 5t}{t}$ for $t \neq 0$, 5π for $t = 0$.

7.23. The value of the integral is $\frac{1}{2}\pi$.

7.25. Boundary values are 0 for all $x \neq 0$, ∞ for $x = 0$ (if one approaches the boundary at right angles).

7.26. $u(x,t) = \dfrac{1}{\sqrt{1+4t}} \exp\left(-\dfrac{x^2}{1+4t}\right) + \dfrac{1}{\sqrt{1+2t}} \exp\left(-\dfrac{x^2}{2+4t}\right)$.

7.29. $u(x,y) \doteq \dfrac{y+1}{x^2+(y+1)^2}$, $y \geq 0$.

7.31. $\cos t$; if $t = 0$, the integral is $\frac{1}{2}$.

7.33. The solution that is attainable by Fourier transformation is $y(t) = (e^{-t} - e^{-2t})H(t)$.

7.35. $\widehat{f}(\omega) = \dfrac{2}{(1+\omega^2)(2 - e^{i\omega} + i\omega)}$.

7.37. (a) $\frac{1}{3}\,\pi e^{-|\omega|/3}$. (b) $\frac{1}{3}\,\pi e^{-|\omega-1|/3}$. (c) $\dfrac{\pi}{6i}\left(e^{-|\omega-1|/3} - e^{-|\omega+1|/3}\right)$.

7.39. (a) $\widehat{f}(\omega) = 4i \cdot \dfrac{\omega \cos\omega - \sin\omega}{\omega^2}$; (c) $\dfrac{\pi}{3}$.

7.41. $\widehat{f}(\omega) = \dfrac{2\sin\pi\omega}{i(1-\omega^2)}$. The integral is $\dfrac{\pi^2}{2}$.

7.43. $\widehat{f}(\omega) = 4\,\dfrac{\sin\omega - \omega\cos\omega}{\omega^3}$. The integrals are $\pi/2$ resp. $\frac{2}{15}\pi$.

7.45. $f(x) = \dfrac{1}{2\pi}\,\dfrac{2(a-1)}{(a-1)^2+x^2}$, $a > 1$.

7.47. $r_{xx} = \frac{1}{2}\,A_1^2 \cos\omega_1 t + \frac{1}{2}\,A_2^2 \cos\omega_2 t$.

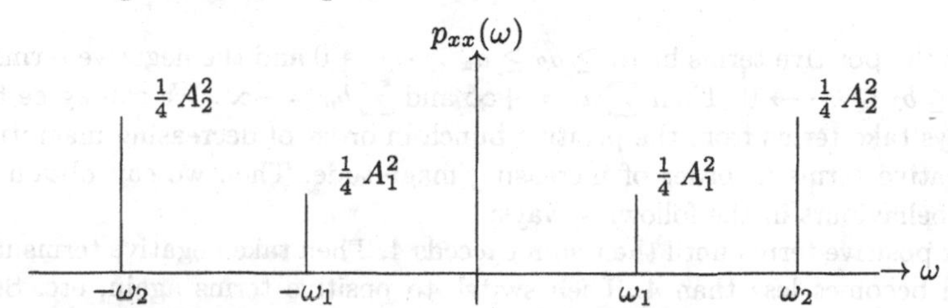

7.49. $\left(1 - \frac{1}{2}x^2\right)e^{-x^2/2}$.

7.51. $\widehat{f}(\omega) = -2i\omega/(1+\omega^2)$, integral $= \frac{1}{2}\,\pi$.

7.53. $f(x) = \frac{4}{3}\,e^{-x}H(x) + \frac{4}{3}e^{x/2}(1 - H(x))$.

7.55. $\pi\left(1 - \dfrac{1}{e}\cosh x\right)$, $|x| < 1$.

Chapter 8

8.1. An antiderivative of $\varphi \in \mathcal{S}$ belongs to $\mathcal{S}$ only if $\int \varphi(x)\,dx = 0$.

8.3. (a) Yes. (b) No (e^x grows to fast as $x \to +\infty$). (c) No (not linear).

8.6. $f'''(x) = 12H(x+1) - 6 - 16\delta(x+1) + 8\delta'(x+1)$.

8.10. $1/(1+i\omega)$, $2\pi e^{\omega}(1 - H(\omega))$ resp. $2\pi i(\delta(\omega) - e^{-\omega}H(\omega))$.

8.13. $f(t) = \delta(t) + H(t)$.

8.18. $\sum_{n\in\mathbf{Z}} \delta(\omega - n)$.

8.19. (a) $E(t) = e^{at}(H(t) - 1)$ if $a > 0$, $E(t) = \frac{1}{2}\operatorname{sgn} t$ if $a = 0$, $E(t) = e^{at}H(t)$ if $a < 0$. (b) $E(t) = (e^{-t} - e^{-2t})H(t)$.

8.20. e^{-x^2} belongs to $\mathcal{S}$ (and thus also to $\mathcal{M}$); $e^{-|x|^5}$ has a discontinuous fifth derivative and belongs to none of the classes; all the others belong to $\mathcal{M}$ but not to $\mathcal{S}$.

8.21. Not $e^{\pm 2x}$ and $e^{2x}H(x)$, but all the others.

8.23. $\psi(x)\delta''(x-a) = \psi''(a)\delta(x-a) - 2\psi'(a)\delta'(x-a) + \psi(a)\delta''(x-a)$.

8.24. $f''(x) = -|\sin x| + 2\sum_{n\in\mathbf{Z}} \delta(x - n\pi)$.

8.25. $n\delta(x - 1/n) - n\delta(x + 1/n) \to -2\delta'(x)$ as $n \to \infty$.

8.27. (a) $i\pi(\delta(\omega + a) - \delta(\omega - a))$. (b) $\pi(\delta(\omega - b) + \delta(\omega + b))$. (c) $i^n\left(\pi\delta^{(n)}(\omega) - \dfrac{i(-1)^n n!}{\omega^{n+1}}\right)$. (d) $2\cos a\omega$. (e) -1.

Chapter 9

9.1. Let the positive terms be $a_1 \geq a_2 \geq a_x \geq \cdots \to 0$ and the negative terms be $b_1 \leq b_2 \leq b_3 \leq \cdots \to 0$. Then $\sum a_n = +\infty$ and $\sum b_n = -\infty$. We can agree that we always take terms from the positive bunch in order of decreasing magnitude, and negative terms in order of increasing magnitude. Then we can obtain the various behaviours in the following ways:
(a) Take positive terms until their sum exceeds 4. Then take negative terms until the sum becomes less than 4. Then switch to positive terms again, etc. Since the terms tend to 0, the sequence of partial sums will oscillate around 4 with diminishing amplitudes and their limit will be 4.
(b) Take negative terms until we get a sum less than -1. Then take one positive term. Then negative terms until we pass -2; one positive term; negative terms past -3; etc.
(c) Take negative terms until we pass -13; then positive terms until we exceed 2003; negative terms again until we pass -13; and carry on like this till the cows come home.
(d) Take positive terms until the sum exceeds 1; then negative terms until we come below -2; then positive terms to pass 3; negative to pass -4; etc.

9.3. 1.

Appendix E
Literature

This list does not attempt to be complete in any way whatsoever. First we mention a few books that cover approximately the same topics as the present volume, and on a similar level.

R. V. CHURCHILL & J. W: BROWN, *Fourier Series and Boundary Value Problems.* McGraw–Hill, New York, 1978.

J. RAY HANNA & JOHN H. ROWLAND, *Fourier Series, Transforms, and Boundary Value Problems.* Wiley, New York, 1990.

P. L. WALKER, *The Theory of Fourier Series and Integrals.* Wiley, Chichester, 1986.

The following books are on a more advanced mathematical level.

THOMAS W. KÖRNER, *Fourier Analysis.* Cambridge University Press; first paperback edition, 1989.

THOMAS W. KÖRNER, *Exercises for Fourier Analysis.* Cambridge University Press, 1993.

These books are excellent reading for the student who wants to go deeper into classical Fourier analysis and its applications. The applications treated cover a wide range: they include matters such as Monte Carlo methods, Brownian motion, linear oscillators, code theory, and the question of the age of the earth. The style is engaging, and the mathematics is 100 percent stringent.

YITZHAK KATZNELSON, *An Introduction to Harmonic Analysis.* Wiley, New York, 1968.

This work goes into generalizations of Fourier analysis that have not been mentioned at all in the present text. It presupposes knowledge of Banach spaces and other parts of functional analysis.

LARS HÖRMANDER, *The Analysis of Linear Partial Differential Operators,* I–IV. Springer-Verlag, Berlin–Heidelberg, 1983–85.

This monumental work is the standard source for distribution theory. It is not an easy read, but it is famous for its depth, breadth, and elegance.

Finally, for the really curious student, we mention a couple of research papers referred to in this text.

LENNART CARLESON, *On convergence and growth of partial sums of Fourier series.* Acta Mathematica 116 (1966), 135–157.

HANS LEWY, *An example of a smooth linear partial differential equation without solution.* Annals of Mathematics (**2**) 66 (1957), 91–107.

Index

Graduate Texts in Mathematics

(continued from page ii)

64 EDWARDS. Fourier Series. Vol. I. 2nd ed.
65 WELLS. Differential Analysis on Complex Manifolds. 2nd ed.
66 WATERHOUSE. Introduction to Affine Group Schemes.
67 SERRE. Local Fields.
68 WEIDMANN. Linear Operators in Hilbert Spaces.
69 LANG. Cyclotomic Fields II.
70 MASSEY. Singular Homology Theory.
71 FARKAS/KRA. Riemann Surfaces. 2nd ed.
72 STILLWELL. Classical Topology and Combinatorial Group Theory. 2nd ed.
73 HUNGERFORD. Algebra.
74 DAVENPORT. Multiplicative Number Theory. 3rd ed.
75 HOCHSCHILD. Basic Theory of Algebraic Groups and Lie Algebras.
76 IITAKA. Algebraic Geometry.
77 HECKE. Lectures on the Theory of Algebraic Numbers.
78 BURRIS/SANKAPPANAVAR. A Course in Universal Algebra.
79 WALTERS. An Introduction to Ergodic Theory.
80 ROBINSON. A Course in the Theory of Groups. 2nd ed.
81 FORSTER. Lectures on Riemann Surfaces.
82 BOTT/TU. Differential Forms in Algebraic Topology.
83 WASHINGTON. Introduction to Cyclotomic Fields. 2nd ed.
84 IRELAND/ROSEN. A Classical Introduction to Modern Number Theory. 2nd ed.
85 EDWARDS. Fourier Series. Vol. II. 2nd ed.
86 VAN LINT. Introduction to Coding Theory. 2nd ed.
87 BROWN. Cohomology of Groups.
88 PIERCE. Associative Algebras.
89 LANG. Introduction to Algebraic and Abelian Functions. 2nd ed.
90 BRØNDSTED. An Introduction to Convex Polytopes.
91 BEARDON. On the Geometry of Discrete Groups.
92 DIESTEL. Sequences and Series in Banach Spaces.
93 DUBROVIN/FOMENKO/NOVIKOV. Modern Geometry—Methods and Applications. Part I. 2nd ed.
94 WARNER. Foundations of Differentiable Manifolds and Lie Groups.
95 SHIRYAEV. Probability. 2nd ed.
96 CONWAY. A Course in Functional Analysis. 2nd ed.
97 KOBLITZ. Introduction to Elliptic Curves and Modular Forms. 2nd ed.
98 BRÖCKER/TOM DIECK. Representations of Compact Lie Groups.
99 GROVE/BENSON. Finite Reflection Groups. 2nd ed.
100 BERG/CHRISTENSEN/RESSEL. Harmonic Analysis on Semigroups: Theory of Positive Definite and Related Functions.
101 EDWARDS. Galois Theory.
102 VARADARAJAN. Lie Groups, Lie Algebras and Their Representations.
103 LANG. Complex Analysis. 3rd ed.
104 DUBROVIN/FOMENKO/NOVIKOV. Modern Geometry—Methods and Applications. Part II.
105 LANG. $SL_2(\mathbf{R})$.
106 SILVERMAN. The Arithmetic of Elliptic Curves.
107 OLVER. Applications of Lie Groups to Differential Equations. 2nd ed.
108 RANGE. Holomorphic Functions and Integral Representations in Several Complex Variables.
109 LEHTO. Univalent Functions and Teichmüller Spaces.
110 LANG. Algebraic Number Theory.
111 HUSEMÖLLER. Elliptic Curves.
112 LANG. Elliptic Functions.
113 KARATZAS/SHREVE. Brownian Motion and Stochastic Calculus. 2nd ed.
114 KOBLITZ. A Course in Number Theory and Cryptography. 2nd ed.
115 BERGER/GOSTIAUX. Differential Geometry: Manifolds, Curves, and Surfaces.
116 KELLEY/SRINIVASAN. Measure and Integral. Vol. I.
117 J.-P. SERRE. Algebraic Groups and Class Fields.
118 PEDERSEN. Analysis Now.
119 ROTMAN. An Introduction to Algebraic Topology.
120 ZIEMER. Weakly Differentiable Functions: Sobolev Spaces and Functions of Bounded Variation.
121 LANG. Cyclotomic Fields I and II. Combined 2nd ed.
122 REMMERT. Theory of Complex Functions. *Readings in Mathematics*
123 EBBINGHAUS/HERMES et al. Numbers. *Readings in Mathematics*

124 DUBROVIN/FOMENKO/NOVIKOV. Modern Geometry—Methods and Applications. Part III
125 BERENSTEIN/GAY. Complex Variables: An Introduction.
126 BOREL. Linear Algebraic Groups. 2nd ed.
127 MASSEY. A Basic Course in Algebraic Topology.
128 RAUCH. Partial Differential Equations.
129 FULTON/HARRIS. Representation Theory: A First Course.
Readings in Mathematics
130 DODSON/POSTON. Tensor Geometry.
131 LAM. A First Course in Noncommutative Rings. 2nd ed.
132 BEARDON. Iteration of Rational Functions.
133 HARRIS. Algebraic Geometry: A First Course.
134 ROMAN. Coding and Information Theory.
135 ROMAN. Advanced Linear Algebra.
136 ADKINS/WEINTRAUB. Algebra: An Approach via Module Theory.
137 AXLER/BOURDON/RAMEY. Harmonic Function Theory. 2nd ed.
138 COHEN. A Course in Computational Algebraic Number Theory.
139 BREDON. Topology and Geometry.
140 AUBIN. Optima and Equilibria. An Introduction to Nonlinear Analysis.
141 BECKER/WEISPFENNING/KREDEL. Gröbner Bases. A Computational Approach to Commutative Algebra.
142 LANG. Real and Functional Analysis. 3rd ed.
143 DOOB. Measure Theory.
144 DENNIS/FARB. Noncommutative Algebra.
145 VICK. Homology Theory. An Introduction to Algebraic Topology. 2nd ed.
146 BRIDGES. Computability: A Mathematical Sketchbook.
147 ROSENBERG. Algebraic *K*-Theory and Its Applications.
148 ROTMAN. An Introduction to the Theory of Groups. 4th ed.
149 RATCLIFFE. Foundations of Hyperbolic Manifolds.
150 EISENBUD. Commutative Algebra with a View Toward Algebraic Geometry.
151 SILVERMAN. Advanced Topics in the Arithmetic of Elliptic Curves.
152 ZIEGLER. Lectures on Polytopes.
153 FULTON. Algebraic Topology: A First Course.
154 BROWN/PEARCY. An Introduction to Analysis.
155 KASSEL. Quantum Groups.
156 KECHRIS. Classical Descriptive Set Theory.
157 MALLIAVIN. Integration and Probability.
158 ROMAN. Field Theory.
159 CONWAY. Functions of One Complex Variable II.
160 LANG. Differential and Riemannian Manifolds.
161 BORWEIN/ERDÉLYI. Polynomials and Polynomial Inequalities.
162 ALPERIN/BELL. Groups and Representations.
163 DIXON/MORTIMER. Permutation Groups.
164 NATHANSON. Additive Number Theory: The Classical Bases.
165 NATHANSON. Additive Number Theory: Inverse Problems and the Geometry of Sumsets.
166 SHARPE. Differential Geometry: Cartan's Generalization of Klein's Erlangen Program.
167 MORANDI. Field and Galois Theory.
168 EWALD. Combinatorial Convexity and Algebraic Geometry.
169 BHATIA. Matrix Analysis.
170 BREDON. Sheaf Theory. 2nd ed.
171 PETERSEN. Riemannian Geometry.
172 REMMERT. Classical Topics in Complex Function Theory.
173 DIESTEL. Graph Theory. 2nd ed.
174 BRIDGES. Foundations of Real and Abstract Analysis.
175 LICKORISH. An Introduction to Knot Theory.
176 LEE. Riemannian Manifolds.
177 NEWMAN. Analytic Number Theory.
178 CLARKE/LEDYAEV/STERN/WOLENSKI. Nonsmooth Analysis and Control Theory.
179 DOUGLAS. Banach Algebra Techniques in Operator Theory. 2nd ed.
180 SRIVASTAVA. A Course on Borel Sets.
181 KRESS. Numerical Analysis.
182 WALTER. Ordinary Differential Equations.

183 MEGGINSON. An Introduction to Banach Space Theory.
184 BOLLOBAS. Modern Graph Theory.
185 COX/LITTLE/O'SHEA. Using Algebraic Geometry.
186 RAMAKRISHNAN/VALENZA. Fourier Analysis on Number Fields.
187 HARRIS/MORRISON. Moduli of Curves.
188 GOLDBLATT. Lectures on the Hyperreals: An Introduction to Nonstandard Analysis.
189 LAM. Lectures on Modules and Rings.
190 ESMONDE/MURTY. Problems in Algebraic Number Theory.
191 LANG. Fundamentals of Differential Geometry.
192 HIRSCH/LACOMBE. Elements of Functional Analysis.
193 COHEN. Advanced Topics in Computational Number Theory.
194 ENGEL/NAGEL. One-Parameter Semigroups for Linear Evolution Equations.
195 NATHANSON. Elementary Methods in Number Theory.
196 OSBORNE. Basic Homological Algebra.
197 EISENBUD/HARRIS. The Geometry of Schemes.
198 ROBERT. A Course in *p*-adic Analysis.
199 HEDENMALM/KORENBLUM/ZHU. Theory of Bergman Spaces.
200 BAO/CHERN/SHEN. An Introduction to Riemann–Finsler Geometry.
201 HINDRY/SILVERMAN. Diophantine Geometry: An Introduction.
202 LEE. Introduction to Topological Manifolds.
203 SAGAN. The Symmetric Group: Representations, Combinatorial Algorithms, and Symmetric Functions.
204 ESCOFIER. Galois Theory.
205 FÉLIX/HALPERIN/THOMAS. Rational Homotopy Theory. 2nd ed.
206 MURTY. Problems in Analytic Number Theory.
Readings in Mathematics
207 GODSIL/ROYLE. Algebraic Graph Theory.
208 CHENEY. Analysis for Applied Mathematics.
209 ARVESON. A Short Course on Spectral Theory.
210 ROSEN. Number Theory in Function Fields.
211 LANG. Algebra. Revised 3rd ed.
212 MATOUŠEK. Lectures on Discrete Geometry.
213 FRITZSCHE/GRAUERT. From Holomorphic Functions to Complex Manifolds.
214 JOST. Partial Differential Equations.
215 GOLDSCHMIDT. Algebraic Functions and Projective Curves.
216 D. SERRE. Matrices: Theory and Applications.
217 MARKER. Model Theory: An Introduction.
218 LEE. Introduction to Smooth Manifolds.
219 MACLACHLAN/REID. The Arithmetic of Hyperbolic 3-Manifolds.
220 NESTRUEV. Smooth Manifolds and Observables.
221 GRÜNBAUM. Convex Polytopes. 2nd ed.
222 HALL. Lie Groups, Lie Algebras, and Representations: An Elementary Introduction.
223 VRETBLAD. Fourier Analysis and Its Applications.